U.S.NRC

United States Nuclear Regulatory Commission

Protecting People and the Environment

NUREG-2124
Volume 3

I0482753

Final Safety Evaluation Report

Related to the Combined Licenses for Vogtle Electric Generating Plant, Units 3 and 4

Volume 3

Docket Nos. 52-025 and 52-026

Office of New Reactors

AVAILABILITY OF REFERENCE MATERIALS
IN NRC PUBLICATIONS

NRC Reference Material

As of November 1999, you may electronically access NUREG-series publications and other NRC records at NRC's Public Electronic Reading Room at http://www.nrc.gov/reading-rm.html. Publicly released records include, to name a few, NUREG-series publications; *Federal Register* notices; applicant, licensee, and vendor documents and correspondence; NRC correspondence and internal memoranda; bulletins and information notices; inspection and investigative reports; licensee event reports; and Commission papers and their attachments.

NRC publications in the NUREG series, NRC regulations, and Title 10, "Energy," in the *Code of Federal Regulations* may also be purchased from one of these two sources.
1. The Superintendent of Documents
   U.S. Government Printing Office
   Mail Stop SSOP
   Washington, DC 20402–0001
   Internet: bookstore.gpo.gov
   Telephone: 202-512-1800
   Fax: 202-512-2250
2. The National Technical Information Service
   Springfield, VA 22161–0002
   www.ntis.gov
   1–800–553–6847 or, locally, 703–605–6000

A single copy of each NRC draft report for comment is available free, to the extent of supply, upon written request as follows:
Address: U.S. Nuclear Regulatory Commission
          Office of Administration
          Publications Branch
          Washington, DC 20555-0001
E-mail: DISTRIBUTION.RESOURCE@NRC.GOV
Facsimile: 301–415–2289

Some publications in the NUREG series that are posted at NRC's Web site address http://www.nrc.gov/reading-rm/doc-collections/nuregs are updated periodically and may differ from the last printed version. Although references to material found on a Web site bear the date the material was accessed, the material available on the date cited may subsequently be removed from the site.

Non-NRC Reference Material

Documents available from public and special technical libraries include all open literature items, such as books, journal articles, transactions, *Federal Register* notices, Federal and State legislation, and congressional reports. Such documents as theses, dissertations, foreign reports and translations, and non-NRC conference proceedings may be purchased from their sponsoring organization.

Copies of industry codes and standards used in a substantive manner in the NRC regulatory process are maintained at—
       The NRC Technical Library
       Two White Flint North
       11545 Rockville Pike
       Rockville, MD 20852–2738

These standards are available in the library for reference use by the public. Codes and standards are usually copyrighted and may be purchased from the originating organization or, if they are American National Standards, from—
       American National Standards Institute
       11 West 42nd Street
       New York, NY 10036–8002
       www.ansi.org
       212–642–4900

Legally binding regulatory requirements are stated only in laws; NRC regulations; licenses, including technical specifications; or orders, not in NUREG-series publications. The views expressed in contractor-prepared publications in this series are not necessarily those of the NRC.

The NUREG series comprises (1) technical and administrative reports and books prepared by the staff (NUREG–XXXX) or agency contractors (NUREG/CR–XXXX), (2) proceedings of conferences (NUREG/CP–XXXX), (3) reports resulting from international agreements (NUREG/IA–XXXX), (4) brochures (NUREG/BR–XXXX), and (5) compilations of legal decisions and orders of the Commission and Atomic and Safety Licensing Boards and of Directors' decisions under Section 2.206 of NRC's regulations (NUREG–0750).

DISCLAIMER: This report was prepared as an account of work sponsored by an agency of the U.S. Government. Neither the U.S. Government nor any agency thereof, nor any employee, makes any warranty, expressed or implied, or assumes any legal liability or responsibility for any third party's use, or the results of such use, of any information, apparatus, product, or process disclosed in this publication, or represents that its use by such third party would not infringe privately owned rights.

United States Nuclear Regulatory Commission

Protecting People and the Environment

NUREG-2124
Volume 3

Final Safety Evaluation Report

Related to the Combined Licenses for Vogtle Electric Generating Plant, Units 3 and 4

Volume 3

Docket Nos. 52-025 and 52-026

Manuscript Completed: August 2011
Date Published: September 2012

Office of New Reactors

ABSTRACT

This final safety evaluation report[1] (FSER) documents the U.S. Nuclear Regulatory Commission (NRC) staff's technical review of the combined license (COL) application submitted by Southern Nuclear Operating Company (SNC or the applicant), for the Vogtle Electric Generating Plant (VEGP) Units 3 and 4. The SER also documents the NRC staff's technical review of the limited work authorization (LWA) activities for which SNC has requested approval.

By letter dated March 28, 2008, SNC, acting on behalf of itself and the proposed owners (Georgia Power Company (GPC), Oglethorpe Power Corporation (an electric membership corporation), Municipal Electric Authority of Georgia, and the City of Dalton, Georgia, an incorporated municipality in the State of Georgia acting by and through its Board of Water, Light and Sinking Fund Commissioners), submitted its application to the NRC for COLs for two AP1000 advanced passive pressurized-water reactors (PWRs) pursuant to the requirements of Sections 103 and 185(b) of the Atomic Energy Act of 1954, as amended; Title 10 of the *Code of Federal Regulations* (10 CFR) Part 52, "Licenses, certifications and approvals for nuclear power plants"; and the associated material licenses under 10 CFR Part 30, "Rules of general applicability to domestic licensing of byproduct material"; 10 CFR Part 40, "Domestic licensing of source material"; and 10 CFR Part 70, "Domestic licensing of special nuclear material." These reactors are identified as VEGP Units 3 and 4, and will be located on the existing VEGP site in Burke County, Georgia.

In October 2009, SNC supplemented its COL application to include a request for an LWA. The LWA, in accordance with 10 CFR 50.10(d), would authorize installation of reinforcing steel, sumps, drain lines, and other embedded items along with placement of concrete for the nuclear island foundation base slab.

The initial application incorporated by reference 10 CFR Part 52, Appendix D, "Design Certification Rule for the AP1000 Design," and the Westinghouse Electric Corporation's (Westinghouse's) application for amendment of the AP1000 design, as described in Revision 16 of the Design Control Document (DCD) (submitted May 26, 2007), as well as Westinghouse Technical Report (TR)-134, APP-GW-GLR-134, "AP1000 DCD Impacts to Support COLA Standardization," Revision 4 (which was submitted on March 18, 2008). The initial application also referenced the VEGP Early Site Permit (ESP) Application, Revision 4, dated March 28, 2008. Subsequent to the initial application, in its submittal dated December 11, 2009, SNC incorporated by reference the VEGP ESP Application, Revision 5, dated December 23, 2008, as approved by the NRC in the VEGP ESP and LWA (ESP-004), dated August 26, 2009. In a letter dated August 6, 2010, SNC incorporated by reference the three amendments issued (on May 21, 2010; June 25, 2010; and July 9, 2010) to the ESP. In a letter dated June 24, 2011(submittal number 8), SNC incorporated by reference AP1000 DCD, Revision 19. The results of the NRC staff's evaluation of the AP1000 DCD are documented in NUREG-1793, "Final Safety Evaluation Report Related to Certification of the AP1000 Standard Design," and its supplements. The results of the NRC staff's evaluation related to the VEGP ESP are documented in NUREG-1923, "Safety Evaluation Report for Early Site Permit (ESP) at the Vogtle Electric Generating Plant (VEGP) ESP Site."

---

[1] This FSER documents the NRC staff's position on all safety issues associated with the combined license application. The Advisory Committee on Reactor Safeguards (ACRS) independently reviewed those aspects of the application that concern safety, as well as the advanced safety evaluation report without open items (an earlier version of this document), and provided the results of its review to the Commission in a report dated January 24, 2011. This report is included as Appendix F to this SER.

This FSER presents the results of the staff's review of information submitted in conjunction with the COL application, except those matters resolved as part of the referenced ESP or design certification rule. In Appendix A to this FSER, the staff has identified certain license conditions and inspections, tests, analyses and acceptance criteria (ITAAC) that the staff recommends the Commission impose, should COLs be issued to the applicant. Appendix A includes the applicable permit conditions and ITAAC from the ESP. Therefore, Appendix A includes COL and ESP conditions, recognizing that should COLs be issued to the applicant, the ESP will be subsumed into the COLs. In addition to the ITAAC in Appendix A, the ITAAC found in the AP1000 DCD, Revision 19 Tier 1 material will also be incorporated into the COLs should COLs be issued to the applicant.

On the basis of the staff's review[2] of the application, as documented in this FSER, the staff recommends that the Commission find the following with respect to the safety aspects of the COL application: 1) the applicable standards and requirements of the Atomic Energy Act and Commission regulations have been met, 2) Required notifications to other agencies or bodies have been duly made, 3) there is reasonable assurance that the facility will be constructed and will operate in conformity with the license, the provisions of the Atomic Energy Act, and the Commission's regulations, 4) the applicant is technically and financially qualified to engage in the activities authorized, and 5) issuance of the license will not be inimical to the common defense and security or to the health and safety of the public.

---

[2] An environmental review was also performed of the COL application and its evaluation and conclusions are documented in NUREG-1947, "Final Supplemental Environmental Impact Statement for Combined Licenses (COLs) for Vogtle Electric Generating Plant Units 3 and 4."

CONTENTS

The chapter and section layout of this SER is consistent with the format of: (1) NUREG-0800, "Standard Review Plan for the Review of Safety Analysis Reports for Nuclear Power Plants (LWR Edition)"; (2) Regulatory Guide 1.206, "Combined License Applications for Nuclear Power Plants"; and (3) the applicant's final safety analysis report. Where applicable, references to other regulatory actions (design certifications, ESPs) are included in the text of the SER.

APPENDICES

FIGURES

TABLES

EXECUTIVE SUMMARY

The U.S. Nuclear Regulatory Commission (NRC) regulations in Title 10 of the *Code of Federal Regulations* (10 CFR) Part 52 include requirements for licensing new nuclear power plants.[3] These regulations include the NRC's requirements for early site permit (ESP), design certification, and combined license (COL) applications. The ESP process (10 CFR Part 52, Subpart A, "Early Site Permits") is intended to address and resolve siting-related issues. The design certification process (10 CFR Part 52, Subpart B, "Standard Design Certifications") provides a means for a vendor to obtain NRC certification of a particular reactor design. Finally, the COL process (10 CFR Part 52, Subpart C, "Combined Licenses") allows an applicant to seek authorization to construct and operate a new nuclear power plant. A COL may reference an ESP, a certified design, both, or neither. As part of demonstrating that all applicable NRC requirements are met, a COL applicant referencing an ESP or certified design must demonstrate compliance with any requirements not already resolved as part of the referenced ESP or design certification before the NRC issues that COL.

This FSER describes the results of a review by the NRC staff of a COL application submitted by Southern Nuclear Operating Company (SNC or the applicant), acting on behalf of itself and the proposed owners (Georgia Power Company (GPC), Oglethorpe Power Corporation (an electric membership corporation), Municipal Electric Authority of Georgia, and the City of Dalton, Georgia, an incorporated municipality in the State of Georgia acting by and through its Board of Water, Light and Sinking Fund Commissioners), for two new reactors to be located at the Vogtle Electric Generating Plant (VEGP) site. The staff's review was to determine the applicant's compliance with the requirements of Subpart C of 10 CFR Part 52, as well as the applicable requirements under 10 CFR Parts 30, 40, and 70 governing the possession and use of applicable source, byproduct, and special nuclear materials. This FSER serves to identify the staff's conclusions with respect to the COL safety review.

The NRC regulations also require an applicant to submit an environmental report pursuant to 10 CFR Part 51, "Environmental protection regulations for domestic licensing and related regulatory functions." The NRC reviews the environmental report as part of the Agency's responsibilities under the National Environmental Policy Act of 1969, as amended. The NRC presents the results of that review in a final environmental impact statement (FEIS), which is a report separate from this FSER. The NRC staff previously prepared an FEIS as part of its review of the VEGP ESP, which is referenced in the VEGP COL application. NUREG-1872, "Final Environmental Impact Statement for an Early Site Permit (ESP) at the Vogtle Electric Generating Plant Site," was issued in August 2008, and can be accessed through the Agencywide Documents Access and Management System (ADAMS) at ML082260190.[4]

---

[3] Applicants may also choose to seek a construction permit (CP) and operating license in accordance with 10 CFR Part 50, "Domestic licensing of production and utilization facilities," instead of using the 10 CFR Part 52 process.

[4] Agencywide Documents Access and Management System (ADAMS) is the NRC's information system that provides access to all image and text documents that the NRC has made public since November 1, 1999, as well as bibliographic records (some with abstracts and full text) that the NRC made public before November 1999. Documents available to the public may be accessed via the Internet at http://www.nrc.gov/reading-rm/adams/web-based.html. Documents may also be viewed by visiting the NRC's Public Document Room at One White Flint North, 11555 Rockville P ke, Rockville, Maryland. Telephone assistance for using web-based ADAMS is available at (800) 397-4209 between 8:30 a.m. and 4:15 p.m., Eastern Time, Monday through Friday, except Federal holidays. The staff is also making this FSER available on the NRC's new reactor licensing public web site at http://www.nrc.gov/reactors/new-reactors/col/vogtle/documents/ser-final.html.

For a COL application that references an ESP, the NRC staff, pursuant to 10 CFR 51.75(c), prepares a supplement to the ESP environmental impact statement (EIS) in accordance with 10 CFR 51.92(e). NRC regulations related to the environmental review of COL applications are in 10 CFR Part 51 and 10 CFR Part 52, Subpart C. Pursuant to 10 CFR 51.50(c)(1), a COL applicant referencing an ESP need not submit information or analyses regarding environmental issues that were resolved in the ESP EIS, except to the extent that the COL applicant has identified new and significant information regarding such issues. In addition, under 10 CFR 52.39, "Finality of early site permit determinations," matters resolved in the ESP proceedings are considered to be resolved in any subsequent proceedings, absent identification of new and significant information. The staff issued a supplement to the ESP EIS, NUREG-1947, "Final Supplemental Environmental Impact Statement for Combined Licenses (COLs) for Vogtle Electric Generating Plant Units 3 and 4," for the COL on March 25, 2011, which can be accessed through ADAMS at ML11076A010.

In a letter dated March 28, 2008, the SNC, acting on behalf of itself and the proposed owners, submitted its application to the NRC for COLs for two AP1000 advanced passive pressurized-water reactors (PWRs) (ADAMS Accession No. ML081050133) to be located at the VEGP site. SNC identified the two units as VEGP Units 3 and 4. The VEGP site is located on a coastal plain bluff on the southwest side of the Savannah River in eastern Burke County, Georgia. The site is approximately 26 miles southeast of Augusta, Georgia, and 100 miles northwest of Savannah, Georgia. Directly across from the site, on the eastern side of the Savannah River, is the U.S. Department of Energy's (DOE's) Savannah River site in Barnwell County, South Carolina. The proposed VEGP Units 3 and 4 would be built on the VEGP site adjacent to two existing nuclear power reactors, VEGP Units 1 and 2, operated by SNC.

In October 2009, SNC supplemented its COL application to include a request for an LWA. The LWA, in accordance with 10 CFR 50.10(d), would authorize installation of reinforcing steel, sumps, drain lines, and other embedded items along with placement of concrete for the nuclear island foundation base slab.

The initial application incorporated by reference 10 CFR Part 52, Appendix D, "Design Certification Rule for the AP1000 Design," and the Westinghouse Electric Corporation's (Westinghouse's) application for amendment of the AP1000 design, as supported by Revision 16 of the Design Control Document (DCD) (submitted May 26, 2007) as well as Westinghouse Technical Report (TR)-134, APP-GW-GLR-134, "AP1000 DCD Impacts to Support COLA Standardization," Revision 4 (which was submitted on March 18, 2008). The initial application also referenced the VEGP Early Site Permit (ESP) Application, Revision 4, dated March 28, 2008. Subsequent to the initial application, in its submittal dated December 11, 2009, SNC incorporated by reference the VEGP ESP Application, Revision 5, dated December 23, 2008, as approved by the NRC in the VEGP ESP and LWA (ESP-004), dated August 26, 2009. In a letter dated August 6, 2010, SNC incorporated by reference the three amendments issued (on May 21, 2010; June 25, 2010; and July 9, 2010) to the ESP. In a letter dated June 24, 2011(submittal number 8), SNC incorporated by reference AP1000 DCD, Revision 19. The results of the NRC staff's evaluation of the AP1000 DCD are documented in NUREG-1793, "Final Safety Evaluation Report Related to Certification of the AP1000 Standard Design," and its supplements. The results of the NRC staff's evaluation related to the VEGP ESP are documented in NUREG-1923, "Safety Evaluation Report for Early Site Permit (ESP) at the Vogtle Electric Generating Plant (VEGP) ESP Site." This FSER presents the results of the staff's review of information submitted in conjunction with the COL application, including any matters that were not already resolved as part of the referenced ESP or the referenced design certification, or subject to resolution in the pending design certification amendment proceeding.

The staff has identified in Appendix A to this FSER certain license conditions, and inspections, tests, analyses and acceptance criteria (ITAAC) that the staff recommends the Commission impose, should COLs be issued to the applicant. Appendix A includes the applicable permit conditions and ITAAC from the ESP. Therefore, Appendix A includes COL and ESP conditions, recognizing that should COLs be issued to the applicant, the ESP will be subsumed into the COLs. In addition to the ITAAC in Appendix A, the ITAAC found in the AP1000 DCD, Revision 19 Tier 1 material will also be incorporated into the COLs should COLs be issued to the applicant.

Inspections conducted by the NRC have verified, where appropriate, the conclusions in this FSER. The inspections focused on selected information in the COL application and its references. The FSER identifies applicable inspection reports as reference documents.

The NRC's Advisory Committee on Reactor Safeguards (ACRS) also reviewed the bases for the conclusions in this report. The ACRS independently reviewed those aspects of the application that concern safety, as well as the advanced safety evaluation report without open items earlier version of this document, and provided the results of its review to the Commission in a report dated January 24, 2011. Appendix F includes a copy of the report by the ACRS on the COL application, as required by 10 CFR 52.87, "Referral to the Advisory Committee on Reactor Safeguards (ACRS)."

ABBREVIATIONS

χ/Q	atmospheric dispersion
A2LA	American Association for Laboratory Accreditation
ac	alternating current
ACI	American Concrete Institute
ACP	access control parts
ACRS	Advisory Committee on Reactor Safeguards
ADAMS	Agencywide Documents Access and Management System
ADS	automatic depressurization system
AE	architect-engineer
AFFF	aqueous film forming foam
ALARA	as low as is reasonably achievable
ALI	annual limits on intake
ALWR	advanced light-water reactor
ANI	American Nuclear Insurers
ANS	American Nuclear Society
ANSI	American National Standards Institute
AOO	anticipated operational occurrence
AOV	air-operated valve
ARS	amplified response spectra
ASCE	American Society of Civil Engineers
ASE	advanced safety evaluation
ASLB	Atomic Safety and Licensing Board
ASME	American Society of Mechanical Engineers
ASTM	American Society for Testing and Materials
ATE	advisory to evacuate
ATWS	anticipated transients without scram
AWWA	American Water Works Association
BBM	Blue Bluff Marl
BCEMA	Burke County Emergency Management Agency
BDBE	beyond-design basis event
BL	Bulletin
BLN	Bellefonte Nuclear Station
BOP	balance of plant
BPV	Boiler & Pressure Vessel Code (ASME BPV Code)
BTP	Branch Technical Position
BWR	boiling-water reactor
C	Celsius
C&C	command & control
CAS	central alarm station
CAV	cumulative absolute velocity
CCS	component cooling water system
CDA	critical digital asset
CDE	committed dose equivalent

CDF	core damage frequency
CDI	conceptual design information
CDM	certified design material
CECC	Central Emergency Control Center
CEUS	Central and Eastern United States
cfm	cubic feet per minute
CFR	*Code of Federal Regulations*
cGy	centiGray
cm	centimeters
CMT	core makeup tank
COL	combined license
COLA	combined license application
CP	construction permit
cpm	counts per minute
CR	control room
CRDM	control rod drive mechanism
CRDS	control rod drive system
CS	containment system
CS	core supports
CSA	control support area
CSC	Communication Support Center
CSDRS	certified seismic design response spectra
CSP	Cyber Security Plan
CST	cyber security team
CTA	critical target area analysis
CVCS	chemical and volume control system
CVS	portions of the chemical and volume control system
CWS	circulating water system
D/Q	dry deposition factor
DAC	derived air concentration
DAS	Diverse Actuation System
DBA	design-basis accident
DBT	design-basis threat
dc	direct current
DC	design certification
DCA	design certification amendment
DCD	design control document
DCP	Design Change Package
DCRA	design-centered review approach
DECT	Digital Enhanced Cordless Telecommunication
DEP	Departure
DG	diesel generator
DHEC	Department of Health and Environmental Control
DHS	Department of Homeland Security
DNBR	departure from nucleate boiling ratio
DOE	Department of Energy
DOT	Department of Transportation
D-RAP	Design Reliability Assurance Program
DTS	demineralized water treatment system
DWS	demineralized water system

EAB	exclusion area boundary
EAL	emergency action level
EAS	Emergency Alert System
ECCS	emergency core cooling system
ED	Emergency Director
EDMG	Extensive Damage Mitigation Guidelines
EIP	emergency implementing procedure
EIS	Environmental Impact Statement
El.	Elevation
ELS	plant lighting system
EMA	Emergency Management Agency
ENC	Emergency News Center
ENN	Emergency Notification Network
ENS	Emergency Notification System
EOC	Emergency Operations Center
EOF	Emergency Operations Facility
EOM	Emergency Offsite Manager
EOP	emergency operating procedure
EP	Emergency Plan
EP	emergency planning
EPA	Environmental Protection Agency
EPAct	Energy Policy Act of 2005
EPC	Engineering, Procurement and Construction
EPI	Emergency Public Information
EPIO	Emergency Public Information Office
EPIP	emergency plan implementing procedures
EP-ITAAC	emergency planning-inspections, tests, analyses, and acceptance criteria
EPM	Emergency Plant Manager
EPOS	Emergency Plant Operations Supervisor
EPRI	Electric Power Research Institute
EPZ	emergency planning zone
EQ	environmental qualification
EQMEL	Environmental Qualification Master Equipment List
ER	Environmental Report
ERDS	Emergency Response Data System
ERF	emergency response facilities
ERO	emergency response officer
ERO	Emergency Response Organization
ESF	engineered safety feature
ESP	Early Site Permit
ESPA	Early Site Permit Application
ESSX	Electric Switch System Exchange
ETE	evacuation time estimate
ETS	Emergency Telecommunications System
F	Fahrenheit
FAA	Federal Aviation Administration
FAC	flow-accelerated corrosion
FBI	Federal Bureau of Investigation
FCEMS	Fairfield County Emergency Medical Services

FD1W	Feeder Ditch 1
FEIS	final environmental impact statement
FEMA	Federal Emergency Management Agency
FERC	Federal Energy Regulatory Commission
FFD	fitness-for-duty
FIFO	first-in-first-out
FIRS	foundation input response spectra
FIV	flow induced vibration
FMCRD	fine motion control rod drive
FMEA	failure mode and effects analysis
fps	feet per second
FPS	fire protection system
FR	*Federal Register*
FRS	floor response spectra
FSAR	final safety analysis report
FSER	final safety evaluation report
ft	feet
FTS	Federal Telecommunications System
GALL	Generic Aging Lessons Learned
GCC	Georgia Transmission Control Center
GDC	General Design Criteria (Criterion)
GEMA	Georgia Emergency Management Agency
GIS	Geographical Information System
GL	Generic Letter
GMRS	ground motion response spectra
GPC	Georgia Power Company
gpm	gallons per minute
GPSC	Georgia Public Service Commission
GSI	Generic Safety Issue
GSM	Global System for Mobile Communications
GSU	generator step-up
GTS	generic technical specification
GWMS	gaseous waste management system
h	hour
HCLPF	high confidence, low probability of failure
HCM	Highway Capacity Manual
HCU	hydraulic control unit
HDPE	high-density polyethylene
HEPA	high efficiency particulate air
HFE	human factors engineering
HICs	high integrity containers
HLD	heavy lift derrick
HP	health physics
HPN	Health Physics Network
HPS	Health Physics Society
HRA	human reliability analysis
HSI	human-system interface
HV	high voltage

HVAC	heating, ventilation, and air conditioning
Hz	Hertz
I&C	instrumentation and controls
IC	initiating conditions
ICM	Interim Compensatory Measures
ICMO	interim compensatory order
iDEN	Integrated Digital Enhanced Network
IDLH	immediate danger to life and health
IEC	International Electrotechnical Commission
IED	Interim Emergency Director
IEEE	Institute of Electrical and Electronic Engineers
IFR	interim findings report
IGSCC	intergranular stress corrosion cracking
IHP	integrated head package
IIS	incore instrumentation system
ILAC	International Laboratory Accreditation Cooperation
in.	inches
INPO	Institute of Nuclear Power Operations
IPEEE	Individual Plant Examination of External Events
IPSAC	Investment Protection Short-Term Availability Control
IPZ	Ingestion Pathway Emergency Planning Zone
IRWST	in-containment refueling water storage tank
ISA	Independent Safety Assessment
ISFSI	independent spent fuel storage installation
ISG	Interim Staff Guidance
ISI	inservice inspection
ISO	International Standardization Organization
ISRS	in-structure response spectra
IST	inservice testing
ITAAC	inspections, tests, analyses, and acceptance criteria
ITP	initial test program
JIC	joint information center
JOG	Joint Owners Group
JTWG	Joint Test Working Group
KI	radio-protective drugs
Kips	kilo pounds
km	kilometers
kPa	kilopascals
kV	kilovolt
kVA	kilovolt amps
kWe	kilowatt electric
LAN	Local Area Network
lb/ft^2	pounds per square foot
LBB	leak-before-break
LCEMS	Lexington County Emergency Medical Services
LCO	limiting condition for operation
LEFM	Leading Flow Edge Meter
LFL	lower flammability limit

LLEA	local law enforcement agency
LLHS	light-load handling system
LLNL	Lawrence Livermore National Laboratory
LLRW	low-level radioactive waste
LOA	Letters of Agreement
LOCA	loss-of-coolant accident
LOLA	loss of large areas
LOOP	loss of offsite power
lpm	liter(s) per minute
LPZ	low population zone
LRF	large release frequency
LSS	low strategic significance
LTOP	low-temperature overpressure protection
LWA	limited work authorization
LWMS	liquid waste management system
LWR	light-water reactor
m	meter(s)
MC	main condenser
MC&A	material control and accounting
MCL	Management Counterpart Link
MCR	main control room
MEAG	Municipal Electric Authority of Georgia
MEI	maximally exposed individual
MERT	Medical Emergency Response Team
mi	mile(s)
MIT	Massachusetts Institute of Technology
MN	Mega Newton
M-O	Mononobe-Okabe
MOU	Memorandum of Understanding
MOV	motor-operated valve
MOX	mixed-oxide
MPA	methoxypropylamine
mph	miles per hour
MR	Maintenance Rule
MRA	Mutual Recognition Arrangement
mrem	millirem
MSD	Mitigative Strategies Description
msl	mean sea level
MSLB	main steam line break
MSSS	main steam supply system
MST	Mitigative Strategies Table
mSv	millisievert
MT	magnetic particle
MUR	measurement uncertainty recapture
MVAR	mega volt amp reactive
MW	megawatt
MWe	megawatts electric
MWt	megawatts thermal

NDCT	natural draft cooling tower
NDL	nuclear data ink
NDQAM	Nuclear Development and Construction Quality Assurance Manual
NEI	Nuclear Energy Institute
NEMA	National Electrical Manufacturers Association
NERC	North American Electric Reliability Corporation
NFPA	National Fire Protection Association
NI	nuclear island
NIRMA	Nuclear Information and Records Management Association
NIST	National Institute of Standards and Technology
NNR	non-nuclear safety
NOV	Notice of Violation
NPIR	Nuclear Plant Interface Requirement
NPPENF	Nuclear Power Plant Emergency Notification
NRC	U.S. Nuclear Regulatory Commission
NRO	Office of New Reactors
NS	nonseismic
NSSS	nuclear steam system supply
NUMARC	Nuclear Management and Resources Council
NVLAP	National Voluntary Laboratory Accreditation Program
NWS	National Weather Service
OBE	operating basis earthquake
OCA	owner controlled area
OCL	Operational Center Local
ODCM	Offsite Dose Calculation Manual
OE	operating experience
OER	operating experience review
OHLHS	overhead heavy-load handling system
OM	Operations and Maintenance (ASME OM Code)
OPC	Oglethorpe Power Corporation
OPRAA	operational phase reliability assurance activity
ORE	occupational radiation exposure
ORM	Onsite Radiation Manager
OSC	Operations Support Center
p.u.	per unit
PA	protected area
PAD	protective action decisions
PAG	protected area guidelines
PAR	protective action recommendations
PAZ	protective action zones
PC	Permit Condition
PCC	Power Coordination Center
PCCAWST	passive containment cooling ancillary water storage tank
PCCWST	passive containment cooling water storage tank
PCP	Process Control Program
PCS	passive containment cooling system
PDC	Personal Digital Cellular
PDP	procedure development program
PE	polyethylene

PGA	peak ground acceleration
PGP	procedures generation package
PM	preventive maintenance
PMCL	Protective Measures Counterpart Link
PMF	probable maximum flood
PMH	probable maximum hurricane
PMP	probable maximum precipitation
PMS	protection and safety monitoring
PMT	probable maximum tsunami
PMWP	probable maximum winter precipitation
PMWS	probable maximum wind storm
PNS	Prompt Notification System
POV	power-operated valve
ppm	parts per million
PRA	probabilistic risk assessment
PRHR	passive residual heat removal
psf	pounds per square foot
PSHA	probabilistic seismic hazard analysis
PSI	preservice inspection
psi	per square inch
psia	pounds per square inch absolute
psig	pounds per square inch gauge
PS-ITAAC	physical security-inspection, test, analysis, and acceptance criteria
PSO	power systems operations
PSP	Physical Security Plan
PSS/E	Power System Simulator for Engineering
P-T	pressure temperature
PT	liquid penetrant
PT&O	plant test and operations
PTLR	pressure-temperature limits report
PTS	pressurized thermal shock
PTS	plant-specific technical specifications
PWR	pressurized-water reactor
PWS	potable water system
PWSCC	primary water stress corrosion cracking
PXS	passive core cooling system
QA	quality assurance
QAPD	Quality Assurance Program description
QAPD	Quality Assurance Program Document
QATR	Quality Assurance Topical Report
QC	quality control
QDF	queue discharge flow
QG	quality group
RAI	request for additional information
RAP	reliability assurance program
RAT	reserve auxiliary transformer
RCCA	rod cluster control assembly
RCL	reactor coolant loop
R-COL	reference combined license

RCP	reactor coolant pump
RCPB	reactor coolant pressure boundary
RCS	reactor coolant system
REAC/TS	Radiation Emergency Assistance Center / Training Site
rem	roentgen equivalent man
REMP	radiological environmental monitoring program
REP	radiological emergency preparedness
RERP	radiological emergency response plan
RET	Radiological Emergency Team
RETS	radiological effluent technical specification
RG	regulatory guide
RIS	Regulatory Issue Summary
RLE	review-level earthquake
RMS	radiation monitoring system
RNS	normal residual heat removal system
RO	reactor operator
RPP	Radiation Protection Program
RPV	reactor pressure vessel
RRS	required response spectrum
RSCL	Reactor Safety Counterpart Link
RTDP	revised thermal design procedure
RT_{NDT}	nil-ductility reference transition temperature
RTNSS	regulatory treatment of nonsafety systems
RTP	rated thermal power
RT_{PTS}	pressurized thermal shock reference temperature
RV	reactor vessel
RVSP	reactor vessel surveillance capsule program
RWS	raw water system
RXS	reactor system
s	second
S&PC	steam and power conversion
SAMG	severe accident management guidance
SAR	safety analysis report
SAS	secondary alarm station
SASSI	system for analysis of soil structure interaction
SAT	systematic approach to training
SBAA	Southern Balancing Authority Area
SBO	station blackout
SC	steel concrete composite
SCBA	self-contained breathing apparatus
SCDPRT	South Carolina Department of Parks, Recreation and Tourism
SCE&G	South Carolina Electric and Gas Company
SCEMD	South Carolina Emergency Management Division
S-COL	subsequent combined license
SCP	Safeguards Contingency Plan
SCSN	South Carolina State Network
SCT	Southern Company Transmission
SE	safety evaluation
SECY	Secretary of the Commission, Office of the Nuclear Regulatory Commission
SER	safety evaluation report

SFP	spent fuel pool
SFS	spent fuel pool cooling system
SG	steam generator
SGI	safeguards information
SGTR	steam generator tube rupture
SMA	seismic margin analysis
SNC	Southern Nuclear Operating Company
SNM	special nuclear material
SOT	station orientation training
SP	Setpoint Program
SPDS	safety parameter display system
SR	surveillance requirement
SREC	standard radiological effluent control
SRM	Staff Requirements Memorandum
SRO	senior reactor operator
SRP	standard review plan
SRSS	square root sum of squares
SSAR	Site Safety Analysis Report
SSCs	structures, systems, and components
SSE	safe shutdown earthquake
SSEP	safety, security and/or emergency preparedness
SSI	soil structure interaction
SS-ITAAC	site-specific inspections, tests, analyses and acceptance criteria
STAC	short-term availability control
STD	Standard
STS	Standard Technical Specification
SUNSI	sensitive unclassified non-safeguard information
SUP	Supplement
Sv	Sievert
SWMS	solid waste management system
SWS	service water system
T&QP	Training and Qualification Plan
TCS	turbine building closed cooling water system
TDMA	Time Division Multiple Access
TEDE	total effective dose equivalent
TG	turbine-generator
TGS	turbine generator system
TLD	thermoluminescent dosimeter
TMI	Three Mile Island
TNT	trinitrotoluene
TR	technical report
TRS	test response spectrum
TS	Technical Specification
TSC	Technical Support Center
TSO	transmission system operator
TSTF	Technical Specification Task Force
TVA	Tennessee Valley Authority
UAT	unit auxiliary transformer
UBC	Uniform Building Code

UFL	upper flammability limit
UFM	ultrasonic flow meter
UHS	ultimate heat sink
UPS	uninterruptible power supply
USACE	U.S. Army Corps of Engineers
USE	upper shelf energy
USGS	United States Geological Society
UT	ultrasonic
V	volt
V&V	verification and validation
VAR	Variance
VBS	nuclear island non-radioactive ventilation system
Vdc	volts direct current
VEGP	Vogtle Electric Generating Plant
VES	main control room emergency habitability system
VFS	containment air filtration system
VHRA	very high radiation area
VOIP	Voice Over Internet Protocol
VPN	Virtual Private Network
WCAP	Westinghouse Commercial Atomic Power
WCS	Waste Control Specialist
WEC	Westinghouse Electric Company
WLS	liquid radioactive waste system
WLS	liquid radwaste system
WWRB	waste water retention basin
WWS	waste water system
YFS	yard fire water system
ZRS	offsite retail power system

20.0 CONCLUSIONS

In accordance with Subpart C, "Combined License," of Title 10 of the Code of Federal Regulations (10 CFR), Part 52, "Licenses, Certifications, and Approvals for Nuclear Power Plants," and 10 CFR Section 50.10, "License required; limited work authorization," the staff of the U.S. Nuclear Regulatory Commission reviewed the combined license (COL) application and associated limited work authorization (LWA) request submitted by Southern Nuclear Operating Company (SNC), for the Vogtle Electric Generating Plant (VEGP) Units 3 and 4. Based on the staff's evaluation documented in this final safety evaluation report (FSER), the staff finds the following with respect to the safety aspects[5] of the COL application:

1) The applicable standards and requirements of the Atomic Energy Act and the Commission's regulations have been met,

2) Required notifications to other agencies or bodies have been duly made,

3) There is reasonable assurance that the facility will be constructed and will operate in conformity with the license, the provisions of the Atomic Energy Act, and the Commission's regulations,

4) The applicant is technically and financially qualified to engage in the activities authorized, and,

5) Issuance of the license will not be inimical to the common defense and security or to the health and safety of the public.

In addition, with respect to the LWA request, the staff also concludes that the applicable standards and the requirements of the Act, and the Commission's regulations applicable to the activities to be conducted under the LWA, have been met. The staff finds that the applicant is technically qualified to engage in the LWA activities authorized, and that issuance of the LWA will provide reasonable assurance of adequate protection to public health and safety and will also not be inimical to the common defense and security. Finally, the staff concludes there are no unresolved safety issues relating to the activities to be conducted under the LWA that would constitute good cause for withholding the authorization.

---

[5] An environmental review of the COL application was also performed, and the staff's evaluation and conclusions are documented in NUREG-1947, "Final Supplemental Environmental Impact Statement for Combined Licenses (COLs) for Vogtle Electric Generating Plant Units 3 and 4," dated March 2011.

APPENDIX A. POST COMBINED LICENSE ACTIVITIES -- LICENSE CONDITIONS, INSPECTIONS, TESTS, ANALYSES, AND ACCEPTANCE CRITERIA, AND FINAL SAFETY ANALYSIS REPORT COMMITMENTS

A.1 License Conditions

The U.S. Nuclear Regulatory Commission's (NRC's) regulations at Title 10 of the *Code of Federal Regulations* (10 CFR) 52.97, "Issuance of combined licenses," requires a combined license (COL) to specify any terms and conditions of the COL the Commission deems appropriate. A license condition is not needed when an existing NRC regulation requires a future regulatory review of a matter to ensure adequate safety during design, construction, inspection activities or operation for a new plant. The staff is proposing that the Commission include the following license conditions, which are set forth below, to control various safety matters. This list also includes applicable early site permit conditions. Therefore, this appendix includes COL and ESP conditions, recognizing that a COL be issued to the applicant, the ESP will be subsumed into the COL.

Proposed License Condition	SER Section	Description
1-1	1.5.5	Subject to the conditions and requirements incorporated herein, the Commission hereby licenses Southern Nuclear Company (SNC): (a) (i) Pursuant to the Act and 10 CFR Part 70, "Domestic licensing of special nuclear material," to receive and possess at any time, special nuclear material as reactor fuel, in accordance with the limitations for storage and amounts required for reactor operation, described in the final safety analysis report (FSAR), as supplemented and amended; (ii) Pursuant to the Act and 10 CFR Part 70, to use special nuclear material as reactor fuel, after a Commission finding under 10 CFR 52.103(g) has been made, in accordance with the limitations for storage and amounts required for reactor operation, and described in the FSAR, as supplemented and amended;

Proposed License Condition	SER Section	Description
		(b) (i) Pursuant to the Act and 10 CFR Parts 30, and 70, to receive, possess, and use, at any time, before a Commission finding under 10 CFR 52.103(g), such byproduct, and special nuclear material as: sealed neutron sources for reactor startup; sealed sources for reactor instrumentation and radiation monitoring equipment, calibration; and fission detectors in amounts as required; (ii) Pursuant to the Act and 10 CFR Parts 30, 40, and 70, to receive, possess, and use, after a Commission finding under 10 CFR 52.103(g), any byproduct, source, and special nuclear material as sealed neutron sources for reactor startup, sealed sources for reactor instrumentation and radiation monitoring equipment, calibration, and as fission detectors in amounts as required;
		(c) (i) Pursuant to the Act and 10 CFR Parts 30, and 70, to receive, possess, and use, before a Commission finding under 10 CFR 52.103 (g), in amounts not exceeding those specified in 10 CFR 30.72, any byproduct, or special nuclear material that is (1) in unsealed form; (2) on foils or plated surfaces, or (3) sealed in glass, for sample analysis or instrument calibration or other activities associated with radioactive apparatus or components; (ii) Pursuant to the Act and 10 CFR Parts 30, 40, and 70, to receive, possess, and use, after a Commission finding under 10 CFR 52.103(g), in amounts as required, any byproduct, source, or special nuclear material without restriction as to chemical or physical form, for sample analysis or instrument calibration or other activity associated with radioactive apparatus or components, but not uranium hexafluoride; and
		(d) Pursuant to the Act and 10 CFR Parts 30 and 70, to possess, but not separate, such byproduct and special nuclear materials as may be produced by the operation of the facility.

Proposed License Condition	SER Section	Description
1-2	1.5.5	Prior to initial receipt of special nuclear materials (SNM) onsite, the licensee shall implement the SNM Material Control and Accounting program. No later than 12 months after issuance of the COL, the licensee shall submit to the Director of Office of New Reactors (NRO) a schedule that supports planning for and conduct of NRC inspections of the SNM Material Control and Accounting program. The schedule shall be updated every 6 months until 12 months before scheduled fuel loading, and every month thereafter until the SNM Material Control and Accounting program has been fully implemented.
1-3	1.5.5	No later than 12 months after issuance of the COL, the licensee shall submit to the Director of NRO a schedule that supports planning for and conduct of NRC inspection of the non-licensed plant staff training program. The schedule shall be updated every 6 months until 12 months before scheduled fuel loading, and every month thereafter until the non-licensed plant staff training program has been fully implemented.
1-4	1.5.5	Prior to initial receipt of SNM on site, the licensee shall implement the SNM physical protection program. No later than 12 months after issuance of the COL, the licensee shall submit to the Director of NRO a schedule that supports planning for and conduct of NRC inspection of the SNM physical protection program. The schedule shall be updated every 6 months until 12 months before scheduled fuel loading, and every month thereafter until the SNM physical protection program has been fully implemented.
1-5	1.5.5	The licensee shall not revise or modify the provisions of Sections 5.3, 5.4, 5.6, 5.9 and 5.10 of the Special Nuclear Material (SNM) Physical Protection Plan until the requirements of 10 CFR 73.55 are implemented.
2-1	2.5.4.5	The licensee shall either remove and replace, or shall improve, the soils directly above the bluff marl for soils under or adjacent to Seismic Category I structures, to eliminate any liquefaction potential.
3-1	3.6.5	Prior to installation of piping and connected components in their final location, the licensee shall complete the as-designed pipe rupture hazards analysis in accordance with the criteria outlined in the AP1000 Design Control Document (DCD), Sections 3.6.1.3.2 and 3.6.2.5.
3-2	3.7.2.5	Prior to initial fuel load, the licensee shall update the seismic interaction review in the AP1000 DCD Section 3.7.3.5 for as-built information. This review must be performed in parallel with the seismic margin evaluation. The review shall be based on as-procured data, as well as the as-constructed condition.

Proposed License Condition	SER Section	Description
3-3	3.7.2.5	Prior to initial fuel load, the licensee shall reconcile the seismic analyses described in Section 3.7.2 of the AP1000 DCD for detailed design changes, such as those due to as-procured or as-built changes in component mass, center of gravity, and support configuration based on as-procured equipment information. The acceptability of deviations must be based on an evaluation consistent with the methods and procedures in Section 3.7 of the AP1000 DCD provided that the amplitude of the seismic floor response spectra (FRS), including the effects due to these deviations, does not exceed the design basis FRS by more than 10 percent.
3-4	3.8.5.5	No later than 12 months after issuance of the COL, the licensee shall submit to the Director of Office of New Reactor (NRO) a schedule that supports planning for and conduct of NRC inspections of the implementation of construction and inspection procedures for steel concrete composite (SC) construction activities for seismic Category I nuclear island modules (including shield building SC) before and after concrete placement, and inspection of such construction before and after concrete placement. The schedule shall be updated every 6 months until 12 months before scheduled fuel loading, and every month thereafter until the procedures have been fully implemented.
3-5	3.9.6.5	Prior to initial fuel load, the licensee shall implement the preservice testing and the motor-operated valve (MOV) testing programs.
3-6	3.9.6.5	No later than 12 months after issuance of the COL, the licensee shall submit to the Director of NRO a schedule that supports planning for and conduct of NRC inspections of the inservice testing program (including preservice and MOV testing). The schedule shall be updated every 6 months until 12 months before scheduled fuel loading, and every month thereafter until the inservice testing program (including preservice and MOV testing) has been fully implemented.
3-7	3.11.5	Prior to initial fuel load, the licensee shall implement the Environmental Qualification Program.
3-8	3.11.5	No later than 12 months after issuance of the COL, the licensee shall submit to the Director of NRO a schedule that supports planning for and conduct of NRC inspections of the Environmental Qualification Program. The schedule shall be updated every 6 months until 12 months before scheduled fuel loading, and every month thereafter until the Environmental Qualification Program has been fully implemented.

Proposed License Condition	SER Section	Description
3-9	3.12.5	Prior to installation of the piping and connected components in their final location, the licensee shall complete the as-designed piping analysis for the piping lines chosen to demonstrate all aspects of the piping design as identified in FSAR Section 3.9.8 and shall inform the Director of NRO of the availability of the piping design information and design reports for the piping packages.
4-1	4.5	Prior to initial fuel load, the licensee shall calculate the instrumentation uncertainties of the actual plant operating instrumentation to confirm that either the design limit departure from nucleate boiling ratio (DNBR) values remain valid or that the safety analysis minimum DNBR bounds the new design limit DNBR values plus DNBR penalties, such as rod bow penalty.
5-1	5.2.4.5	No later than 12 months after issuance of the COL, the licensee shall submit to the Director of the Office of New Reactors (NRO) a schedule that supports planning for and conduct of NRC inspections of the preservice inspection (PSI)/inservice inspection (ISI) programs (including the augmented ISI program). The schedule shall be updated every 6 months until 12 months before scheduled fuel loading, and every month thereafter until either the PSI/ISI programs (including the augmented ISI program) have been fully implemented or the plant has been placed in commercial service, whichever comes first.
5-2	5.3.2.5	The licensee shall implement the Reactor Vessel (RV) Material Surveillance program prior to initial criticality.
5-3	5.3.2.5	No later than 12 months after issuance of the COL, the licensee shall submit to the Director of NRO a schedule that supports planning for and conduct of NRC inspections of the RV Material Surveillance program. The schedule shall be updated every 6 months until 12 months before scheduled fuel loading, and every month thereafter until the RV Material Surveillance program has been fully implemented.
5-4	5.3.3.5	Prior to initial fuel load, the licensee shall update the pressure temperature (P-T) limits using the pressure temperature limits report (PTLR) methodologies approved in the AP1000 DCD using the plant-specific material properties or confirm that the RV material properties meet the specifications and use the Westinghouse generic PTLR curves.
5-5	5.3.4.5	Prior to initial fuel load, the licensee shall complete verification of plant-specific belt line material properties consistent with the requirements in FSAR Section 5.3.3.1 and FSAR Tables 5.3-1 and 5.3-3. The verification shall include a pressurized thermal shock (PTS) evaluation based on as-procured RV material data and the projected neutron fluence for the plant design objective of 60 years. This evaluation report shall be submitted for an NRC confirmatory review at least 18 months prior to initial fuel load.

Proposed License Condition	SER Section	Description
5-6	5.4.5	No later than 12 months after issuance of the COL, the licensee shall submit to the Director of NRO a schedule that supports planning for and conduct of NRC inspections of the steam generator (SG) PSI/ISI program. The schedule shall be updated every 6 months until 12 months before scheduled fuel loading, and every month thereafter until either the SG PSI/ISI program has been fully implemented or the plant has been placed in commercial service, whichever comes first.
6-1	6.2.5	The licensee shall implement the containment leakage rate testing program prior to initial fuel load.
6-2	6.2.5	No later than 12 months after issuance of the COL, the licensee shall submit to the Director of the Office of New Reactors (NRO) a schedule that supports planning for and conduct of NRC inspections of the containment leakage rate testing program. The schedule shall be updated every 6 months until 12 months before scheduled fuel loading, and every month thereafter until the containment leakage rate testing program has been fully implemented.
6-3	6.6.5	No later than 12 months after issuance of the COL, the licensee shall submit to the Director of the NRO a schedule that supports planning for and conduct of NRC inspections of the PSI and ISI programs. The schedule shall be updated every 6 months until 12 months before scheduled fuel loading, and every month thereafter until either the PSI and ISI programs have been fully implemented or the plant has been placed in commercial service, whichever comes first.
9-1	9.1.2.5	Prior to initial fuel load, the licensee shall implement the spent fuel rack Metamic Coupon Monitoring Program. No later than 12 months after issuance of the COL, the licensee shall submit to the Director of the Office of New Reactors (NRO) a schedule that supports planning for and conduct of NRC inspections of the spent fuel rack Metamic Coupon Monitoring Program. The schedule shall be updated every 6 months until 12 months before scheduled fuel loading, and every month thereafter until the spent fuel rack Metamic Coupon Monitoring Program has been fully implemented.

Proposed License Condition	SER Section	Description
9-2	9.5.1.5	The licensee shall implement the Fire Protection (FP) Program or portions of the FP Program identified below on or before the associated milestones identified below: 1. Applicable portions of the FP Program – prior to initial receipt of byproduct, source, or special nuclear materials onsite (excluding Exempt Quantities as described in 10 CFR 30.18). 2. Applicable portions of the FP Program – prior to initial receipt of fuel onsite. 3. FP Program – prior to initial fuel load.
9-3	9.5.1.5	No later than 12 months after issuance of the COL, the licensee shall submit to the Director of NRO a schedule that supports planning for and conduct of NRC inspections of the FP Program. The schedule shall be updated every 6 months until 12 months before scheduled fuel loading, and every month thereafter until the FP Program has been fully implemented.
10-1	10.1.5	Prior to initial fuel load, the licensee shall implement the flow accelerated corrosion (FAC) program including construction phase activities. No later than 12 months after issuance of the COL, the licensee shall submit to the Director of the Office of New Reactors (NRO) a schedule that supports planning for and conduct of NRC inspections of the FAC program implementation including construction phase activities. The schedule shall be updated every 6 months until 12 months before scheduled fuel loading, and every month thereafter until the FAC program has been fully implemented.
10-2	10.2.5	Prior to initial fuel load, the licensee shall implement a turbine maintenance and inspection program, which will be consistent with the maintenance and inspection program plan activities and inspection intervals identified in FSAR Section 10.2.3.6. No later than 12 months after issuance of the COL, the licensee shall submit to the Director of NRO a schedule that supports planning for and conduct of NRC inspections of the turbine maintenance and inspection program. The schedule shall be updated every 6 months until 12 months before scheduled fuel loading, and every month thereafter until the turbine maintenance and inspection program has been fully implemented.
11-1	11.4.5	Prior to initial fuel load, the licensee shall implement an operational program for process and effluent monitoring and sampling. The program shall include the subprogram and documents for a Process Control Program.

Proposed License Condition	SER Section	Description
11-2	11.4.5	No later than 12 months after issuance of the COL, the licensee shall submit to the Director of the Office of New Reactors (NRO) a schedule that supports planning for and conduct of NRC inspections of the operational program for process and effluent monitoring and sampling (including process control program). The schedule shall be updated every 6 months until 12 months before scheduled fuel loading, and every month thereafter until the operational program for process and effluent monitoring and sampling (including process control program) has been fully implemented.
11-3	11.5.5	Prior to initial fuel load, the licensee shall implement an operational program for process and effluent monitoring and sampling. The program shall include the following subprograms and documents: a. Radiological Effluent Technical Specifications/Standard Radiological Effluent Controls b. Offsite Dose Calculation Manual c. Radiological Environmental Monitoring Program
11-4	11.5.5	No later than 12 months after issuance of the COL, the licensee shall submit to the Director of NRO a schedule that supports planning for and conduct of NRC inspections of the operational program for process and effluent monitoring and sampling (including Radiological Effluent Technical Specifications/Standard Radiological Effluent Controls, Offsite Dose Calculation Manual, and Radiological Environmental Monitoring Program). The schedule shall be updated every 6 months until 12 months before scheduled fuel loading, and every month thereafter until the above operational program has been fully implemented.
12-1	12.5.5	The licensee shall implement the Radiation Protection Program (RPP) including the as low as is reasonably achievable (ALARA) principle (or applicable portions thereof) on or before the associated milestones identified below: – Receipt of Materials – Prior to initial receipt of byproduct, source, or special nuclear materials onsite (excluding exempt quantities as described in 10 CFR 30.18, "Exempt quantities) – Fuel Receipt – Prior to initial receipt of fuel onsite – Fuel Loading – Prior to initial fuel load – Waste Shipment – Prior to initial radioactive waste shipment

Proposed License Condition	SER Section	Description
12-2	12.5.5	No later than 12 months after issuance of the COL, the licensee shall submit to the Director of the Office of New Reactors a schedule that supports planning for and conduct of NRC inspections of the operational program (RPP). The schedule shall be updated every 6 months until 12 months before scheduled fuel loading, and every month thereafter until this operational program has been fully implemented.
13-1	13.2.5	The licensee shall implement the Reactor Operator Training Program at least 18 months prior to the scheduled date of initial fuel load.
13-2	13.2.5	No later than 12 months after issuance of the COL, the licensee shall submit to the Director of the Office of New Reactors (NRO) a schedule that supports planning for and conduct of NRC inspections of the operational programs (the Non-Licensed Plant Staff Training Program (required in accordance with 10 CFR 50.120), Reactor Operator Training Program, and Reactor Operator Requalification Program). The schedule shall be updated every 6 months until 12 months before scheduled fuel loading, and every month thereafter until these operational programs have been fully implemented.
13-3	13.3.5	The licensee shall submit a fully developed set of plant-specific emergency action levels (EALs) for VEGP Units 3 and 4 in accordance with Nuclear Energy Institute (NEI) 07-01, "Methodology for Development of Emergency Action Levels Advanced Passive Light Water Reactors," Revision 0, with no deviations. The EALs shall have been discussed and agreed upon with State and local officials. These fully developed EALs shall be submitted to the NRC for confirmation at least 180 days prior to initial fuel load.
13-4	13.3.5	No later than 12 months after issuance of the COL, the licensee shall submit to the Director of NRO a schedule that supports planning for and conduct of NRC inspections of the emergency planning (EP) program implementation. The schedule shall be updated every 6 months until 12 months before scheduled fuel loading, and every month thereafter until the EP operational program has been fully implemented.
13-5	13.6.5	No later than 12 months after issuance of the COL, the licensee shall submit to the Director of NRO a schedule that supports planning for and conduct of NRC inspections of the physical security programs. The schedule shall be updated every 6 months until 12 months before scheduled fuel loading, and every month thereafter until the physical security program has been fully implemented.

Proposed License Condition	SER Section	Description
13-6	13.7.5	No later than 12 months after issuance of the COL, the licensee shall submit to the Director of NRO a schedule that supports planning for and conduct of NRC inspections of the fitness for duty (FFD) operational program. The schedule shall be updated every 6 months until 12 months before scheduled fuel loading, and every month thereafter until the FFD operational program has been fully implemented.
13-7	13.8.5	No later than 12 months after issuance of the COL, the licensee shall submit to the Director of NRO a schedule that supports planning for and conduct of NRC inspections of the cyber security program implementation. The schedule shall be updated every 6 months until 12 months before scheduled fuel loading, and every month thereafter until the cyber security program has been fully implemented.
14-1	14.2.3.5	No later than 12 months after issuance of the COL, the licensee shall submit to the Director of the Office of New Reactors (NRO) a schedule that supports planning for and conduct of NRC inspections of the approved preoperational and startup procedures (including the site-specific startup administration manual.) The schedule shall be updated every 6 months until the approved preoperational and startup procedures have been implemented. Prior to initiating the initial test program, the approved preoperational and startup procedures (including the site-specific startup administration manual) shall be available.
14-2	14.2.3.5	Within one month of a change, any changes to the Initial Startup Test Program described in Chapter 14 of the VEGP COL FSAR made in accordance with the provisions of 10 CFR 50.59, "Changes, tests and experiments," or Section VIII, "Processes for Changes and Departures of Appendix D, "Design Certification Rule for the AP1000 Design," to 10 CFR Part 52, "Licenses, certifications, and approvals for nuclear power plants," shall be reported in accordance with 10 CFR 50.59(d).
14-3	14.2.5.5	<u>First-Plant-Only and First-Three-Plant-Only Testing</u> The licensee shall notify the Director of the NRO, in writing, when it determines that it has completed the design-specific testing identified below and confirmed that the test results are within the range of acceptable values predicted or otherwise confirm that the tested systems perform their specific functions in accordance with the FSAR: a. The licensee shall perform "first plant only" tests. b. The licensee shall perform "first three plants" tests.

Proposed License Condition	SER Section	Description
14-4	14.2.8.5	The licensee shall implement the initial test program (applicable portions) on or before the associated milestones identified below: 1. Construction Testing - Prior to initial construction testing 2. Preoperational Testing - Prior to initial preoperational testing 3. Startup Testing - Prior to initial fuel load
14-5	14.2.8.5	No later than 12 months after issuance of the COL, the licensee shall submit to the Director of NRO a schedule that supports planning for and conduct of NRC inspections of the operational program (initial test program). The schedule shall be updated every 6 months until 12 months before scheduled fuel loading, and every month thereafter until this operational program (ITP) has been fully implemented.
14-6	14.2.8.5	Pre-operational Testing Following completion of pre-operational testing, the licensee shall review and evaluate individual test results and confirm the test results are within the range of acceptable values predicted or otherwise confirm that the tested systems perform their specific functions in accordance with the FSAR. Pre-critical and Criticality Testing 1. Following completion of pre-critical and criticality testing, the licensee shall review and evaluate individual test results and confirm the test results are within the range of acceptable values predicted or otherwise confirm that the tested systems perform their specific functions in accordance with the FSAR. 2. The licensee shall provide written notification to the Director of the NRO upon completion of pre-critical and criticality testing. Upon submission of this notification, the licensee is authorized to perform low-power testing as described in the FSAR and operate the facility at reactor steady-state core power levels, not in excess of 170 megawatts thermal (5-percent power), in accordance with the conditions specified herein. Low-Power (<5% Rated Thermal Power) Testing 1. Following completion of low-power testing (<5% RTP), the licensee shall review and evaluate individual test results and confirm that the test results are within the range of acceptable values predicted or otherwise confirm that the

Proposed License Condition	SER Section	Description
		tested systems perform their specific functions in accordance with the FSAR. 2. The licensee shall provide written notification to the Director of the NRO upon completion of low power testing. Upon submission of this notification, the licensee is authorized to perform power ascension testing as described in the FSAR and operate the facility at reactor steady-state core power levels, not in excess of 3400 megawatts thermal (100 percent power), in accordance with the conditions specified herein. At-Power (5%-100% RTP) Testing 1. Following completion of at-power testing (at or above 5% RTP up to and including testing at 100% RTP), the licensee shall review and evaluate individual test results and confirm that the results are within the range of acceptable values predicted or otherwise confirm that the tested systems perform their specific functions in accordance with the FSAR. 2. The licensee shall provide written notification to the Director of NRO upon completion of the at-power testing.
15-1	15.0.5	No later than 12 months after issuance of the COL, the licensee shall submit to the Director of the Office of New Reactors a schedule that supports planning for and conduct of NRC inspections of license calculations for power calorimetric uncertainty and administrative controls to implement maintenance and contingency activities related to the power calorimetric uncertainty instrumentation. The schedule shall be updated every 6 months until 12 months before scheduled fuel loading, and every month thereafter until the license condition has been fully implemented. This schedule shall address: • The availability of documented instrumentation uncertainties to calculate a power calorimetric uncertainty (prior to initial fuel load). • The availability of administrative controls to implement maintenance and contingency activities related to the power calorimetric uncertainty instrumentation (prior to initial fuel load).

Proposed License Condition	SER Section	Description
17-1	17.6.5	No later than 12 months after issuance of the COL, the licensee shall submit to the Director of the Office of New Reactors a schedule that supports planning for and conduct of NRC inspections of the Maintenance Rule (MR) program. The schedule shall be updated every 6 months until 12 months before scheduled fuel loading, and every month thereafter until the MR program has been fully implemented.
19-1	19.59.5	The licensee shall review differences between the as-built plant and the design used as the basis for the AP1000 seismic margin analysis prior to initial fuel load. The licensee shall perform a verification walkdown to identify differences between the as-built plant and the design. The licensee shall evaluate any differences and shall modify the seismic margin analysis as necessary to account for the plant-specific design and any design changes or departures from the certified design. The licensee shall compare the as-built structure, system, and component (SSC) high confidence, low probability of failures (HCLPFs) to those assumed in the AP1000 seismic margin evaluation prior to initial fuel load. The licensee shall evaluate deviations from the HCLPF values or assumptions in the seismic margin evaluation due to the as-built configuration and final analysis to determine if vulnerabilities have been introduced.
19-2	19.59.5	The licensee shall review differences between the as-built plant and the design used as the basis for the AP1000 probabilistic risk assessment (PRA) and Table 19.59-18 prior to initial fuel load. The plant-specific PRA-based insight differences shall be evaluated and the plant-specific PRA model modified as necessary to account for the plant-specific design and any design changes or departure from the certification PRA.
19-3	19.59.5	The licensee shall review differences between the as-built plant and the design used as the basis for the AP1000 internal fire and internal flood analysis prior to initial fuel load. The licensee shall evaluate the plant-specific internal fire and internal flood analyses and shall modify the analyses as necessary to account for the plant-specific design and any design changes or departures from the certified design.
19-4	19.59.5	Prior to startup testing, the license shall implement the site-specific severe accident management guidelines. No later than 12 months after issuance of the COL, the licensee shall submit to the Director of the Office of New Reactors (NRO) a schedule that supports planning for and conduct of NRC inspections of the implementation of site-specific severe accident management guidelines. The schedule shall be updated every 6 months until 12 months before scheduled fuel loading, and every month thereafter until the site-specific severe accident management guidelines have been fully implemented.

Proposed License Condition	SER Section	Description
19-5	19.59.5	Prior to initial fuel load, the licensee shall perform a thermal lag assessment of the as-built equipment listed in Tables 6b and 6c in Attachment A of APP-GW-GLR-069, "Equipment Survivability Assessment," to provide additional assurance that this equipment can perform its severe accident functions during environmental conditions resulting from hydrogen burns associated with severe accidents. This assessment is required only for equipment used for severe accident mitigation that has not been tested at severe accident conditions. The license shall assess the ability of the as-built equipment to perform during accident hydrogen burns using the environment enveloping method or the test based thermal analysis method described in Electric Power Research Institute (EPRI) NP-4354, "Large Scale Hydrogen Burn Equipment Experiments."
19.A-1	19.A.5	Prior to initial fuel load, the licensee shall implement the operational and programmatic elements of its mitigative strategies for responding to a LOLA event developed in accordance with 10 CFR 50.54(hh)(2). No later than 12 months after issuance of the COL, the licensee shall submit to the Director of the Office of New Reactors a schedule that supports planning for and conduct of NRC inspection of the operational and programmatic elements of responding to an event associated with a loss of large areas of the plant due to explosions or fires. The schedule shall be updated every 6 months until 12 months before scheduled fuel loading, and every month thereafter until these operational and programmatic elements have been fully implemented. The licensee shall maintain the guidance and strategies developed in accordance with10 CFR 50.54(hh)(2).

A.2   Inspections, Tests, Analyses, and Acceptance Criteria

The staff has identified the certain ITAAC that it will recommend the Commission impose with respect to a COL issued to the applicant. The list also includes the applicable ESP ITAAC from the ESP. The following is a list of those ITAAC. In addition to the ITAAC contained in this list, the ITAAC found in the AP1000 DCD, Revision 19 Tier 1 material will also be incorporated into the COL should a COL be issued to the applicant.

1.   The licensee shall perform and satisfy the backfill ITAAC defined in SER Table 2.5-1, "Backfill ITAAC."

Table 2.5-1 Backfill ITAAC

Design Requirement	Inspections, Tests, Analyses	Acceptance Criteria
Backfill material under Seismic Category 1 structures is installed to meet a minimum of 95 percent modified Proctor compaction.	Required testing will be performed during placement of the backfill materials.	A report exists that documents that the backfill material under Seismic Category 1structures meets the minimum 95 percent modified Proctor compaction
Backfill shear wave velocity is greater than or equal to 1,000 fps at the depth of the NI foundation and below.	Field shear wave velocity measurements will be performed when backfill placement is at the elevation of the bottom of the Nuclear Island foundation and at finish grade.	A report exists and documents that the as built backfill shear wave velocity at the NI foundation depth and below is greater than or equal to 1,000 fps.

2. The licensee shall perform and satisfy the pipe rupture hazards analysis ITAAC defined in SER Table 3.6-1, "Pipe Rupture Hazards Analysis ITAAC."

Table 3.6-1. Pipe Rupture Hazards Analysis ITAAC

Design Commitment	Inspections, Tests, Analyses	Acceptance Criteria
Systems, structures, and components (SSCs) that are required to be functional during and following a design basis event shall be protected against or qualified to withstand the dynamic and environmental effects associated with analyses of postulated failures in high and moderate energy piping.	Inspection of the as-designed pipe rupture hazard analysis report will be conducted. The report documents the analyses to determine where protection features are necessary to mitigate the consequence of a pipe break. Pipe break events involving high-energy fluid systems are analyzed for the effects of pipe whip, jet impingement, flooding, room pressurization, and temperature effects. Pipe break events involving moderate-energy fluid systems are analyzed for wetting from spray, flooding, and other environmental effects, as appropriate.	An as-designed pipe rupture hazard analysis report exists and concludes that the analysis performed for high and moderate energy piping confirms the protection of SSCs required to be functional during and following a design basis event.

3. The licensee shall perform and satisfy the waterproof membrane ITAAC defined in SER Table 3.8-1, "Waterproof Membrane ITAAC."

Table 3.8-1. Waterproof Membrane ITAAC

Design Commitment	Inspections, Tests, Analyses	Acceptance Criteria
The friction coefficient to resist sliding is 0.7 or higher.	Testing will be performed to confirm that the mudmat-waterproofing-mudmat interface beneath the nuclear island basemat has a minimum coefficient of friction to resist sliding of 0.7.	A report exists and documents that the as-built waterproof system (mudmat-waterproofing-mudmat interface) has a minimum coefficient of friction of 0.7 as demonstrated through material qualification testing.

4. The licensee shall perform and satisfy the piping design analysis ITAAC in SER Table 3.12-1, "Piping Design ITAAC."

Table 3.12-1. Piping Design ITAAC

Design Commitment	Inspections, Tests, Analyses	Acceptance Criteria
The American Society of Mechanical Engineers (ASME) Code, Section III piping is designed in accordance with the ASME Code, Section III requirements.	Inspection of the ASME Code Design Reports (NCA-3550) and required documents will be conducted for the set of lines chosen to demonstrate compliance.	The ASME Code Design Report(s) (NCA-3550) (certified, when required by the ASME Code) exist and conclude that the design of the piping for lines chosen to demonstrate all aspects of the piping design complies with the requirements of the ASME Code section.

5. The licensee shall perform and satisfy the ITAAC defined in Table 8.2A-1, "Offsite Power System."

Table 8.2A-1. Offsite Power System

Design Commitment	Inspections, Tests, and Analyses	Acceptance Criteria
1. A minimum of one offsite circuit supplies electric power from the transmission network to the interface with the onsite alternating current (ac) power system.	Inspections of the as-built offsite circuit will be performed.	At least one offsite circuit is provided from the transmission switchyard interface to the interface with the onsite ac power system.
2. Each offsite power circuit interfacing with the onsite ac power system is adequately rated to supply assumed loads during normal, abnormal and accident conditions.	Analyses of the offsite power system will be performed to evaluate the as-built ratings of each offsite circuit interfacing with the onsite ac power system against the load assumptions.	A report exists and concludes that each as-built offsite circuit is rated to supply the load assumptions during normal, abnormal and accident conditions.
3. During steady state operation, each offsite power source is capable of supplying required voltage to the interface with the onsite ac power system that will support operation of assumed loads during normal, abnormal and accident conditions.	Analyses of the as-built offsite circuit will be performed to evaluate the capability of each offsite circuit to supply the voltage requirements at the interface with the onsite ac power system.	A report exists and concludes that during steady state operation each as-built offsite circuit is capable of supplying the voltage at the interface with the onsite ac power system that will support operation of assumed loads during normal, abnormal and accident conditions.

Table 8.2A-1. Offsite Power System

Design Commitment	Inspections, Tests, and Analyses	Acceptance Criteria
4. During steady state operation, each offsite circuit is capable of supplying required frequency to the interface with the onsite ac power system that will support operation of assumed loads during normal, abnormal and accident conditions.	Analyses of the as-built offsite circuit will be performed to evaluate the capability of each offsite circuit to supply the frequency requirements at the interface with the onsite ac power system.	A report exists and concludes that during steady state operation each as-built offsite circuit is capable of supplying the frequency at the interface with onsite ac power system that will support operation of assumed loads during normal, abnormal and accident conditions.
5. The fault current contribution of each offsite circuit is compatible with the interrupting capability of the onsite short circuit interrupting devices.	Analyses of the as-built offsite circuit will be performed to evaluate the fault current contribution of each offsite circuit at the interface with the onsite ac power system.	A report exists and concludes the short circuit contribution of each as-built offsite circuit at the interface with the onsite ac power system is compatible with the interrupting capability of the onsite fault current interrupting devices.
6. The reactor coolant pumps continue to receive power from either the main generator or the grid for a minimum of 3 seconds following a turbine trip.	Analyses of the as-built offsite power system will be performed to confirm that power will be available to the reactor coolant pumps for a minimum of 3 seconds following a turbine trip when the buses powering the reactor coolant pumps are aligned to either the unit auxiliary transformers (UATs) or the reserve auxiliary transformers (RATs).	A report exists and concludes that voltage at the high-side of the generator stepup transformer (GSU), and the RATs, does not drop more than 0.15 per unit (pu) from the pre-trip steady-state voltage for a minimum of 3 seconds following a turbine trip when the buses powering the reactor coolant pumps are aligned to either the UATs or the RATs.

6. The licensee shall perform, and satisfy the acceptance criteria of the EP ITAAC set forth in SER Tables 13.3-1 and 13.3-2.

Table 13.3-1. VEGP Unit 3 ITAAC

Planning Standard	EP Program Elements (From NUREG-0654/FEMA-REP-1)	Inspections, Tests, Analyses	Acceptance Criteria
1.0 Emergency Classification System			
10 CFR 50.47(b)(4) – A standard emergency classification and action level scheme, the bases of which include facility system and effluent parameters, is in use by the nuclear facility licensee, and State and local plans call for reliance on information provided by facility licensees for determinations of minimum initial offsite response measures.	1.1 An emergency classification and emergency action level (EAL) scheme must be established by the licensee. The specific instruments, parameters, or equipment status shall be shown for establishing each emergency class, in the in-plant emergency procedures. The plan shall identify the parameter values and equipment status for each emergency class. [D.1]	1.1.1 An inspection of the control room, technical support center (TSC), and emergency operations facility (EOF) will be performed to verify that the displays for retrieving system and effluent parameters specified in Table Annex V2 D.2-1, "Hot Initiating Condition Matrix, Modes 1, 2, 3, and 4"; Table V2 D.2-2, "Cold Initiating Condition Matrix, Modes 5, 6, and De-fueled", are installed and perform their intended functions; and that emergency implementing procedures (EIPs) have been completed. 1.1.2 An analysis of the EAL technical bases will be performed to verify as-built, site-specific implementation of the EAL scheme.	1.1.1 The parameters specified in Table Annex V2 H-1, "Post Accident Monitoring Variables," are retrievable in the control room, TSC, and EOF. The ranges of values of these parameters that can be displayed encompass the values specified in the emergency classification and EAL scheme. 1.1.2 The EAL scheme is consistent with Regulatory Guide (RG) 1.101, "Emergency Planning and Preparedness for Nuclear Power Reactors."

Table 13.3-1. VEGP Unit 3 ITAAC

Planning Standard	EP Program Elements (From NUREG-0654/FEMA-REP-1)	Inspections, Tests, Analyses	Acceptance Criteria
3.0 Emergency Communications			
10 CFR 50.47(b)(6) – Provisions exist for prompt communications among principal response organizations to emergency personnel and to the public.	3.1 The means exists for communications between the control room, operations support center (OSC), TSC, EOF, principal State and local emergency operations centers (EOCs), and radiological field monitoring teams. [F.1.d]	3.1 A test will be performed of the communications capabilities between the control room, OSC, TSC and EOF, and to the State and local EOCs, and radiological field monitoring teams.	3.1 Communications are established between the control room, OSC, TSC, and EOF. Communications are established between the control room, TSC, and the Georgia Emergency Management Agency (GEMA) Operation Center; Burke County EOC; SRS Operations Center; South Carolina Warning Point; and Aiken, Allendale, and Barnwell County Dispatchers. Communications are established between the TSC and radiological monitoring teams.
	3.2 The means exists for communications from the control room, TSC, and EOF to the NRC headquarters and regional office EOC, including establishment of the Emergency Response Data System (ERDS) between the onsite computer system and the NRC Operations Center. [F.1.f]	3.2 A test will be performed of the communications capabilities from the control room, TSC and EOF to the NRC, including ERDS.	3.2 Communications are established from the control room, TSC, and EOF to the NRC headquarters and regional office EOCs, and an access port for the ERDS is provided.
5.0 Emergency Facilities and Equipment			
10 CFR 50.47(b)(8) – Adequate emergency facilities and equipment to support the emergency response are provided and maintained.	5.1 The licensee has established a technical support center (TSC) and an onsite operations support center (OSC). [H.1]	5.1 An inspection of the as-built TSC and OSC will be performed, including a test of the capabilities.	5.1.1 The TSC has at least 2,175 square feet of floor space.

Table 13.3-1. VEGP Unit 3 ITAAC

Planning Standard	EP Program Elements (From NUREG-0654/FEMA-REP-1)	Inspections, Tests, Analyses	Acceptance Criteria
			5.1.2 Communication equipment is installed in the TSC and OSC, and voice transmission and reception are accomplished.
			5.1.3 The plant parameters listed in Table Annex V2 H-1, "Post Accident Monitoring Values," can be retrieved and displayed in the TSC.
			5.1.4 The TSC is located within the protected area, and no major security barriers exist between the TSC and the control room.
			5.1.5 The OSC is located adjacent to the passage from the annex building to the control room.
			5.1.6 The TSC ventilation system includes a high-efficiency particulate air (HEPA) and charcoal filter, and radiation monitors are installed.
			5.1.7 A reliable and backup electrical power supply is available for the TSC.
			5.1.8 Controls and displays exist in the TSC to control and monitor the status of the TSC ventilation system including heating and cooling, and the activation of the HEPA and charcoal filter system upon detection of high radiation in the TSC.

Table 13.3-1. VEGP Unit 3 ITAAC

Planning Standard	EP Program Elements (From NUREG-0654/FEMA-REP-1)	Inspections, Tests, Analyses	Acceptance Criteria
	5.2 The licensee has established an EOF. [H.2]	5.2 An inspection of the EOF will be performed, including a test of the capabilities.	5.2.1 Voice transmission and reception are accomplished between the EOF and the control room. 5.2.2 The plant parameters listed in Table Annex V2 H-1, "Post Accident Monitoring Values," can be retrieved and displayed in the EOF.
6.0 Accident Assessment			
10 CFR 50.47(b)(9) – Adequate methods, systems, and equipment for assessing and monitoring actual or potential offsite consequences of a radiological emergency condition are in use.	6.1 The means exists to provide initial and continuing radiological assessment throughout the course of an accident. [I.2]	6.1 A test of the emergency plan will be conducted by performing a drill to verify the capability to perform accident assessment.	6.1 Using selected monitoring parameters listed in Table Annex V2 H-1 of the VEGP emergency plan, simulated degraded plant conditions are assessed and protective actions are initiated in accordance with the following criteria: A. *Accident Assessment and Classification* 1. Demonstrate the ability to identify initiating conditions, determine EAL parameters, and correctly classify the emergency throughout the drill.

Table 13.3-1. VEGP Unit 3 ITAAC

Planning Standard	EP Program Elements (From NUREG-0654/FEMA-REP-1)	Inspections, Tests, Analyses	Acceptance Criteria
			B. *Radiological Assessment and Control*
			1. Demonstrate the ability to obtain onsite radiological surveys and samples.
			2. Demonstrate the ability to continuously monitor and control radiation exposure to emergency workers.
			3. Demonstrate the ability to assemble and deploy field monitoring teams within 60 minutes from the decision to do so.
			4. Demonstrate the ability to satisfactorily collect and disseminate field team data.
			5. Demonstrate the ability to develop dose projections.
			6. Demonstrate the ability to make the decision whether to issue radio-protective drugs (KI) to emergency workers.
			7. Demonstrate the ability to develop appropriate protective action recommendations (PARs) and notify appropriate authorities within 15 minutes of development.

Table 13.3-1.   VEGP Unit 3 ITAAC

Planning Standard	EP Program Elements (From NUREG-0654/FEMA-REP-1)	Inspections, Tests, Analyses	Acceptance Criteria
	6.2 The means exists to determine the source term of releases of radioactive material within plant systems, and the magnitude of the release of radioactive materials based on plant system parameters and effluent monitors. [I.3]	6.2 An analysis of the EIPs and the Offsite Dose Calculation Manual (ODCM) will be completed to verify ability to determine the source term and magnitude of releases.	6.2 The EIPs and ODCM correctly calculate source terms and magnitudes of postulated releases.
	6.3 The means exists to continuously assess the impact of the release of radioactive materials to the environment, accounting for the relationship between effluent monitor readings, and onsite and offsite exposures and contamination for various meteorological conditions. [I.4]	6.3 An analysis of the EIPs and the ODCM will be completed to verify the relationship between effluent monitor readings, and onsite and offsite exposures and contamination.	6.3 The EIPs and ODCM calculate the relationship between effluent monitor readings, and onsite and offsite exposures and contamination.
	6.4 The means exists to acquire and evaluate meteorological information. [I.5]	6.4 A test will be performed to verify the ability to access meteorological information in the TSC and control room.	6.4 The following parameters are displayed in the TSC and control room: • Wind speed (at 10 and 60 meters) • Wind direction (at 10 and 60 meters) • Standard deviation of horizontal wind direction (at 10 meters) • Vertical temperature difference (between 10 and 60 meters) • Ambient temperature (at 10 meters) • Precipitation (at the tower base)

Table 13.3-1. VEGP Unit 3 ITAAC

Planning Standard	EP Program Elements (From NUREG-0654/FEMA-REP-1)	Inspections, Tests, Analyses	Acceptance Criteria
	6.5 The means exists to make rapid assessments of actual or potential magnitude and locations of any radiological hazards through liquid or gaseous release pathways, including activation, notification means, field team composition, transportation, communication, monitoring equipment, and estimated deployment times. [I.8]	6.5 A test will be performed of the capabilities to make rapid assessment of actual or potential radiological hazards through liquid or gaseous release pathways.	6.5 Demonstrate the capability to make rapid assessment of actual or potential magnitude and locations of any radiological hazards through liquid or gaseous release pathways.
	6.6 The means exists to estimate integrated dose from the projected and actual dose rates, and for comparing these estimates with the Environmental Protection Agency's (EPA's) protective action guides (PAGs). [I.10]	6.6 An analysis of the methodology contained in the EIPs for estimating dose and preparing PARs, and in the ODCM will be performed to verify the ability to estimate an integrated dose from projected and actual dose rates.	6.6 The EIPs and ODCM estimate an integrated dose.

Table 13.3-1. VEGP Unit 3 ITAAC

Planning Standard	EP Program Elements (From NUREG-0654/FEMA-REP-1)	Inspections, Tests, Analyses	Acceptance Criteria
7.0 Protective Response			
10 CFR 50.47(b)(10) – A range of protective actions has been developed for the plume exposure pathway emergency planning zone (EPZ) for emergency workers and the public. In developing this range of actions, consideration has been given to evacuation, sheltering, and, as a supplement to these, the prophylactic use of potassium iodide (KI), as appropriate. Guidelines for the choice of protective actions during an emergency, consistent with Federal guidance, are developed and in place, and protective actions for the ingestion exposure pathway EPZ appropriate to the locale have been developed.	7.1 The means exists to warn and advise onsite individuals of an emergency, including those in areas controlled by the operator, including: • Employees not having emergency assignments • Visitors • Contractor and construction personnel • Other persons who may be in the public access areas, on or passing through the site, or within the owner controlled area [J.1]	7.1 A test of the onsite warning and communication capability EIPs including protective action guidelines, assembly and accountability, and site dismissal will be performed during a drill.	7.1.1 Demonstrate the capability to direct and control emergency operations. 7.1.2 Demonstrate the ability to transfer emergency direction from the control room (simulator) to the technical support center (TSC) within 30 minutes from activation. 7.1.3 Demonstrate the ability to prepare for around-the-clock staffing requirements. 7.1.4 Demonstrate the ability to perform assembly and accountability for all onsite individuals within 30 minutes of an emergency requiring protected area assembly and accountability. 7.1.5 Demonstrate the ability to perform site dismissal.
8.0 Exercises and Drills			
10 CFR 50.47(b)(14) – Periodic exercises are (will be) conducted to evaluate major portions of emergency response capabilities, periodic drills are (will be) conducted to develop and maintain key skills, and deficiencies identified as a result of exercises or drills are (will be) corrected.	8.1 The licensee conducts a full participation exercise to evaluate major portions of emergency response capabilities, which includes participation by each State and local agency within the plume exposure emergency planning zone (EPZ), and each State within the ingestion pathway EPZ. [N.1]	8.1 A full participation exercise (test) will be conducted within the specified time periods of 10 CFR Part 50, Appendix E.	8.1.1 The exercise is completed within the specified time periods of Appendix E to 10 CFR Part 50, onsite exercise objectives listed below have have been met and there are no uncorrected onsite exercise deficiencies. A. *Accident Assessment and Classification*

Table 13.3-1. VEGP Unit 3 ITAAC

Planning Standard	EP Program Elements (From NUREG-0654/FEMA-REP-1)	Inspections, Tests, Analyses	Acceptance Criteria
			1. Demonstrate the ability to identify initiating conditions, determine EAL parameters, and correctly classify the emergency throughout the exercise
			Standard Criteria:
			a. Determine the correct highest emergency classification level based on events which were in progress, considering past events and their impact on the current conditions, within 15 minutes from the time the initiating condition(s) or EAL is identified.
			B. *Notifications*
			1. Demonstrate the ability to alert, notify, and mobilize site emergency response personnel.
			Standard Criteria:
			a. Complete the designated checklist and perform the announcement within 5 minutes of the initial event classification for an Alert or higher.
			b. Activate the emergency recall system within 5 minutes of the initial event classification for an Alert or higher.

Table 13.3-1. VEGP Unit 3 ITAAC

Planning Standard	EP Program Elements (From NUREG-0654/FEMA-REP-1)	Inspections, Tests, Analyses	Acceptance Criteria
			2. Demonstrate the ability to notify responsible State and local government agencies within 15 minutes and the NRC within 60 minutes after declaring an emergency.

Standard Criteria:

a. Transmit information using the designated checklist, in accordance with approved EIPs, within 15 minutes of event classification.

b. Transmit information using the designated checklist, in accordance with approved EIPs, within 60 minutes of last transmittal for a follow-up notification to State and local authorities.

c. Transmit information using the designated checklist within 60 minutes of event classification for an initial notification of the NRC.

3. Demonstrate the ability to warn or advise onsite individuals of emergency conditions.

Standard Criteria:

a. Initiate notification of onsite individuals (via plant page or telephone), using the designated checklist within 15 minutes of notification. |

Table 13.3-1. VEGP Unit 3 ITAAC

Planning Standard	EP Program Elements (From NUREG-0654/FEMA-REP-1)	Inspections, Tests, Analyses	Acceptance Criteria
			4. Demonstrate the capability of the Prompt Notification System (PNS), for the public, to operate properly when required. Standard Criteria: a. Ninety percent of the sirens operate properly, as indicated by the Whelen feedback system. b. A National Oceanic and Atmospheric Administration (NOAA) tone alert radio is activated. C. *Emergency Response* 1. Demonstrate the capability to direct and control emergency operations. Standard Criteria: a. Command and control is demonstrated by the control room in the early phase of the emergency and the TSC within 60 minutes from TSC activation. 2. Demonstrate the ability to transfer emergency direction from the control room (simulator) to the TSC within 30 minutes from activation. Standard Criteria:

Table 13.3-1. VEGP Unit 3 ITAAC

Planning Standard	EP Program Elements (From NUREG-0654/FEMA-REP-1)	Inspections, Tests, Analyses	Acceptance Criteria
			a. Briefings were conducted prior to turnover responsibility. Personnel document transfer of duties.
			3. Demonstrate the ability to prepare for around-the-clock staffing requirements.
			Standard Criteria:
			a. Complete 24-hour staff assignments.
			4. Demonstrate the ability to perform assembly and accountability for all onsite individuals within 30 minutes of an emergency requiring protected area assembly and accountability.
			Standard Criteria:
			a. Protected area personnel assembly and accountability completed within 30 minutes of the Alert or higher emergency declaration via public address announcement.
			D. *Emergency Response Facilities*
			1. Demonstrate activation of the OSC, and full functional operation of the TSC and EOF within 60 minutes of activation.
			Standard Criteria:

Table 13.3-1. VEGP Unit 3 ITAAC

Planning Standard	EP Program Elements (From NUREG-0654/FEMA-REP-1)	Inspections, Tests, Analyses	Acceptance Criteria
			a. The TSC, OSC, and EOF are activated within about 60 minutes of the initial notification.
			2. Demonstrate the adequacy of equipment, security provisions, and habitability precautions for the TSC, OSC, EOF, and emergency news center (ENC), as appropriate.
			Standard Criteria:
			a. Demonstrate the adequacy of the emergency equipment in the emergency response facilities, including availability and general consistency with EIPs.
			b. The Security Shift Captain implements and follows applicable EIPs.
			c. The Health Physics Supervisor (TSC) implements the designated checklist if an onsite or offsite release has occurred.
			d. Demonstrate the capability of TSC and EOF equipment and data displays to clearly identify and reflect the affected unit.
			3. Demonstrate the adequacy of communications for all emergency support resources.

Table 13.3-1. VEGP Unit 3 ITAAC

Planning Standard	EP Program Elements (From NUREG-0654/FEMA-REP-1)	Inspections, Tests, Analyses	Acceptance Criteria
			Standard Criteria:
			a. Emergency response communications listed in EIPs are available and operational.
			b. Communications systems are tested in accordance with TSC, OSC, and EOF activation checklists.
			c. Emergency response facility personnel are able to operate all specified communication systems.
			d. Clear primary and backup communications links are established and maintained for the duration of the exercise.
			E. *Radiological Assessment and Control*
			1. Demonstrate the ability to obtain onsite radiological surveys and samples.
			Standard Criteria:
			a. Health Physics Technicians demonstrate the ability to obtain appropriate instruments (range and type) and take surveys.
			b. Airborne samples are taken when the conditions indicate the need for the information.

Table 13.3-1. VEGP Unit 3 ITAAC

Planning Standard	EP Program Elements (From NUREG-0654/FEMA-REP-1)	Inspections, Tests, Analyses	Acceptance Criteria
			2. Demonstrate the ability to continuously monitor and control radiation exposure to emergency workers. Standard Criteria: a. Emergency workers are issued self-reading dosimeters when radiation levels require, and exposures are controlled to 10 CFR Part 20 limits (unless the Emergency Director authorizes emergency limits). b. Exposure records are available, either from the as low as is reasonably achievable (ALARA) computer or a hard copy dose report. c. Emergency workers include Security and personnel within all emergency facilities. 3. Demonstrate the ability to assemble and deploy field monitoring teams within 60 minutes from the decision to do so. Standard Criteria: a. One field monitoring team is ready to be deployed within 60 minutes of being requested from the OSC, and no later than 90 minutes from the declaration of

Table 13.3-1. VEGP Unit 3 ITAAC

Planning Standard	EP Program Elements (From NUREG-0654/FEMA-REP-1)	Inspections, Tests, Analyses	Acceptance Criteria
			an Alert or higher emergency. 4. Demonstrate the ability to satisfactorily collect and disseminate field team data. Standard Criteria: a. Field team data to be collected is dose rate or counts per minute (cpm) from the plume, both open and closed window, and air sample (gross/net cpm) for particulate and iodine, if applicable. b. Satisfactory data dissemination is from the field team to the Dose Assessment Supervisor, via the field team communicator and field team coordinator. 5. Demonstrate the ability to develop dose projections. Standard Criteria: a. The on-shift Health Physics/Chemistry Shared Foreman or Dose Assessment Supervisor performs timely and accurate dose projections, in accordance with EIPs. 6. Demonstrate the ability to make the decision whether to issue radioprotective drugs (KI) to emergency workers.

Table 13.3-1. VEGP Unit 3 ITAAC

Planning Standard	EP Program Elements (From NUREG-0654/FEMA-REP-1)	Inspections, Tests, Analyses	Acceptance Criteria
			Standard Criteria:
			a. KI is taken (simulated) if the estimated dose to the thyroid will exceed 25 rem committed dose equivalent (CDE).
			7. Demonstrate the ability to develop appropriate PARs and notify appropriate authorities within 15 minutes of development.
			Standard Criteria:
			a. Total effective dose equivalent (TEDE) and CDE dose projections from the dose assessment computer code are compared to EIPs.
			b. PARs are developed within 15 minutes of data availability.
			c. PARs are transmitted to responsible State and local government agencies via voice or fax within 15 minutes of PAR development.
			F. *Public Information*
			1. Demonstrate the capability to develop and disseminate clear, accurate, and timely information to the news media, in accordance with EIPs.

Table 13.3-1. VEGP Unit 3 ITAAC

Planning Standard	EP Program Elements (From NUREG-0654/FEMA-REP-1)	Inspections, Tests, Analyses	Acceptance Criteria
			Standard Criteria:
			a. Media information (e.g., press releases, press briefings, electronic media) is made available within 60 minutes of notification of the on-call media representative.
			b. Follow-up information is provided, at a minimum, within 60 minutes of an emergency classification or PAR change.
			2. Demonstrate the capability to establish and effectively operate rumor control in a coordinated fashion.
			Standard Criteria:
			a. Calls are answered in a timely manner with the correct information, in accordance with EIPs.
			b. Calls are returned or forwarded, as appropriate, to demonstrate responsiveness.
			c. Rumors are identified and addressed.
			G. *Evaluation*
			1. Demonstrate the ability to conduct a post-exercise critique, to determine areas requiring improvement and corrective action.

Table 13.3-1. VEGP Unit 3 ITAAC

Planning Standard	EP Program Elements (From NUREG-0654/FEMA-REP-1)	Inspections, Tests, Analyses	Acceptance Criteria
			Standard Criteria:
			a. An exercise time line is developed, followed by an evaluation of the objectives.
			b. Significant problems in achieving the objectives are discussed to ensure understanding of why objectives were not fully achieved.
			c. Recommendations for improvement in non-objective areas are discussed.
			8.1.2 Onsite emergency response personnel are mobilized in sufficient number to fill the emergency positions identified in emergency plan Section B, "VEGP Emergency Organization," and they successfully perform their assigned responsibilities as outlined in Acceptance Criterion 8.1.1.D, "Emergency Response Facilities."
			8.1.3 The exercise is completed within the specified time periods of Appendix E to 10 CFR Part 50, offsite exercise objectives have been met, and there are no uncorrected offsite deficiencies, or a license condition requires offsite deficiencies to be corrected prior to operation above 5 percent of rated power.

Table 13.3-1. VEGP Unit 3 ITAAC

Planning Standard	EP Program Elements (From NUREG-0654/FEMA-REP-1)	Inspections, Tests, Analyses	Acceptance Criteria
9.0 Implementing Procedures			
10 CFR Part 50, Appendix E, Section V – No less than 180 days prior to the scheduled issuance of an operating license for a nuclear power reactor or a license to possess nuclear material, the applicant's detailed implementing procedures for its emergency plan shall be submitted to the Commission.	9.1 The licensee has submitted detailed implementing procedures for its emergency plan no less than 180 days prior to fuel load.	9.1 An inspection of the submittal letter will be performed.	9.1 The licensee has submitted detailed EIPs for the onsite emergency plan no less than 180 days prior to fuel load.

Table 13.3-2. VEGP Unit 4 ITAAC

Planning Standard	EP Program Elements (From NUREG-0654/FEMA-REP-1)	Inspections, Tests, Analyses	Acceptance Criteria
1.0 Emergency Classification System			
10 CFR 50.47(b)(4) – A standard emergency classification and action level scheme, the bases of which include facility system and effluent parameters, is in use by the nuclear facility licensee, and State and local plans call for reliance on information provided by facility licensees for determinations of minimum initial offsite response measures.	1.1 An emergency classification and EAL scheme must be established by the licensee. The specific instruments, parameters, or equipment status shall be shown for establishing each emergency class, in the in-plant emergency procedures. The plan shall identify the parameter values and equipment status for each emergency class. [D.1]	1.1.1 An inspection of the control room will be performed to verify that the displays for retrieving system and effluent parameters specified in Table Annex V2 D.2-1, "Hot Initiating Condition Matrix, Modes 1, 2, 3, and 4"; Table V2 D.2-2, "Cold Initiating Condition Matrix, Modes 5, 6, and De-fueled"; are installed and perform their intended functions; and that EIPs have been completed.	1.1.1 The parameters specified in Table Annex V2 H-1, "Post Accident Monitoring Variables," are retrievable in the control room. The ranges of values of these parameters that can be displayed encompass the values specified in the emergency classification and EAL scheme.
		1.1.2 An analysis of the EAL technical bases will be performed to verify as-built, site-specific implementation of the EAL scheme.	1.1.2 The EAL scheme is consistent with RG 1.101, "Emergency Planning and Preparedness for Nuclear Power Reactors."
3.0 Emergency Communications			
10 CFR 50.47(b)(6) – Provisions exist for prompt communications among principal response organizations to emergency personnel and to the public.	3.1 The means exists for communications between the control room, OSC, TSC, and EOF. [F.1.d]	3.1 A test will be performed of the communications capabilities between the control room, OSC, TSC and EOF, and to the State and local EOCs.	3.1 Communications are established between the control room, OSC, TSC, and EOF. Communications are established between the control room, Georgia Emergency Management Agency (GEMA) Operation Center; Burke County EOC; SRS Operations Center; South Carolina Warning Point; and Aiken, Allendale, and Barnwell County Dispatchers.

Table 13.3-2. VEGP Unit 4 ITAAC

Planning Standard	EP Program Elements (From NUREG-0654/FEMA-REP-1)	Inspections, Tests, Analyses	Acceptance Criteria
	3.2 The means exists for communications from the control room to the NRC headquarters and regional office EOC. [F.1.f]	3.2 A test will be performed of the communications capabilities from the control room, TSC and EOF to the NRC, including ERDS.	3.2 Communications are established from the control room, TSC, and EOF, to the NRC headquarters and regional office EOCs, and an access port for the ERDS is provided.
5.0 Emergency Facilities and Equipment			
10 CFR 50.47(b)(8) – Adequate emergency facilities and equipment to support the emergency response are provided and maintained.	5.1 The licensee has established an onsite OSC. [H.1]	5.1 An inspection of the as-built OSC will be performed, including a test of the capabilities.	5.1.1 Communication equipment is installed in the OSC, and voice transmission and reception are accomplished. 5.1.2 The plant parameters listed in Table Annex V2 H-1, "Post Accident Monitoring Values," can be retrieved and displayed in the TSC. 5.1.3 The OSC is located adjacent to the passage from the annex building to the control room.
	5.2 The licensee has established an EOF. [H.2]	5.2 An inspection of the EOF will be performed, including a test of the capabilities.	5.2.1 Voice transmission and reception are accomplished between the EOF and the control room. 5.2.2 The plant parameters listed in Table Annex V2 H-1, "Post Accident Monitoring Values," can be retrieved and displayed in the EOF.

Table 13.3-2. VEGP Unit 4 ITAAC

Planning Standard	EP Program Elements (From NUREG-0654/FEMA-REP-1)	Inspections, Tests, Analyses	Acceptance Criteria
6.0 Accident Assessment			
10 CFR 50.47(b)(9) – Adequate methods, systems, and equipment for assessing and monitoring actual or potential offsite consequences of a radiological emergency condition are in use.	6.1 The means exists to provide initial and continuing radiological assessment throughout the course of an accident. [I.2]	6.1 A test of the emergency plan will be conducted by performing a drill to verify the capability to perform accident assessment.	6.1 Using selected monitoring parameters listed in Table Annex V2 H-1 of the VEGP emergency plan, simulated degraded plant conditions are assessed and protective actions are initiated in accordance with the following criteria: A. *Accident Assessment and Classification* 1. Demonstrate the ability to identify initiating conditions, determine EAL parameters, and correctly classify the emergency throughout the drill. B. *Radiological Assessment and Control* 1. Demonstrate the ability to obtain onsite radiological surveys and samples. 2. Demonstrate the ability to continuously monitor and control radiation exposure to emergency workers. 3. Demonstrate the ability to assemble and deploy field monitoring teams within 60 minutes from the decision to do so. 4. Demonstrate the ability to satisfactorily collect and

Table 13.3-2. VEGP Unit 4 ITAAC

Planning Standard	EP Program Elements (From NUREG-0654/FEMA-REP-1)	Inspections, Tests, Analyses	Acceptance Criteria
			disseminate field team data. 5. Demonstrate the ability to develop dose projections. 6. Demonstrate the ability to make the decision whether to issue radio-protective drugs (KI) to emergency workers. 7. Demonstrate the ability to develop appropriate PARs and notify appropriate authorities within 15 minutes of development.
	6.2 The means exists to determine the source term of releases of radioactive material within plant systems, and the magnitude of the release of radioactive materials based on plant system parameters and effluent monitors. [I.3]	6.2 An analysis of the EIPs and the ODCM will be completed to verify ability to determine the source term and magnitude of releases.	6.2 The EIPs and ODCM correctly calculate source terms and magnitudes of postulated releases.
	6.3 The means exists to continuously assess the impact of the release of radioactive materials to the environment, accounting for the relationship between effluent monitor readings, and onsite and offsite exposures and contamination for various meteorological conditions. [I.4]	6.3 An analysis of the EIPs and the ODCM will be completed to verify the relationship between effluent monitor readings, and onsite and offsite exposures and contamination.	6.3 The EIPs and ODCM calculate the relationship between effluent monitor readings, and onsite and offsite exposures and contamination.
	6.4 The means exists to acquire and evaluate meteorological information. [I.5]	6.4 A test will be performed to verify the ability to access meteorological information in the TSC and control room.	6.4 The following parameters are displayed in the TSC and control room: • Wind speed (at 10 and 60 meters)

Table 13.3-2. VEGP Unit 4 ITAAC

Planning Standard	EP Program Elements (From NUREG-0654/FEMA-REP-1)	Inspections, Tests, Analyses	Acceptance Criteria
			• Wind direction (at 10 and 60 meters)
			• Standard deviation of horizontal wind direction (at 10 meters)
			• Vertical temperature difference (between 10 and 60 meters)
			• Ambient temperature (at 10 meters)
			• Dew-point temperature (at 10 meters)
			• Precipitation (at the tower base)
	6.5 The means exists to make rapid assessments of actual or potential magnitude and locations of any radiological hazards through liquid or gaseous release pathways, including activation, notification means, field team composition, transportation, communication, monitoring equipment, and estimated deployment times. [I.8]	6.5 A test will be performed of the capabilities to make rapid assessments of actual or potential radiological hazards through liquid or gaseous release pathways.	6.5 Demonstrate the capability to make rapid assessment of actual or potential magnitude and locations of any radiological hazards through liquid or gaseous release pathways.
	6.6 The means exists to estimate integrated dose from the projected and actual dose rates, and for comparing these estimates with the EPA PAGs. [I.10]	6.6 An analysis of the methodology contained in the EIPs for estimating dose and preparing PARs, and in the ODCM will be performed to verify the ability to estimate an integrated dose from projected and actual dose rates.	6.6 The EIPs and ODCM estimate an integrated dose.

Table 13.3-2. VEGP Unit 4 ITAAC

Planning Standard	EP Program Elements (From NUREG-0654/FEMA-REP-1)	Inspections, Tests, Analyses	Acceptance Criteria
7.0 Protective Response			
10 CFR 50.47(b)(10) – A range of protective actions has been developed for the plume exposure pathway EPZ for emergency workers and the public. In developing this range of actions, consideration has been given to evacuation, sheltering, and, as a supplement to these, the prophylactic use of potassium iodide (KI), as appropriate. Guidelines for the choice of protective actions during an emergency, consistent with Federal guidance, are developed and in place, and protective actions for the ingestion exposure pathway EPZ appropriate to the locale have been developed.	7.1 The means exists to warn and advise onsite individuals of an emergency, including those in areas controlled by the operator, including: • Employees not having emergency assignments • Visitors • Contractor and construction personnel • Other persons who may be in the public access areas, on or passing through the site, or within the owner controlled area [J.1]	7.1 A test of the onsite warning and communication capability emergency implementing procedures (EIPs) including protective action guidelines, assembly and accountability, and site dismissal will be performed during a drill.	7.1.1 Demonstrate the capability to direct and control emergency operations. 7.1.2 Demonstrate the ability to transfer emergency direction from the control room (simulator) to the TSC within 30 minutes of activation. 7.1.3 Demonstrate the ability to prepare for around-the-clock staffing requirements. 7.1.4 Demonstrate the ability to perform assembly and accountability for all onsite individuals within 30 minutes of an emergency requiring protected area assembly and accountability. 7.1.5 Demonstrate the ability to perform site dismissal.
8.0 Exercises and Drills			
10 CFR 50.47(b)(14) – Periodic exercises are (will be) conducted to evaluate major portions of emergency response capabilities, periodic drills are (will be) conducted to develop and maintain key skills, and deficiencies identified as a result of exercises or drills are (will be) corrected.	8.1 The licensee conducts a limited participation exercise to evaluate portions of emergency response capabilities, which includes participation by each State and local agency within the plume exposure EPZ that have not been tested in a previous exercise. [N.1]	8.1 A limited participation exercise (test) will be conducted within the specified time periods of 10 CFR Part 50, Appendix E.	8.1.1 The exercise is completed within the specified time periods of Appendix E to 10 CFR Part 50, onsite exercise objectives listed below have been met and there are no uncorrected onsite exercise deficiencies. A. *Accident Assessment and Classification*

Table 13.3-2. VEGP Unit 4 ITAAC

Planning Standard	EP Program Elements (From NUREG-0654/FEMA-REP-1)	Inspections, Tests, Analyses	Acceptance Criteria
			1. Demonstrate the ability to identify initiating conditions, determine EAL parameters, and correctly classify the emergency throughout the exercise Standard Criteria: a. Determine the correct highest emergency classification level based on events, which were in progress, considering past events and their impact on the current conditions, within 15 minutes from the time the initiating condition(s) or EAL is identified. B. *Notifications* 1. Demonstrate the ability to alert, notify, and mobilize site emergency response personnel. Standard Criteria: a. Complete the designated checklist and perform the announcement within 5 minutes of the initial event classification for an Alert or higher. b. Activate the emergency recall system within 5 minutes of the initial event classification for an Alert or higher.

Table 13.3-2. VEGP Unit 4 ITAAC

Planning Standard	EP Program Elements (From NUREG-0654/FEMA-REP-1)	Inspections, Tests, Analyses	Acceptance Criteria
			2. Demonstrate the ability to notify responsible State and local government agencies within 15 minutes and the NRC within 60 minutes after declaring an emergency. Standard Criteria: a. Transmit information using the designated checklist, in accordance with approved EIPs, within 15 minutes of event classification. b. Transmit information using the designated checklist, in accordance with approved EIPs, within 60 minutes of last transmittal for a follow-up notification to State and local authorities. c. Transmit information using the designated checklist within 60 minutes of event classification for an initial notification of the NRC. 3. Demonstrate the ability to warn or advise onsite individuals of emergency conditions. Standard Criteria: a. Initiate notification of onsite individuals (via plant page or telephone) using the designated checklist, within 15 minutes of notification.

Table 13.3-2. VEGP Unit 4 ITAAC

Planning Standard	EP Program Elements (From NUREG-0654/FEMA-REP-1)	Inspections, Tests, Analyses	Acceptance Criteria
			C. *Emergency Response*
			1. Demonstrate the capability to direct and control emergency operations.
			Standard Criteria:
			a. Command and control is demonstrated by the control room in the early phase of the emergency and by the TSC within 60 minutes from activation.
			2. Demonstrate the ability to transfer emergency direction from the control room (simulator) to the TSC within 30 minutes from activation.
			Standard Criteria:
			a. Briefings were conducted prior to turnover responsibility. Personnel document transfer of duties.
			3. Demonstrate the ability to prepare for around-the-clock staffing requirements.
			Standard Criteria:
			a. Complete 24-hour staff assignments.

Table 13.3-2. VEGP Unit 4 ITAAC

Planning Standard	EP Program Elements (From NUREG-0654/FEMA-REP-1)	Inspections, Tests, Analyses	Acceptance Criteria
			4. Demonstrate the ability to perform assembly and accountability for all onsite individuals within 30 minutes of an emergency requiring protected area assembly and accountability.
			Standard Criteria:
			a. Protected area personnel assembly and accountability completed within 30 minutes of the Alert or higher emergency declaration via public address announcement.
			D. *Emergency Response Facilities*
			1. Demonstrate timely activation of the OSC.
			Standard Criteria:
			a. The OSC is activated within about 60 minutes of the initial notification.
			2. Demonstrate the adequacy of equipment, security provisions, and habitability precautions for the OSC, as appropriate.
			Standard Criteria:
			a. Demonstrate the adequacy of the emergency equipment in the emergency response facilities,

Table 13.3-2. VEGP Unit 4 ITAAC

Planning Standard	EP Program Elements (From NUREG-0654/FEMA-REP-1)	Inspections, Tests, Analyses	Acceptance Criteria
			including availability and general consistency with EIPs.
			b. The Security Shift Captain implements and follows applicable EIPs.
			c. The Health Physics Supervisor (TSC) implements the designated checklist if an onsite or offsite release has occurred.
			3. Demonstrate the adequacy of communications for all emergency support resources.
			Standard Criteria:
			a. Emergency response communications listed in EIPs are available and operational.
			b. Communications systems are tested in accordance with OSC activation checklist.
			c. Emergency response facility personnel are able to operate all specified communication systems.
			d. Clear primary and backup communications links are established and maintained for the duration of the exercise.
			E. *Radiological Assessment and Control*

Table 13.3-2. VEGP Unit 4 ITAAC

Planning Standard	EP Program Elements (From NUREG-0654/FEMA-REP-1)	Inspections, Tests, Analyses	Acceptance Criteria
			1. Demonstrate the ability to obtain onsite radiological surveys and samples. Standard Criteria: a. Health Physics Technicians demonstrate the ability to obtain appropriate instruments (range and type) and take surveys. b. Airborne samples are taken when the conditions indicate the need for the information. 2. Demonstrate the ability to continuously monitor and control radiation exposure to emergency workers. Standard Criteria: a. Emergency workers are issued self-reading dosimeters when radiation levels require, and exposures are controlled to 10 CFR Part 20 limits (unless the Emergency Director authorizes emergency limits). b. Exposure records are available, either from the ALARA computer or a hard copy dose report. c. Emergency workers include Security and personnel within all

Table 13.3-2. VEGP Unit 4 ITAAC

Planning Standard	EP Program Elements (From NUREG-0654/FEMA-REP-1)	Inspections, Tests, Analyses	Acceptance Criteria
			emergency facilities.
			3. Demonstrate the ability to assemble and deploy field monitoring teams within 60 minutes from the decision to do so.
			Standard Criteria:
			a. One field monitoring team is ready to be deployed within 60 minutes of being requested from the OSC, and no later than 90 minutes from the declaration of an Alert or higher emergency.
			4. Demonstrate the ability to satisfactorily collect and disseminate field team data.
			Standard Criteria:
			a. Field team data to be collected is dose rate or counts per minute (cpm) from the plume, both open and closed window, and air sample (gross/net cpm) for particulate and iodine, if applicable.
			b. Satisfactory data dissemination is from the field team to the Dose Assessment Supervisor, via the field team communicator and field team coordinator.
			5. Demonstrate the ability to develop dose projections.

Table 13.3-2. VEGP Unit 4 ITAAC

Planning Standard	EP Program Elements (From NUREG-0654/FEMA-REP-1)	Inspections, Tests, Analyses	Acceptance Criteria
			Standard Criteria:
			a. The on-shift Health Physics/Chemistry Shared Foreman or Dose Assessment Supervisor performs timely and accurate dose projections, in accordance with EIPs.
			6. Demonstrate the ability to develop appropriate PARs and notify appropriate authorities within 15 minutes of development.
			Standard Criteria:
			a. Total effective dose equivalent (TEDE) and CDE dose projections from the dose assessment computer code are compared to EIPs.
			b. PARs are developed within 15 minutes of data availability.
			c. PARs are transmitted to responsible State and local government agencies via voice or fax within 15 minutes of PAR development.
			8.1.2 Onsite emergency response personnel are mobilized in sufficient number to fill the emergency positions identified in the emergency plan, Section B, "VEGP

Table 13.3-2. VEGP Unit 4 ITAAC

Planning Standard	EP Program Elements (From NUREG-0654/FEMA-REP-1)	Inspections, Tests, Analyses	Acceptance Criteria
			Emergency Organization," and successfully perform their assigned responsibilities as outlined in Acceptance Criterion 8.1.1.D, "Emergency Response Facilities." 8.1.3 The exercise is completed within the specified time periods of Appendix E to 10 CFR Part 50, offsite exercise objectives have been met, and there are no uncorrected offsite deficiencies, or a license condition requires offsite deficiencies to be corrected prior to operation above 5 percent of rated power.
9.0 Implementing Procedures			
10 CFR Part 50, Appendix E, Section V – No less than 180 days prior to the scheduled issuance of an operating license for a nuclear power reactor or a license to possess nuclear material, the applicant's detailed implementing procedures for its emergency plans shall be submitted to the Commission.	9.1 The licensee has submitted detailed implementing procedures for its emergency plan no less than 180 days prior to fuel load.	9.1 An inspection of the submittal letter will be performed.	9.1 The licensee has submitted detailed EIPs for the onsite emergency plan no less than 180 days prior to fuel load.

7. The licensee shall perform and satisfy the ITAAC defined in Table 13.6A-1, "Site-Specific Physical Security Inspections, Tests, Analyses, and Acceptance Criteria."

Table 13.6A-1. Site-Specific Physical Security Inspections, Tests, Analyses and Acceptance Criteria

Design Commitment	Inspections, Tests, and Analyses	Acceptance Criteria
1. The external walls, doors, ceiling, and floors in the location within which the last access control function for access to the protected area is performed are bullet-resistant to at least Underwriters Laboratory Ballistic Standard 752, level 4.	Type test, analysis, or a combination of type test and analysis will be performed for the external walls, doors, ceilings, and floors in the location within which the last access control function for access to the protected area is performed.	The external walls, doors, ceilings, and floors in the location within which the last access control function for access to the protected area is performed are bullet-resistant to at least Underwriters Laboratory Ballistic Standard 752, level 4.
2. Physical barriers for the protected area perimeter are not part of vital area barriers.	An inspection of the protected area perimeter barrier will be performed.	Physical barriers at the perimeter of the protected area are separated from any other barrier designated as a vital area barrier.
3. a) Isolation zones exist in outdoor areas adjacent to the physical barrier at the perimeter of the protected area that allows 20 feet of observation on either side of the barrier. Where permanent buildings do not allow a 20-foot observation distance on the inside of the protected area, the building walls are immediately adjacent to, or an integral part of, the protected area barrier.	Inspections will be performed of the isolation zones in outdoor areas adjacent to the physical barrier at the perimeter of the protected area.	Isolation zones exist in outdoor areas adjacent to the physical barrier at the perimeter of the protected area and allow 20 feet of observation and assessment of the activities of people on either side of the barrier. Where permanent buildings do not allow a 20-foot observation and assessment distance on the inside of the protected area, the building walls are immediately adjacent to, or an integral part of, the protected area barrier and the 20-foot observation and assessment distance does not apply.
b) The isolation zones are monitored with intrusion detection	Inspections will be performed of the intrusion detection equipment within	The isolation zones are equipped with intrusion detection equipment that

Table 13.6A-1. Site-Specific Physical Security Inspections, Tests, Analyses and Acceptance Criteria

Design Commitment	Inspections, Tests, and Analyses	Acceptance Criteria
equipment that provides the capability to detect and assess unauthorized persons.	the isolation zones.	provides the capability to detect and assess unauthorized persons.
4. The intrusion detection and assessment equipment at the protected area perimeter: a) detects penetration or attempted penetration of the protected area barrier and concurrently alarms in both the Central Alarm Station and Secondary Alarm Station; and b) remains operable from an uninterruptible power supply in the event of the loss of normal power.	Tests, inspections or a combination of tests and inspections of the intrusion detection and assessment equipment at the protected area perimeter and its uninterruptible power supply will be performed.	The intrusion detection and assessment equipment at the protected area perimeter: a) detects penetration or attempted penetration of the protected area barrier and concurrently alarms in the Central Alarm Station and Secondary Alarm Station; and b) remains operable from an uninterruptible power supply in the event of the loss of normal power.
5. Access control points are established to: a) control personnel and vehicle access into the protected area. b) detect firearms, explosives, and incendiary devices at the protected area personnel access points.	Tests, inspections, or combination of tests and inspections of installed systems and equipment at the access control points to the protected area will be performed.	The access control points for the protected area: a) are configured to control personnel and vehicle access. b) include detection equipment that is capable of detecting firearms, incendiary devices, and explosives at the protected area personnel access points.

Table 13.6A-1. Site-Specific Physical Security Inspections, Tests, Analyses and Acceptance Criteria

Design Commitment	Inspections, Tests, and Analyses	Acceptance Criteria
6. An access control system with numbered picture badges is installed for use by individuals who are authorized access to protected areas and vital areas without escort.	A test of the access control system with numbered picture badges will be performed.	The access authorization system with numbered picture badges can identify and authorize protected area and vital area access only to those personnel with unescorted access authorization.
7. Access to vital equipment physical barriers requires passage through the protected area perimeter barrier.	Inspection will be performed to confirm that access to vital equipment physical barriers requires passage through the protected area perimeter barrier.	Vital equipment is located within a protected area such that access to vital equipment physical barriers requires passage through the protected area perimeter barrier.
8. a) Penetrations through the protected area barrier are secured and monitored. b) Unattended openings (such as underground pathways) that intersect the protected area boundary or vital area boundary will be protected by a physical barrier and monitored by intrusion detection equipment or provided surveillance at a frequency sufficient to detect exploitation.	Inspections will be performed of penetrations through the protected area barrier. Inspections will be performed of unattended openings that intersect the protected area boundary or vital area boundary.	Penetrations and openings through the protected area barrier are secured and monitored. Unattended openings (such as underground pathways) that intersect the protected area boundary or vital area boundary are protected by a physical barrier and monitored by intrusion detection equipment or provided surveillance at a frequency sufficient to detect exploitation.
9. Emergency exits through the protected area perimeter are alarmed and secured with locking devices to allow for emergency egress.	Tests, inspections, or a combination of tests and inspections of emergency exits through the protected area perimeter will be performed.	Emergency exits through the protected area perimeter are alarmed and secured by locking devices that allow prompt egress during an emergency.

8. The licensee shall perform and satisfy the plant calorimetric uncertainty and plant instrumentation performance analysis ITAAC defined in SER Table 15.0-1, "Power Calorimetric Uncertainty Methodology."

Table 15.0-1. Power Calorimetric Uncertainty Methodology

Design Commitment	Inspections, Tests, Analyses	Acceptance Criteria
4. The plant calorimetric uncertainty and plant instrumentation performance is bounded by the 1 percent calorimetric uncertainty value assumed for the initial reactor power in the safety analysis.	Inspection will be performed of the plant operating instrumentation installed for feedwater flow measurement, its associated power calorimetric uncertainty calculation, and the calculated calorimetric values.	a) the as-built system takes input for feedwater flow measurement from a Caldon [Cameron] LEFM CheckPlus™ System; b) the power calorimetric uncertainty calculation documented for that instrumentation is based on an NRC-accepted Westinghouse methodology and the uncertainty values for that instrumentation are not lower than those for the actual installed instrumentation; and c) the calculated calorimetric power uncertainty measure values are bounded by the 1 percent uncertainty value assumed for the initial reactor power in the safety analysis.

A.3   Final Safety Analysis Report (FSAR) Commitments

The following FSAR commitments are identified as the responsibility of the licensee:

SER Section	Description
1.4.5	A site-specific construction plan and startup schedule will be provided after issuance of the COL.
5.2.5.5	Prior to initial fuel load, the operating procedures, that include identifying, monitoring, trending, and managing the prolonged low-level RCS leakage, will be developed.
6.4.5	FSAR Commitment 6.4-1. The licensee's CR operator training program shall address the following: • Regulatory Position C.5, "Emergency Planning," of RG 1.78 • Regulatory Position 2.5, "Hazardous Chemicals," of RG 1.196 • Regulatory Position 2.2.1, "Comparison of System Design, Configuration, and Operation with Licensing Basis," of RG 1.196 • Regulatory Position 2.7.1, "Periodic Evaluations and Maintenance," of RG 1.196
9.1.4.5	The light load handling program, including system inspections, will be implemented prior to receipt of fuel onsite.
9.1.5.5	The overhead heavy-load handling program, including system inspections, will be implemented prior to receipt of fuel onsite.

Appendix B

Chronology of the Combined License Application for Vogtle Electric Generating Plant Units 3 and 4

This appendix contains a chronological listing of routine licensing correspondence between the staff of the U.S. Nuclear Regulatory Commission (NRC) regarding the review of the Vogtle Electric Generating Plant, Units 3 and 4 plant design under Docket Nos. 052-000025 and 052-000026

Document Date	Accession Number	Title & Estimated Page Count	Document Type	Author Affiliation	Addressee Affiliation	Docket Number
02/09/2007	ML070430110	Site Visit to Vogtle to Observe Combined License Pre-Application Subsurface Investigation Activities (Project No. 755.) 8 Page(s)	Letter	NRC/RGN-II/DCI/CIB1	Southern Nuclear Operating Co, Inc	05200011 05200025 05200026 PROJ0755
04/20/2007	ML071130348	Vogtle Combined License Application Vendors. 3 Page(s)	Letter	Southern Nuclear Operating Co, Inc	NRC/Document Control Desk NRC/NRO	05200011 05200025 05200026 PROJ0755
05/30/2007	ML071520072	Vogtle Early Site Permit Application, Response to Regulatory Issue Summary 2007-08. 11 Page(s)	Letter	Southern Nuclear Operating Co, Inc	NRC/Document Control Desk NRC/NRO	05200011 05200025 05200026 PROJ0755
01/11/2008	ML080110651	Site Visit to Vogtle to Observe Test Pad Activities to Support Early Site Permit, Limited Work Authorization and Combined License Pre-application Activities (Project No. 755). 9 Page(s)	Letter Trip Report	NRC/RGN-II/DCI/CIB1	Southern Nuclear Operating Co, Inc	05200025 05200026 PROJ0755
02/08/2008	ML080430082	Vogtle Combined License Application, Response to Regulatory Issue Summary 2008-01. 4 Page(s)	Letter	Southern Nuclear Operating Co, Inc	NRC/Document Control Desk NRC/NRO	05200025 05200026 PROJ0755

Appendix B

Chronology of the Combined License Application for Vogtle Electric Generating Plant Units 3 and 4

This appendix contains a chronological listing of routine licensing correspondence between the staff of the U.S. Nuclear Regulatory Commission (NRC) regarding the review of the Vogtle Electric Generating Plant, Units 3 and 4 plant design under Docket Nos. 052-000025 and 052-000026

Document Date	Accession Number	Title & Estimated Page Count	Document Type	Author Affiliation	Addressee Affiliation	Docket Number
03/26/2008	ML080860341	04/10/2008 - Notice of Meeting with Southern Nuclear to Discuss Vogtle, Unit 3 & 4, COL Application on Technical Content. 8 Page(s)	Meeting Agenda Meeting Notice Memoranda	NRC/NRO/DNRL/AP1000B 1	NRC/NRO/DNRL/AP1000B 1	05200025 05200026
03/28/2008	ML080910423	Vogtle, Units 3 and 4 - COL Application Physical Security Plan. 3 Page(s)	Letter	Southern Nuclear Operating Co, Inc	NRC/Document Control Desk NRC/NRO	05200011 05200025 05200026 PROJ0755
03/28/2008	ML080920632	Vogtle, Units 3 and 4 - Combined License Application Part 1, Appendix 1A - Estimated Total Construction Costs. 6 Page(s)	Letter	Southern Nuclear Operating Co, Inc	NRC/Document Control Desk NRC/NRO	05200025 05200026 PROJ0755
03/28/2008	ML080920633	Vogtle, Units 3 and 4, COL Application Part 1, Appendix 1A, Contract and Financial Information. 4 Page(s)	License-Application for Combined License (COLA)	Southern Nuclear Operating Co, Inc	NRC/NRO	05200025 05200026 PROJ0755
03/28/2008	ML081050133	Application for Combined Licenses for Vogtle Electric Generating Plant Units 3 and 4. 6 Page(s)	Letter License-Application for Combined License (COLA)	Southern Nuclear Operating Co, Inc	NRC/Document Control Desk NRC/NRO	05200025 05200026 PROJ0755
03/31/2008	ML091000566	Vogtle COL SER/OI Section 2.3.1: Snowfall Data from the National Climatic Data Centers's Snow Climatology Web Site. 21 Page(s)	Graphics incl Charts and Tables	NRC/NRO		05200025 05200026
04/03/2008	ML080930338	Request for Federal Emergency	Letter	NRC/NSIR/DPR/RIOB	US Dept of Homeland	05200025

Appendix B

Chronology of the Combined License Application for Vogtle Electric Generating Plant Units 3 and 4

This appendix contains a chronological listing of routine licensing correspondence between the staff of the U.S. Nuclear Regulatory Commission (NRC) regarding the review of the Vogtle Electric Generating Plant, Units 3 and 4 plant design under Docket Nos. 052-000025 and 052-000026

Document Date	Accession Number	Title & Estimated Page Count	Document Type	Author Affiliation	Addressee Affiliation	Docket Number
		Management Agency Review of a Combined License Application from Southern Nuclear Operating Company for the Vogtle Site. 43 Page(s)			Security US Federal Emergency Mgmt Agency (FEMA)	05200026 PROJ0755
04/04/2008	ML080940668	4/17/2008 - Forthcoming Meeting with Southern Nuclear Operation Company (SNC) Regarding Vogtle Units 3 and 4 COL Application Meeting on Technical Content. 8 Page(s)	Meeting Agenda Meeting Notice	NRC/NRO/DNRL/AP1000B 1	NRC/NRO/DNRL/AP1000B 1	05200025 05200026 PROJ0755
04/11/2008	ML081350247	Letter Re: PNNL Technical Assistance in Support of "New Reactors Review of the Evacuation Time Estimate Analysis for Grand Gulf 3 COLA and Vogtle COL Environmental Review" Under Q-4007. 3 Page(s)	Letter	NRC/NRO/DNRL	US Environmental Protection Agency (EPA)	05200024 05200025 05200026 PROJ0755
04/15/2008	ML081060578	Press Release-08-076: Vogtle Application for New Reactors Available on NRC Website. 1 Page(s)	Press Release	NRC/OPA		05200025 05200026 PROJ0755
04/15/2008	ML081060225	Press Releas-08-075: Deadline Extended Until April 23 to Pre-Register for Public comment Sessions on Vogtle Early Site Permit. 2 Page(s)	Press Release	NRC/OPA		05200025 05200026 PROJ0755

Appendix B

Chronology of the Combined License Application for Vogtle Electric Generating Plant Units 3 and 4

This appendix contains a chronological listing of routine licensing correspondence between the staff of the U.S. Nuclear Regulatory Commission (NRC) regarding the review of the Vogtle Electric Generating Plant, Units 3 and 4 plant design under Docket Nos. 052-000025 and 052-000026

Document Date	Accession Number	Title & Estimated Page Count	Document Type	Author Affiliation	Addressee Affiliation	Docket Number
04/17/2008	ML081360628	04/17/2008 Presentation Slides, "Vogtle 3 & 4 COL Application Overview," for Southern Nuclear - Vogtle Meeting. 88 Page(s)	Meeting Briefing Package/Handouts Slides and Viewgraphs	Southern Nuclear Operating Co, Inc	NRC/NRO	05200025 05200026
04/24/2008	ML081060305	Acknowledgment of Receipt of The Combined License Application for Vogtle Electric Generating Plant Unit 3 and 4 and Associated Federal Register Notice. 8 Page(s)	Federal Register Notice Letter	NRC/NRO/DNRL/AP1000B 1	Southern Nuclear Operating Co, Inc	05200011 05200025 05200026 PROJ0755
04/24/2008	ML081190060	Acceptance Review Results for the Vogtle Electric Generating Plant Unit 3 and 4 Combined License Application (Project No. 0755). 28 Page(s)	Graphics incl Charts and Tables Memoranda	NRC/NRO/DSRA	NRC/NRO/DNRL/AP1000B 1	05200012 05200013 05200025 05200026 PROJ0755
04/25/2008	ML081190319	Vogtle Combined License - 10 CFR 50.46 Report for the AP1000 Standard Plant Design. 2 Page(s)	Letter	Southern Nuclear Operating Co, Inc	NRC/Document Control Desk NRC/NRO	05200025 05200026 PROJ0755

Appendix B

Chronology of the Combined License Application for Vogtle Electric Generating Plant Units 3 and 4

This appendix contains a chronological listing of routine licensing correspondence between the staff of the U.S. Nuclear Regulatory Commission (NRC) regarding the review of the Vogtle Electric Generating Plant, Units 3 and 4 plant design under Docket Nos. 052-000025 and 052-000026

Document Date	Accession Number	Title & Estimated Page Count	Document Type	Author Affiliation	Addressee Affiliation	Docket Number
04/29/2008	ML081190160	Acceptance Review Results for the Vogtle Electric Generating Plant Units 3 and 4 Project Combined License Application. 19 Page(s)	Memoranda	NRC/NRO/DSER/HEB	NRC/NRO/DNRL/AP1000B 1	05200025 05200026 PROJ0755
05/02/2008	ML081220976	Acceptance Review Results for the Vogtle Electric Generating Plant Units 3 & 4 Combined License Application Final Safety Analysis Report (TAC RD3008). 4 Page(s)	Memoranda	NRC/NRO/DCIP/QVB1	NRC/NRO/DNRL/AP1000B 1	05200025 05200026 PROJ0755
05/05/2008	ML081230272	Acceptance Review Results for the Vogtle Electric Generating Plant, Units 3 & 4, Combined License Application - Chapter 7. 3 Page(s)	Memoranda	NRC/NRO/DE/ICEEB1	NRC/NRO/DNRL/AP1000B 1	05200025 05200026 PROJ0755
05/06/2008	ML081220526	Acceptance Review Results for the Vogtle Electric Generating Plant, Units 3 & 4, Combined License Application. 7 Page(s)	Memoranda	NRC/NRO/DSRA/BPB1	NRC/NRO/DNRL/AP1000B 1	05200025 05200026 PROJ0755
05/06/2008	ML081230117	Acceptance Review Results for FSAR Chapters 11 and 12 of the Southern Nuclear Operating Company Vogtle Electric Generating Plants Units 3 and 4 (PROJ755) Combined License Application (COLA). 18 Page(s)	Memoranda	NRC/NRO/DCIP/HPB	NRC/NRO/DNRL/AP1000B 1	05200025 05200026 PROJ0755

Appendix B

Chronology of the Combined License Application for Vogtle Electric Generating Plant Units 3 and 4

This appendix contains a chronological listing of routine licensing correspondence between the staff of the U.S. Nuclear Regulatory Commission (NRC) regarding the review of the Vogtle Electric Generating Plant, Units 3 and 4 plant design under Docket Nos. 052-000025 and 052-000026

Document Date	Accession Number	Title & Estimated Page Count	Document Type	Author Affiliation	Addressee Affiliation	Docket Number
05/06/2008	ML081260458	Acceptance Review Results for Vogtle Electric Generating Plant Units 3 and 4 Combined License Application. 21 Page(s)	Memoranda	NRC/NRO/DE/EMB1 NRC/NRO/DE/EMB2	NRC/NRO/DNRL/AP1000B 1	05200025 05200026 PROJ0755
05/07/2008	Ml081260291	Acceptance Review Results for the Vogtle Electric Generating Plant, Units 3 and 4 Combined License Application – Chapter 7. 3 Page(s)	Graphics incl Charts and Tables Memoranda	NRC/NRO/DE/CIPTB1	NRC/NRO/DNRL/AP1000B 1	05200011 05200025 05200026 PROJ0755
05/06/2008	ML081260675	Acceptance Review Results for the Vogtle Electric Generating Plant, Units 3 and 4 Combined License Application Review (PROJECT 755). 3 Page(s)	Memoranda	NRC/NRO/DE	NRC/NRO/DNRL/AP1000B 1	05200025 05200026 PROJ0755
05/07/2008	ML081270689	Acceptance Review Results For The Vogtle Units 3 & 4 Combined License Application. 3 Page(s)	Memoranda	NRC/NSIR/DPR/DDEP/LIB	NRC/NRO/DNRL/AP1000B 1	05200011 05200025 05200026 PROJ0755
05/07/2008	ML081280507	Acceptance Review of Vogtle Units 3 and 4 Combined License Application Final Safety Analysis Report. 3 Page(s)	Acceptance Review Letter Memoranda	NRC/NRO/DSER/SACB	NRC/NRO/DNRL/AP1000B 1	05200025 05200026 PROJ0755
05/08/2008	ML081270174	Acceptance Review Results for the Vogtle Electric Generating Plant Combined License Application (TAC No. 3028).	Memoranda	NRC/NRO/DSER/EPB1	NRC/NRO/DNRL/AP1000B 1	05200025 05200026 PROJ0755

Chronology of the Combined License Application for Vogtle Electric Generating Plant Units 3 and 4

This appendix contains a chronological listing of routine licensing correspondence between the staff of the U.S. Nuclear Regulatory Commission (NRC) regarding the review of the Vogtle Electric Generating Plant, Units 3 and 4 plant design under Docket Nos. 052-000025 and 052-000026

Document Date	Accession Number	Title & Estimated Page Count	Document Type	Author Affiliation	Addressee Affiliation	Docket Number
		7 Page(s)				
05/08/2008	ML081280575	Acceptance Review Results for Vogtle AP-1000 Combined License Application (Subsequent COLA) (Units 3 and 4). 23 Page(s)	Acceptance Review Letter Memoranda	NRC/NRO/DE/SEB1	NRC/NRO/DNRL/AP1000B 1	05200025 05200026 PROJ0755
05/08/2008	ML081290404	Acceptance Review Results for the Vogtle Electric Generating Plant Units 3 and 4 Combined License Application. 10 Page(s)	Memoranda	NRC/NRO/DSRA/CVB1	NRC/NRO/DNRL/AP1000B 1	05200025 05200026 PROJ0755
05/08/2008	ML081290574	Acceptance Review Results for the Vogtle Nuclear Site Units 3 and 4 Combined License Application. 5 Page(s)	Memoranda	NRC/NRO/DSER	NRC/NRO/DNRL/AP1000B 1	05200025 05200026 PROJ0755
05/08/2008	ML081330598	Acceptance Review Results for the Vogtle Electric Generating Plant, Units 3 and 4 Combined License Application. 6 Page(s)	Memoranda	NRC/NRO/DSRA/RSNPCR B	NRC/NRO/DNRL/AP1000B 1	05200025 05200026 PROJ0755
05/09/2008	ML081290764	Acceptance Review Results for Vogtle S-COLA. 4 Page(s)	Acceptance Review Letter Memoranda	NRC/NRO/DCIP/TSB	NRC/NRO/DNRL/AP1000B 1	05200025 05200026 PROJ0755
05/12/2008	ML081330224	SPCV CTH Vogtle SCOLA Acceptance Review Notes 3. 2 Page(s)	Acceptance Review Letter	NRC/NRO/DSRA		05200025 05200026 PROJ0755

Appendix B

Chronology of the Combined License Application for Vogtle Electric Generating Plant Units 3 and 4

This appendix contains a chronological listing of routine licensing correspondence between the staff of the U.S. Nuclear Regulatory Commission (NRC) regarding the review of the Vogtle Electric Generating Plant, Units 3 and 4 plant design under Docket Nos. 052-000025 and 052-000026

Document Date	Accession Number	Title & Estimated Page Count	Document Type	Author Affiliation	Addressee Affiliation	Docket Number
05/12/2008	ML091170687	2008/05/12-Limited Appearance Statement of Tom Clements on Vogtle, Units 3 & 4 COL. 11 Page(s)	Legal-Limited Appearance Statement	- No Known Affiliation	NRC/SECY/RAS	05200025 05200026
05/19/2008	ML081400086	Results of Acceptance Review for the Vogtle Electric Generating Plant, Units 3 & 4 Combined Operating License Application. 2 Page(s)	Memoranda	NRC/NRR/ADRO/DPR/FPR B	NRC/NRO/DNRL/AP1000B 1	05200025 05200026 PROJ0755
05/20/2008	ML081290740	04/17/2008 - Summary of Meeting Regarding the Vogtle Units 3 and 4 Combined License Application. 8 Page(s)	Meeting Summary	NRC/NRO/DNRL/AP1000B 1		05200025 05200026
05/30/2008	ML081480138	Acceptance Review For The Vogtle Electric Generating Plant Units 3 And 4 Combined License Application. 10 Page(s)	Federal Register Notice Letter	NRC/NRO/DNRL/AP1000B 1	Southern Nuclear Operating Co, Inc	05200025 05200026
06/04/2008	ML081500246	Vogtle Combined License Application - Safety Review Schedule. 3 Page(s)	Letter	NRC/NSIR/DPR/DDEP/RIO B	US Federal Emergency Mgmt Agency (FEMA)	05200025 05200026
06/04/2008	ML081500305	Enclosure: Project Details: AP1000 Vogtle SCOL, First Three Phases of the Preliminary Review Schedule. 1 Page(s)	Graphics incl Charts and Tables	NRC/NSIR		05200025 05200026
07/16/2008	ML081500677	Notice of Intent to Prepare an Environmental Impact Statement	Letter	NRC/NRO/DSER	Southern Nuclear Operating Co, Inc	05200025 05200026

Appendix B

Chronology of the Combined License Application for Vogtle Electric Generating Plant Units 3 and 4

This appendix contains a chronological listing of routine licensing correspondence between the staff of the U.S. Nuclear Regulatory Commission (NRC) regarding the review of the Vogtle Electric Generating Plant, Units 3 and 4 plant design under Docket Nos. 052-000025 and 052-000026

Document Date	Accession Number	Title & Estimated Page Count	Document Type	Author Affiliation	Addressee Affiliation	Docket Number
		and Conduct Scoping Related to the Combined License Application for the Vogtle Electric Generating Plant. 11 Page(s)				
06/09/2008	ML081610588	Press Release-08-113: NRC Accepts Application for New Reactors at Vogtle. 1 Page(s)	Press Release	NRC/OPA		05200025 05200026
06/06/2008	ML081630150	Vogtle, Units 3 and 4, Combined License Application 10 CFR 2.101 Affidavit. 5 Page(s)	Legal-Affidavit Letter	Southern Nuclear Operating Co, Inc	NRC/Document Control Desk NRC/NRO	05200025 05200026
06/09/2008	ML091050326	2008/06/09 Vogtle COL Review - SNC Letter AR-08-0865 transmitting VEGP Units 3&4 Combined License Application 10 CFR 2.101 Affidavit 7 Page(s)	E-Mail	- No Known Affiliation	NRC/NRO/DNRL/NWE1	05200025 05200026
06/11/2008	ML081640477	Ltr. Eric Pupe - Vogtle Units 3 and 4 Combined License Application. 2 Page(s)	Letter	NRC/NSIR/DSP/DDRSR/R SPLB	US Dept of Homeland Security, Office of Infrastructure Protection	05200025 05200026
06/11/2008	ML081770650	FRN - Southern Nuclear Operating Company; Acceptance for Docketing of an Application for Combined License for Vogtle, Units 3 and 4. 2 Page(s)	Federal Register Notice	NRC/NRO/DNRL/AP1000B 1		05200025 05200026
06/27/2008	ML081780539	Vogtle Electric Generating Plant	Letter	NRC/NRO/DNRL/AP1000B	Southern Nuclear Operating	05200025

Chronology of the Combined License Application for Vogtle Electric Generating Plant Units 3 and 4

This appendix contains a chronological listing of routine licensing correspondence between the staff of the U.S. Nuclear Regulatory Commission (NRC) regarding the review of the Vogtle Electric Generating Plant, Units 3 and 4 plant design under Docket Nos. 052-000025 and 052-000026

Document Date	Accession Number	Title & Estimated Page Count	Document Type	Author Affiliation	Addressee Affiliation	Docket Number
07/03/2008	ML081850263	Units 3 and 4 Combined License Application Review Schedule. 8 Page(s)		2	Co, Inc	05200026
07/17/2008	ML081900234	07/17/2008 - Notice of Public Outreach Meeting on Vogtle Electric Generating Plant, Units 3 & 4 COLA to Discuss Details of the Safety and Environmental Reviews, and Public Participation in NRC Processes. 7 Page(s)	Meeting Notice Memoranda	NRC/NRO/DNRL/AP1000B 1	NRC/NRO/DNRL/AP1000B 1	05200025 05200026
07/08/2008	ML081780805	Acceptance Review Results for the Vogtle Electric Generating Plant, Units 3 & 4, Subsequent Combined License Application. 3 Page(s)	Memoranda	NRC/NRO/DSRA/SFPB	NRC/NRO/DNRL/AP1000B 1	05200025 05200026
07/09/2008	ML082100092	Application by Southern Nuclear Operating Company for a Combined License for Units 3 and 4 at the Vogtle Electric Generating Plant. 9 Page(s)	Letter	NRC/NRO/DSER	Burke County, GA	05200025 05200026 PROJ0755
07/16/2008	ML082490543	Vogtle LR DSEIS Comments from Radiation and Public Health Project. 7 Page(s)	E-Mail	Radiation & Public Health Project	NRC/ADM/DAS/RDEB	05200025 05200026
07/16/2008		Comment (5) of Joseph J. Mangano on Behalf of Radiation and Public Health Project on the	General FR Notice Comment Letter	Radiation & Public Health Project	NRC/ADM/DAS/RDEB	05200025 05200026

Appendix B

Chronology of the Combined License Application for Vogtle Electric Generating Plant Units 3 and 4

This appendix contains a chronological listing of routine licensing correspondence between the staff of the U.S. Nuclear Regulatory Commission (NRC) regarding the review of the Vogtle Electric Generating Plant, Units 3 and 4 plant design under Docket Nos. 052-000025 and 052-000026

Document Date	Accession Number	Title & Estimated Page Count	Document Type	Author Affiliation	Addressee Affiliation	Docket Number
		Environmental Impact Statement Opposing for the Proposed New Nuclear Reactors at the Alvin Vogtle Plant. 6 Page(s)				
07/31/2008	ML082130375	Confirmation Closed Meeting Vogtle Electric Generating Plant, Units 3 and 4. 6 Page(s)	Letter Meeting Agenda Meeting Notice	NRC/RGN-II/DCP/CPB3	Southern Nuclear Operating Co, Inc	05200025 05200026
08/08/2008	ML081910396	Final Environmental Impact Statement for an Early Site Permit (ESP) at the Vogtle Electric Generating Plant Site (TAC MD3010). 17 Page(s)	Environmental Impact Statement Letter	NRC/NRO/DSER	US Environmental Protection Agency (EPA)	05200025 05200026
08/01/2008	ML082140699	SCDNR Comments on Vogtle License Renewal DSEIS. 3 Page(s)	Letter	State of SC, Dept of Natural Resources	NRC/NRR/ADRO/DLR	05000424 05000425 05200011 05200025 05200026 PROJ0755
08/06/2008	ML082280763	2008/08/06 Vogtle COL Review - test 2 Page(s)	E-Mail	NRC/NRO	NRC/NRO/DNRL/NWE1	05200025 05200026
08/11/2008	ML082190977	Meeting Summary of 07/17/2008 Public Outreach Meeting to Discuss the Review of the Vogtle Combined License Application. 7 Page(s)	Meeting Summary	NRC/NRO/DNRL/AP1000B1		05200025 05200026

Chronology of the Combined License Application for Vogtle Electric Generating Plant Units 3 and 4

This appendix contains a chronological listing of routine licensing correspondence between the staff of the U.S. Nuclear Regulatory Commission (NRC) regarding the review of the Vogtle Electric Generating Plant, Units 3 and 4 plant design under Docket Nos. 052-000025 and 052-000026

Document Date	Accession Number	Title & Estimated Page Count	Document Type	Author Affiliation	Addressee Affiliation	Docket Number
08/13/2008	ML081910113	Notice of Availability of the Final Environmental Impact Statement for an Early Site Permit at the Vogtle Electric Generating Plant Site (Tac No. MD3010). 18 Page(s)	Federal Register Notice Letter	NRC/NRO/DSER	Southern Nuclear Operating Co, Inc	05200011 05200025 05200026
08/13/2008	ML081920332	Changes In The Processing of Requests For Additional Information - The Vogtle Electric Generating Plant Units 3 And 4 Combined License Application. 6 Page(s)	Letter Request for Additional Information (RAI)	NRC/NRO/DNRL/AP1000B1	Southern Nuclear Operating Co, Inc	05200025 05200026
08/14/2008	ML082280539	Site Visit to Vogtle to Observe Erection of a Mechanically Stabilized Earth (MSE) Demonstration retaining Wall to Support Early Site Permit, Limited Work Authorization, and Combined License Activities. 8 Page(s)	Letter Trip Report	NRC/RGN-II/DCI/CIB2	Southern Nuclear Operating Co, Inc	05200011 05200025 05200026
08/19/2008	ML082320837	Request for Additional Information Letter No. 001 Related to SRP Section 02.03.04 for the Vogtle Electric Generating Plant Units 3 and 4 Combined License Application. 4 Page(s)	Letter Request for Additional Information (RAI)	NRC/NRO/DNRL/AP1000B1	Southern Nuclear Operating Co, Inc	05200025 05200026
08/19/2008	ML082321224	2008/08/19 Vogtle COL Review - RAI Letter No. 001 Related SRP	E-Mail	NRC/NRO	NRC/NRO/DNRL/NWE1	05200025 05200026

Appendix B

Chronology of the Combined License Application for Vogtle Electric Generating Plant Units 3 and 4

This appendix contains a chronological listing of routine licensing correspondence between the staff of the U.S. Nuclear Regulatory Commission (NRC) regarding the review of the Vogtle Electric Generating Plant, Units 3 and 4 plant design under Docket Nos. 052-000025 and 052-000026

Document Date	Accession Number	Title & Estimated Page Count	Document Type	Author Affiliation	Addressee Affiliation	Docket Number
		Section 2.3.4 for Vogtle Units 3 and 4 6 Page(s)				
08/22/2008	ML082330172	Preliminary Safety Evaluation Report for Combined License Application for Vogtle, Units 3 and 4. 34 Page(s)	Memoranda Safety Evaluation Report	NRC/NRO/DSER/SACB	NRC/NRO/DNRL/AP1000B 1	05200025 05200026
08/29/2008	ML082421046	2008/08/29 Vogtle RAI for SER - Request for Additional Information Letter No. 001 Related To SRP Sections 02.03.01 and 02.03.05 for the Vogtle 3 and 4 Combined License Application 7 Page(s)	E-Mail	NRC/NRO	NRC/NRO/DNRL/NWE1	05200025 05200026
08/29/2008	ML082421137	2008/08/29 Vogtle RAI for SER - Request for Additional Information Letter No. 002 Related To SRP Sections 02.03.01 and 02.03.05 for the Vogtle 3 and 4 Combined License Application 7 Page(s)	E-Mail Request for Additional Information (RAI)	NRC/NRO	NRC/NRO/DNRL/NWE1	05200025 05200026
08/29/2008	ML082421141	2008/08/29 Vogtle COL Review - RAI Letter No. 002 Related SRP Sections 2.3.1 and 2.3.5 for Vogtle Units 3 and 4 9 Page(s)	E-Mail	NRC/NRO	NRC/NRO/DNRL/NWE1	05200025 05200026
08/31/2008	ML082260203	NUREG-1872, Vol. 2, "Final Environmental Impact Statement for	Environmental Impact Statement	NRC/NRO/DSER/EPB1		05200025 05200026

Appendix B

Chronology of the Combined License Application for Vogtle Electric Generating Plant Units 3 and 4

This appendix contains a chronological listing of routine licensing correspondence between the staff of the U.S. Nuclear Regulatory Commission (NRC) regarding the review of the Vogtle Electric Generating Plant, Units 3 and 4 plant design under Docket Nos. 052-000025 and 052-000026

Document Date	Accession Number	Title & Estimated Page Count	Document Type	Author Affiliation	Addressee Affiliation	Docket Number
		an Early Site Permit (ESP) at the Vogtle Electric Generating Plant Site," Final Report, Appendix F. 112 Page(s)	NUREG			
08/31/2008	ML092460461	Redacted - Confirmation Closed Meeting Vogtle Electric Generating Plant, Unit 3 and 4. 6 Page(s)	Letter Meeting Notice	NRC/RGN-II/DCP/CPB3	Southern Nuclear Operating Co, Inc	05200025 05200026
09/08/2008	ML082520409	Draft Final Safety Evaluation Report for Vogtle for Sections 2.5.4 (Excluding 2.5.4.10). 92 Page(s)	Memoranda	NRC/NRO/DSER/RGS1 NRC/NRO/DSER/RGS2	NRC/NRO/DNRL/AP1000B 1	05200025 05200026
09/10/2008	ML082590003	Joseph M. Farley, Units 1 and 2, Edwin I. Hatch, Units 1 and 2, Vogtle, Units 1, 2, 3 and 4 - Management Organization Change. 2 Page(s)	Letter	Southern Nuclear Operating Co, Inc	NRC/Document Control Desk NRC/NRO	05000321 05000348 05000364 05000366 05000424 05000425 05200011 05200025 05200026
09/11/2008	ML082590051	Vogtle, Units 3 & 4 Combined License Application, Response to Request for Additional Information No. 02.03.04-1. 5 Page(s)	Letter	Southern Nuclear Operating Co, Inc	NRC/Document Control Desk NRC/NRO	05200025 05200026

Appendix B

Chronology of the Combined License Application for Vogtle Electric Generating Plant Units 3 and 4

This appendix contains a chronological listing of routine licensing correspondence between the staff of the U.S. Nuclear Regulatory Commission (NRC) regarding the review of the Vogtle Electric Generating Plant, Units 3 and 4 plant design under Docket Nos. 052-000025 and 052-000026

Document Date	Accession Number	Title & Estimated Page Count	Document Type	Author Affiliation	Addressee Affiliation	Docket Number
09/11/2008	ML082590052	Vogtle, Units 3 and 4 – Combined License Application, Departure Report Update. 3 Page(s)	Letter	Southern Nuclear Operating Co, Inc	NRC/Document Control Desk NRC/NRO	05200025 05200026
09/11/2008	ML082590333	Vogtle, Units 3 and 4 Combined License Application - Submittal of Hydrology-Related Model Input Files. 13 Page(s)	Letter	Southern Nuclear Operating Co, Inc	NRC/Document Control Desk NRC/NRO	05200025 05200026
09/11/2008	ML091000565	Vogtle COL SER/OI Section 2.3.4: ARCON96 Atmospheric Dispersion Model Runs of Control Room X/Q Values. 41 Page(s)	- No Document Type Applies Final Safety Analysis Report (FSAR) Graphics incl Charts and Tables Letter	NRC/NRO		05200025 05200026
09/11/2008	ML091050324	2008/09/11 Vogtle COL Review - SNC Letter AR-08-1408 transmitting VEGP COLA Departure Report Update 5 Page(s)	E-Mail	- No Known Affiliation	NRC/NRO/DNRL/NWE1	05200025 05200026
09/14/2008	ML082590094	2008/09/14 Vogtle COL Review - Draft RAI 568 related to SRP Section: 14.02 - Initial Plant Test Program for Vogtle S-COL Units 3 and 4 3 Page(s)	E-Mail	NRC/NRO	NRC/NRO/DNRL/NWE1	05200025 05200026
09/15/2008	ML082680154	2008/09/15-Vogtle ESP E-mail	Legal-Hearing File	NRC/NRO	- No Known Affiliation	05200025

Appendix B

Chronology of the Combined License Application for Vogtle Electric Electric Generating Plant Units 3 and 4

This appendix contains a chronological listing of routine licensing correspondence between the staff of the U.S. Nuclear Regulatory Commission (NRC) regarding the review of the Vogtle Electric Generating Plant, Units 3 and 4 plant design under Docket Nos. 052-000025 and 052-000026

Document Date	Accession Number	Title & Estimated Page Count	Document Type	Author Affiliation	Addressee Affiliation	Docket Number
		(REDACTED) re: Robust Redhorse, and Other Ball-Busting Suckers. 3 Page(s)	(For Informal Hearings)			05200026
09/16/2008	ML082600493	Press Release-08-172: NRC Announces Opportunity to Participate in Hearing on New Reactor Application for Vogtle Site. 2 Page(s)	Press Release	NRC/OPA		05200025 05200026
09/18/2008	ML082660545	Vogtle, Units 3 & 4 Combined License Application Response to Request for Additional Information Letter No. 002. 6 Page(s)	Letter	Southern Nuclear Operating Co, Inc	NRC/Document Control Desk NRC/NRO	05200025 05200026
09/18/2008	ML082660546	Southern Nuclear Operating Co., Review of Bellefonte Response to Requests for Additional Information No. 01-05 for Applicability to Vogtle, Units 3 & 4 Combined License Application. 3 Page(s)	Letter	Southern Nuclear Operating Co, Inc	NRC/Document Control Desk NRC/NRO	05200025 05200026
09/23/2008	ML082670197	9/16/08 FRN - Official Notice Regarding Hearing Opportunity to Petition for Leave to Intervene Vogtle Units 3 & 4 5 Page(s)	Federal Register Notice	NRC/NRO/DNRL/AP1000B 1		05200025 05200026
09/25/2008	ML082682042	Draft Final Safety Evaluation Report for Vogtle for Sections 2.5.4.7 and 2.5.4.10. 10 Page(s)	Memoranda Safety Evaluation Report, Draft	NRC/NRO/DSER/GGEB1 NRC/NRO/DSER/GGEB2	NRC/NRO/DNRL/AP1000B 1	05200025 05200026

Appendix B

Chronology of the Combined License Application for Vogtle Electric Generating Plant Units 3 and 4

This appendix contains a chronological listing of routine licensing correspondence between the staff of the U.S. Nuclear Regulatory Commission (NRC) regarding the review of the Vogtle Electric Generating Plant, Units 3 and 4 plant design under Docket Nos. 052-000025 and 052-000026

Document Date	Accession Number	Title & Estimated Page Count	Document Type	Author Affiliation	Addressee Affiliation	Docket Number
09/25/2008	ML082682050	SGI Letter for DHS Consultation at Vogtle. 2 Page(s)	Letter	NRC/NSIR	Southern Nuclear Operating Co, Inc	05200025 05200026
09/26/2008	ML082690115	FSAR Section 6.4 Preliminary Safety Evaluation Report Input for Vogtle Units 3 & 4 Combined License Application. 8 Page(s)	Final Safety Analysis Report (FSAR) Memoranda	NRC/NRO/DSER/RSAC	NRC/NRO/DNRL/AP1000B 1	05200025 05200026
09/26/2008	ML082690630	Safety Evaluation Report (SER) on Hydrology for the Vogtle Early Site Permit Application. 36 Page(s)	Memoranda Safety Evaluation	NRC/NRO/DSER/HEB	NRC/NRO/DNRL/AP1000B 1	05200025 05200026
09/30/2008	ML082740307	2008/09/30 Vogtle COL Review - Draft RAIs 1241, 1242 and 1247 Related SRP Section 9.2.5, 9.2.2 and 9.2.1 for Vogtle Units 3 and 4 2 Page(s)	E-Mail	NRC/NRO	NRC/NRO/DNRL/NWE1	05200025 05200026
09/30/2008	ML082740309	2008/09/30 Vogtle COL Review - RESEND: Draft RAIs 1241, 1242 and 1247 Related SRP Section 9.2.5, 9.2.2 and 9.2.1 for Vogtle Units 3 and 4 7 Page(s)	E-Mail	NRC/NRO	NRC/NRO/DNRL/NWE1	05200025 05200026

Appendix B

Chronology of the Combined License Application for Vogtle Electric Generating Plant Units 3 and 4

This appendix contains a chronological listing of routine licensing correspondence between the staff of the U.S. Nuclear Regulatory Commission (NRC) regarding the review of the Vogtle Electric Generating Plant, Units 3 and 4 plant design under Docket Nos. 052-000025 and 052-000026

Document Date	Accession Number	Title & Estimated Page Count	Document Type	Author Affiliation	Addressee Affiliation	Docket Number
10/01/2008	ML082750234	2008/10/01 Vogtle COL Review - RE: RESEND: Draft RAIs 1241, 1242 and 1247 Related SRP Section 9.2.5, 9.2.2 and 9.2.1 for Vogtle Units 3 and 4 2 Page(s)	E-Mail	NRC/NRO	NRC/NRO/DNRL/NWE1	05200025 05200026
10/01/2008	ML082810052	Bellefonte, Response to Request for Additional Information No. 03.02.01-02 for Applicability to the Vogtle Electric Generating Plant, Units 3 & 4 Combined License Application. 3 Page(s)	Letter	Southern Nuclear Operating Co, Inc	NRC/Document Control Desk NRC/NRO	05200014 05200015 05200025 05200026
10/01/2008	ML082810053	Review of Bellefonte Response to Request for Additional Information No. 01-07 for Applicability to the Vogtle Electric Generating Plant Units 3 and 4 Combined License Application. 3 Page(s)	Letter	Southern Nuclear Operating Co, Inc	NRC/Document Control Desk NRC/NRO	05200014 05200015 05200025 05200026
10/02/2008	ML082770551	2008/10/02 Vogtle COL Review - RE: RESEND: Draft RAIs 1241, 1242 and 1247 Related SRP Section 9.2.5, 9.2.2 and 9.2.1 for Vogtle Units 3 and 4 2 Page(s)	E-Mail	- No Known Affiliation	NRC/NRO/DNRL/NWE1	05200025 05200026
10/03/2008	ML082770190	2008/10/03 Vogtle COL Review - Draft RAIs 1444 Related SRP Section 7.5 for Vogtle Units 3 and 4	E-Mail	NRC/NRO	NRC/NRO/DNRL/NWE1	05200025 05200026

Appendix B

Chronology of the Combined License Application for Vogtle Electric Generating Plant Units 3 and 4

This appendix contains a chronological listing of routine licensing correspondence between the staff of the U.S. Nuclear Regulatory Commission (NRC) regarding the review of the Vogtle Electric Generating Plant, Units 3 and 4 plant design under Docket Nos. 052-000025 and 052-000026

Document Date	Accession Number	Title & Estimated Page Count	Document Type	Author Affiliation	Addressee Affiliation	Docket Number
		3 Page(s)				
10/06/2008	ML082800255	2008/10/06 Vogtle RAI for SER - Request for Additional Information Letter No. 003 Related to SRP Section 09.02.01 for the Vogtle Units 3 and 4 Combined License Application 7 Page(s)	E-Mail Request for Additional Information (RAI)	NRC/NRO	NRC/NRO/DNRL/NWE1	05200025 05200026
10/06/2008	ML082800285	2008/10/06 Vogtle COL Review - RAI Letter No. 003 Related SRP Section 9.2.1 for Vogtle Units 3 and 4 9 Page(s)	E-Mail	NRC/NRO	NRC/NRO/DNRL/NWE1	05200025 05200026
10/06/2008	ML082800303	Enclosure 2 - Vogtle Site Tour. 27 Page(s)	Meeting Agenda Meeting Briefing Package/Handouts Slides and Viewgraphs	Southern Co Services	NRC/NRO	05200025 05200026
10/06/2008	ML082800361	2008/10/06 Vogtle RAI for SER - Request for Additional Information Letter No. 004 Related to SRP Section 09.02.05 for the Vogtle Units 3 and 4 Combined License Application 6 Page(s)	E-Mail Request for Additional Information (RAI)	NRC/NRO	NRC/NRO/DNRL/NWE1	05200025 05200026
10/06/2008	ML082800382	2008/10/06 Vogtle RAI for SER - Request for Additional Information Letter No. 005 Related to SRP Section 09.05.02 for the Vogtle	E-Mail Request for Additional Information (RAI)	NRC/NRO	NRC/NRO/DNRL/NWE1	05200025 05200026

Appendix B

Chronology of the Combined License Application for Vogtle Electric Generating Plant Units 3 and 4

This appendix contains a chronological listing of routine licensing correspondence between the staff of the U.S. Nuclear Regulatory Commission (NRC) regarding the review of the Vogtle Electric Generating Plant, Units 3 and 4 plant design under Docket Nos. 052-000025 and 052-000026

Document Date	Accession Number	Title & Estimated Page Count	Document Type	Author Affiliation	Addressee Affiliation	Docket Number
		Units 3 and 4 Combined License Application 7 Page(s)				
10/06/2008	ML082800384	2008/10/06 Vogtle COL Review - RAI Letter No. 004 Related SRP Section 9.2.5 for Vogtle Units 3 and 4 8 Page(s)	E-Mail	NRC/NRO	NRC/NRO/DNRL/NWE1	05200025 05200026
10/06/2008	ML082800390	2008/10/06 Vogtle COL Review - RAI Letter No. 005 Related SRP Section 9.5.2 for Vogtle Units 3 and 4 9 Page(s)	E-Mail	NRC/NRO	NRC/NRO/DNRL/NWE1	05200025 05200026
10/08/2008	ML082820306	2008/10/08 Vogtle COL Review - RE: Update RE: RAI Calls and Schedule 3 Page(s)	E-Mail	NRC/NRO	NRC/NRO/DNRL/NWE1	05200025 05200026
10/08/2008	ML091770466	2008/10/08 Vogtle COL Review - Draft RAI (#1260) Related to SRP Section 15.00.003 for Vogtle Units 3 and 4 3 Page(s)	E-Mail	NRC/NRO	NRC/NRO/DNRL/NWE1	05200025 05200026
10/09/2008	ML082831718	2008/10/09 Vogtle COL Review - Draft RAIs 1432 and 1433 Related SRP Section 2.3.1 and 2.3.5 for Vogtle Units 3 and 4 4 Page(s)	E-Mail	NRC/NRO	NRC/NRO/DNRL/NWE1	05200025 05200026
10/09/2008	ML082880094	Vogtle, Units 3 and 4, Combined License Application Response to	Letter	Southern Nuclear Operating Co, Inc	NRC/Document Control Desk	05200025 05200026

Appendix B

Chronology of the Combined License Application for Vogtle Electric Generating Plant Units 3 and 4

This appendix contains a chronological listing of routine licensing correspondence between the staff of the U.S. Nuclear Regulatory Commission (NRC) regarding the review of the Vogtle Electric Generating Plant, Units 3 and 4 plant design under Docket Nos. 052-000025 and 052-000026

Document Date	Accession Number	Title & Estimated Page Count	Document Type	Author Affiliation	Addressee Affiliation	Docket Number
		Questions Raised in Combined License Public Meeting. 3 Page(s)			NRC/NRO	
10/09/2008	ML091050321	2008/10/09 Vogtle COL Review - SNC Letter AR-08-1430 transmitting VEGP 3&4 COLA Response to Questions Raised in COL Public Meeting 5 Page(s)	E-Mail	- No Known Affiliation	NRC/NRO/DNRL/NWE1	05200025 05200026
10/10/2008	ML082840402	2008/10/10 Vogtle COL Review - Telcon Summary – Telcon with Vogtle 10/06/2008 2 Page(s)	E-Mail	NRC/NRO	NRC/NRO/DNRL/NWE1	05200025 05200026
10/14/2008	ML082620184	Vogtle Site Audit Trip Report for August 2008. 9 Page(s)	Memoranda Trip Report	NRC/NRO/DSER/EPB1	NRC/NRO/DSER/EPB1	05200025 05200026
10/14/2008	ML092151040	2008/10/14 Vogtle COL Review - FW: Met Tower Information in the Vogtle ESP 2 Page(s)	E-Mail	NRC/NRO	NRC/NRO/DNRL/NWE1	05200025 05200026
10/15/2008	ML082800251	Trip Report for the Vogtle, Units 3 and 4 Combined License Application Department of Homeland Security Consultation. 3 Page(s)	Memoranda Trip Report	NRC/NRO/DNRL/EPB2	NRC/NRO/DNRL/AP1000B 1	05200025 05200026
10/16/2008	ML082900188	2008/10/16 Vogtle RAI for SER - Request for Additional Information Letter No. 006 Related to SRP Section 07.05 for the Vogtle Units 3	E-Mail Request for Additional Information (RAI)	NRC/NRO	NRC/NRO/DNRL/NWE1	05200025 05200026

Appendix B

Chronology of the Combined License Application for Vogtle Electric Generating Plant Units 3 and 4

This appendix contains a chronological listing of routine licensing correspondence between the staff of the U.S. Nuclear Regulatory Commission (NRC) regarding the review of the Vogtle Electric Generating Plant, Units 3 and 4 plant design under Docket Nos. 052-000025 and 052-000026

Document Date	Accession Number	Title & Estimated Page Count	Document Type	Author Affiliation	Addressee Affiliation	Docket Number
		and 4 Combined License Application 6 Page(s)				
10/16/2008	ML082900726	2008/10/16 Vogtle COL Review - RAI Letter No. 006 Related SRP Section 7.5 for Vogtle Units 3 and 4 8 Page(s)	E-Mail	NRC/NRO	NRC/NRO/DNRL/NWE1	05200025 05200026
10/17/2008	ML082910046	2008/10/17 Vogtle RAI for SER – Request for Information Letter No. 007 Related to SRP Section 15.00.03 for the Vogtle Electric Generating Plant Units 3 and 4 combined License Application 6 Page(s)	Request for Additional Information (RAI)	NRC/NRO/DNRL	NRC/NRO/DNRL/NWE1	05200025 05200026
10/17/2008	ML082910177	2008/10/17 Vogtle COL Review - Telcon Summary - Telcon with Vogtle 10/14/2008 2 Page(s)	E-Mail	NRC/NRO	NRC/NRO/DNRL/NWE1	05200025 05200026
10/17/2008	ML082910203	2008/10/17 Vogtle COL Review - Telcon Summary - Telcon with Vogtle 10/14/2008 2 Page(s)	E-Mail	NRC/NRO	NRC/NRO/DNRL/NWE1	05200025 05200026
10/17/2008	ML082911086	2008/10/17 Vogtle COL Review - Draft RAI 1453 Related SRP Section 2.5.4 for Vogtle Units 3 and 4 3 Page(s)	E-Mail	NRC/NRO	NRC/NRO/DNRL/NWE1	05200025 05200026
10/17/2008	ML092151041	2008/10/17 Vogtle COL Review - RAI letter No. 007 related to SRP	E-Mail	NRC/NRO	NRC/NRO/DNRL/NWE1	05200025 05200026

Appendix B

Chronology of the Combined License Application for Vogtle Electric Generating Plant Units 3 and 4

This appendix contains a chronological listing of routine licensing correspondence between the staff of the U.S. Nuclear Regulatory Commission (NRC) regarding the review of the Vogtle Electric Generating Plant, Units 3 and 4 plant design under Docket Nos. 052-000025 and 052-000026

Document Date	Accession Number	Title & Estimated Page Count	Document Type	Author Affiliation	Addressee Affiliation	Docket Number
		section 15.00.03 for Vogtle Units 3 and 4 8 Page(s)				
10/20/2008	ML082940631	2008/10/20 Vogtle RAI for SER - Request for Additional Information Letter No. 008 Related to SRP Sections 02.03.01 and 02.03.05 for the Vogtle 3 and 4 Combined License Application 6 Page(s)	E-Mail Request for Additional Information (RAI)	NRC/NRO	NRC/NRO/DNRL/NWE1	05200025 05200026
10/20/2008	ML082940632	2008/10/20 Vogtle COL Review - RAI Letter No. 008 Related SRP Sections 2.3.1 and 2.3.5 for Vogtle Units 3 and 4 8 Page(s)	E-Mail	NRC/NRO	NRC/NRO/DNRL/NWE1	05200025 05200026
10/23/2008	ML082970557	Request for Extension of Vogtle, Units 3 and 4 DHS Consultation Report. 2 Page(s)	Letter	NRC/NSIR/DSP/DDRSR/R SPLB	US Dept of Homeland Security	05200025 05200026
10/24/2008	ML093570257	Federal Emergency Management Agency Interim Finding Report on the Adequacy of Offsite Emergency Plans for Vogtle Combined License Application. 1 Page(s)	Letter	US Federal Emergency Mgmt Agency (FEMA)	NRC/NSIR/DPR	05200025 05200026
10/29/2008	ML083100678	FEMA E-Mail and 10-24-08 Letter - Interim Findings Report re: Vogtle COL Application (Offsite Emergency Plans).	E-Mail	NRC/NSIR	NRC/NSIR	05200025 05200026

Appendix B

Chronology of the Combined License Application for Vogtle Electric Generating Plant Units 3 and 4

This appendix contains a chronological listing of routine licensing correspondence between the staff of the U.S. Nuclear Regulatory Commission (NRC) regarding the review of the Vogtle Electric Generating Plant, Units 3 and 4 plant design under Docket Nos. 052-000025 and 052-000026

Document Date	Accession Number	Title & Estimated Page Count	Document Type	Author Affiliation	Addressee Affiliation	Docket Number
10/30/2008	ML083030384	FSAR Sections 3.51.5 and 3.5.1.6 Preliminary Safety Evaluation Report Input for Vogtle Unit 3 and 4 Combined License Application. 7 Page(s)	Memoranda Safety Evaluation Report	NRC/NRO/DSER/RSAC	NRC/NRO/DNRL/AP1000B 1	05200025 05200026
11/02/2008	ML092151086	2008/11/02 Vogtle COL Review - FW: Vogtle Info 4 Page(s)	E-Mail	NRC/NRO	NRC/NRO/DNRL/NWE1	05200025 05200026
11/03/2008	ML092151089	2008/11/03 Vogtle COL Review - Draft RAI # 1351 related to SRP section 6.4 for Vogtle units 3 and 4 3 Page(s)	E-Mail	NRC/NRO	NRC/NRO/DNRL/NWE1	05200025 05200026
11/03/2008	ML092151091	2008/11/03 Vogtle COL Review - Draft RAIs # 1527 and 1528 related to SRP section 11.02 and 11.03 respectively for Vogtle units 3 and 4 4 Page(s)	E-Mail	NRC/NRO	NRC/NRO/DNRL/NWE1	05200025 05200026
11/04/2008	ML083090425	2008/11/04 Vogtle RAI for SER - Request for Additional Information Letter No. 009 Related to SRP Section 09.02.02 for the Vogtle Units 3 and 4 Combined License Application 6 Page(s)	E-Mail Request for Additional Information (RAI)	NRC/NRO	NRC/NRO/DNRL/NWE1	05200025 05200026

Appendix B

Chronology of the Combined License Application for Vogtle Electric Generating Plant Units 3 and 4

This appendix contains a chronological listing of routine licensing correspondence between the staff of the U.S. Nuclear Regulatory Commission (NRC) regarding the review of the Vogtle Electric Generating Plant, Units 3 and 4 plant design under Docket Nos. 052-000025 and 052-000026

Document Date	Accession Number	Title & Estimated Page Count	Document Type	Author Affiliation	Addressee Affiliation	Docket Number
11/04/2008	ML083090458	2008/11/04 Vogtle COL Review - RAI Letter No. 009 Related SRP Section 9.2.2 for Vogtle Units 3 and 4 8 Page(s)	E-Mail	NRC/NRO	NRC/NRO/DNRL/NWE1	05200025 05200026
11/04/2008	ML083110369	Vogtle, Units 3 and 4 - Response to Request for Additional Information Letter No. 003. 12 Page(s)	Letter	Southern Nuclear Operating Co, Inc	NRC/Document Control Desk NRC/NRO	05200025 05200026
11/04/2008	ML083110370	Vogtle, Units 3 and 4 Combined License Application, Response to Request for Additional Information Letter No. 005. 8 Page(s)	Letter	Southern Nuclear Operating Co, Inc	NRC/Document Control Desk NRC/NRO	05200025 05200026
11/04/2008	ML083110470	Vogtle, Units 3 and 4 - Response to Request for Additional Information Letter No. 004 re: Involving SWS Cooling Towers. 7 Page(s)	Letter	Southern Nuclear Operating Co, Inc	NRC/Document Control Desk NRC/NRO	05200025 05200026
11/07/2008	ML092151093	2008/11/07 Vogtle COL Review - FW: Draft RAIs # 1527 and 1528 related to SRP section 11.02 and 11.03 respectively for Vogtle units 3 and 4 4 Page(s)	E-Mail	NRC/NRO	NRC/NRO/DNRL/NWE1	05200025 05200026
11/07/2008	ML092151095	2008/11/07 Vogtle COL Review - Draft RAI # 1526 related to Section 19 for Vogtle Units 3 and 4 3 Page(s)	E-Mail	NRC/NRO	NRC/NRO/DNRL/NWE1	05200025 05200026

Chronology of the Combined License Application for Vogtle Electric Generating Plant Units 3 and 4

This appendix contains a chronological listing of routine licensing correspondence between the staff of the U.S. Nuclear Regulatory Commission (NRC) regarding the review of the Vogtle Electric Generating Plant, Units 3 and 4 plant design under Docket Nos. 052-000025 and 052-000026

Document Date	Accession Number	Title & Estimated Page Count	Document Type	Author Affiliation	Addressee Affiliation	Docket Number
11/13/2008	ML083180069	2008/11/13 Vogtle RAI for SER – Request for Information Letter No. 010 Related to SRP Section 11.02 for the Vogtle Electric Generating Plant Untis 3 and 4 Combined License Application 6 Page(s)	Request for Additional Information (RAI)	NRC/NRO	NRC/NRO/DNRL/NWE1	05200025 05200026
11/13/2008	ML083180070	2008/11/13 Vogtle RAI for SER - Request for Additional Information Letter No 011 Related to SRP Section 11.03 for the Vogtle Electric Generating Plant Units 3 and 4 Combiend License Application 6 Page(s)	Request for Additional Information (RAI)	NRC/NRO	NRC/NRO/DNRL/NWE1	05200025 05200026
11/13/2008	ML083180071	2008/11/13 Vogtle RAI for SER – Request for Additional Information Letter No. 012 Related to SRP Section 6.4 for the Vogtle Electric Generating Plant Untis 3 and 4 Combined License Application 6 Page(s)	Request for Additional Information (RAI)	NRC/NRO	NRC/NRO/DNRL/NWE1	05200025 05200026
11/13/2008	ML083180932	2008/11/13 Vogtle RAI for SER - Request for Additional Information Letter No. 013 Related to SRP Section 02.05.04 for the Vogtle 3 and 4 Combined License Application 6 Page(s)	E-Mail Request for Additional Information (RAI)	NRC/NRO	NRC/NRO/DNRL/NWE1	05200025 05200026
11/13/2008	ML083181080	2008/11/13 Vogtle COL Review - RAI Letter No. 013 Related SRP	E-Mail	NRC/NRO	NRC/NRO/DNRL/NWE1	05200025 05200026

Appendix B

Chronology of the Combined License Application for Vogtle Electric Generating Plant Units 3 and 4

This appendix contains a chronological listing of routine licensing correspondence between the staff of the U.S. Nuclear Regulatory Commission (NRC) regarding the review of the Vogtle Electric Generating Plant, Units 3 and 4 plant design under Docket Nos. 052-000025 and 052-000026

Document Date	Accession Number	Title & Estimated Page Count	Document Type	Author Affiliation	Addressee Affiliation	Docket Number
11/13/2008	ML083181255	Section 2.5.4 for Vogtle Units 3 and 4 8 Page(s)				
11/13/2008	ML083181255	2008/11/13 Vogtle COL Review - Draft RAIs 1531 and 1532 Related SRP Section 2.4.13 for Vogtle Units 3 and 4 4 Page(s)	E-Mail	NRC/NRO	NRC/NRO/DNRL/NWE1	05200025 05200026
11/13/2008	ML092151096	2008/11/13 Vogtle COL Review - Summary of Telephone call with Applicant (Vogtle) November 12, 2008 2 Page(s)	E-Mail	NRC/NRO	NRC/NRO/DNRL/NWE1	05200025 05200026
11/13/2008	ML092151097	2008/11/13 Vogtle COL Review - RAI letter Number 010 related to SRP section 11.02 for Vogtle Units 3 and 4 8 Page(s)	E-Mail	NRC/NRO	NRC/NRO/DNRL/NWE1	05200025 05200026
11/13/2008	ML092151099	2008/11/13 Vogtle COL Review - RAI letter number 011 related to SRP section 11.03 for Vogtle units 3 and 4 8 Page(s)	E-Mail	NRC/NRO	NRC/NRO/DNRL/NWE1	05200025 05200026
11/13/2008	ML092151100	2008/11/13 Vogtle COL Review - RAI letter number 012 related to SRP section 6.4 for Vogtle units 3 and 4 8 Page(s)	E-Mail	NRC/NRO	NRC/NRO/DNRL/NWE1	05200025 05200026
11/13/2008	ML092151101	2008/11/13 Vogtle COL Review -	E-Mail	NRC/NRO	NRC/NRO/DNRL/NWE1	05200025

Appendix B

Chronology of the Combined License Application for Vogtle Electric Generating Plant Units 3 and 4

This appendix contains a chronological listing of routine licensing correspondence between the staff of the U.S. Nuclear Regulatory Commission (NRC) regarding the review of the Vogtle Electric Generating Plant, Units 3 and 4 plant design under Docket Nos. 052-000025 and 052-000026

Document Date	Accession Number	Title & Estimated Page Count	Document Type	Author Affiliation	Addressee Affiliation	Docket Number
		FW: Draft RAIs # 1527 and 1528 related to SRP section 11.02 and 11.03 respectively for Vogtle units 3 and 4 5 Page(s)				05200026
11/13/2008	ML092151102	2008/11/13 Vogtle COL Review - FW: Participants in the Ch6 draft RAI call on 11/12/08 2 Page(s)	E-Mail	NRC/NRO	NRC/NRO/DNRL/NWE1	05200025 05200026
11/14/2008	ML083210003	2008/11/14-Vogtle - Letter to Colonel Edward Kertis. 2 Page(s)	Legal-Correspondence	NRC/OGC	US Dept of the Army, Corps of Engineers US Dept of the Army, Corps of Engineers, Savannah District	05200025 05200026
11/14/2008	ML083220125	2008/11/14-Vogtle - Attachment 2 - Anticipated Questions for Testimony Re: Savannah River Federal Navigation Channel and Barge Transportation. 4 Page(s)	Legal-Correspondence	NRC/OGC		05200025 05200026
11/14/2008	ML083230221	Vogtle, Units 3 and 4 Combined License Application - Response to Request for Additional Information Letter No. 007. 8 Page(s)	Letter	Southern Nuclear Operating Co, Inc	NRC/Document Control Desk NRC/NRO	05200025 05200026
11/14/2008	ML083230224	Vogtle, Units 3 and 4 - Response to Request for Additional Information Letter No. 006. 9 Page(s)	Letter	Southern Nuclear Operating Co, Inc	NRC/Document Control Desk NRC/NRO	05200025 05200026

Appendix B

Chronology of the Combined License Application for Vogtle Electric Generating Plant Units 3 and 4

This appendix contains a chronological listing of routine licensing correspondence between the staff of the U.S. Nuclear Regulatory Commission (NRC) regarding the review of the Vogtle Electric Generating Plant, Units 3 and 4 plant design under Docket Nos. 052-000025 and 052-000026

Document Date	Accession Number	Title & Estimated Page Count	Document Type	Author Affiliation	Addressee Affiliation	Docket Number
11/17/2008	ML083240585	2008/11/17 - Petition Request to Intervene in the Vogtle COL Proceeding, submitted by Atlanta Women's Action for New Directions, et al. 24 Page(s)	Legal-Petition for Rulemaking	Emory Univ School of Law	NRC/SECY	05200025 05200026
11/18/2008	ML083250482	Vogtle, Units 3 and 4 Combined License Application Response to Request for Additional Information Letter No. 008. 11 Page(s)	Letter	Southern Nuclear Operating Co, Inc	NRC/Document Control Desk NRC/NRO	05200025 05200026
11/19/2008	ML092151136	2008/11/19 Vogtle COL Review - DRAFT - RAI 1486 - SRP section 17.5 - Vogtle Units 3 and 4 Combined License Application 5 Page(s)	E-Mail	NRC/NRO	NRC/NRO/DNRL/NWE1	05200025 05200026
11/20/2008	ML083250361	2008/11/20 Vogtle COL Review - Draft RAIs 1396 and Draft RAI 1400 Related SRP Section 9.5.1 for Vogtle Units 3 and 4 4 Page(s)	E-Mail	NRC/NRO	NRC/NRO/DNRL/NWE1	05200025 05200026
11/21/2008	ML083260749	2008/11/21-Request for Hearing with Respect to the Combined License Application of the Southern Nuclear Operating Company, for the Vogtle Electric Generating Plant, Units 3 and 4. 1 Page(s)	Legal-Order	NRC/SECY	NRC/ASLBP	05200025 05200026
11/25/2008	ML083300157	2008/11/25 Vogtle RAI for SER –	E-Mail	NRC/NRO	NRC/NRO/DNRL/NWE1	05200025

Appendix B

Chronology of the Combined License Application for Vogtle Electric Generating Plant Units 3 and 4

This appendix contains a chronological listing of routine licensing correspondence between the staff of the U.S. Nuclear Regulatory Commission (NRC) regarding the review of the Vogtle Electric Generating Plant, Units 3 and 4 plant design under Docket Nos. 052-000025 and 052-000026

Document Date	Accession Number	Title & Estimated Page Count	Document Type	Author Affiliation	Addressee Affiliation	Docket Number
		Request for Additonal Information Letter No. 014 Related to SRP Section 17.5 for the Vogtle Electric Generating Plant 7 Page(s)	Request for Additional Information (RAI)			05200026
11/25/2008	ML092151137	2008/11/25 Vogtle COL Review – Request for Additional Information Letter No. 014 Related to SRP Section 17.5 for the Vogtle Untis 3 and 4 Combined License Application 9 Page(s)	E-Mail	NRC/NRO	NRC/NRO/DNRL/NWE1	05200025 05200026
11/25/2008	ML092151138	2008/11/25 Vogtle COL Review - FW: Request for Additional Information Letter No. 014 Related to SRP Section 17.5 for the Vogtle Untis 3 and 4 Combined License Application 9 Page(s)	E-Mail	NRC/NRO	NRC/NRO/DNRL/NWE1	05200025 05200026
11/26/2008	ML083310028	2008/11/26 Vogtle RAI for SER - Request for Additional Information Letter No. 015 Related to SRP Section 09.05.01 for the Vogtle Units 3 and 4 Combined License Application 6 Page(s)	E-Mail Request for Additional Information (RAI)	NRC/NRO	NRC/NRO/DNRL/NWE1	05200025 05200026
11/26/2008	ML083310029	2008/11/26 Vogtle RAI for SER - Request for Additional Information Letter No. 016 Related to SRP Section 09.05.01 for the Vogtle	E-Mail Request for Additional Information (RAI)	NRC/NRO	NRC/NRO/DNRL/NWE1	05200025 05200026

Chronology of the Combined License Application for Vogtle Electric Generating Plant Units 3 and 4

This appendix contains a chronological listing of routine licensing correspondence between the staff of the U.S. Nuclear Regulatory Commission (NRC) regarding the review of the Vogtle Electric Generating Plant, Units 3 and 4 plant design under Docket Nos. 052-000025 and 052-000026

Document Date	Accession Number	Title & Estimated Page Count	Document Type	Author Affiliation	Addressee Affiliation	Docket Number
		Units 3 and 4 Combined License Application 6 Page(s)				
11/26/2008	ML083310122	2008/11/26 Vogtle COL Review - RAI Letter No. 015 and 016 Related to SRP Section 9.5.1 for Vogtle Units 3 and 4 14 Page(s)	E-Mail	NRC/NRO	NRC/NRO/DNRL/NWE1	05200025 05200026
12/05/2008	ML083310313	Safety Evaluation Report Input Regarding the Vogtle Electric Generating Plant Combined License Application, Final Safety Analysis Report Chapter 17. 27 Page(s)	Memoranda Safety Evaluation Report	NRC/NRO/DCIP/CQVP	NRC/NRO/DNRL/NWE1	05200025 05200026
11/26/2008	ML083310349	2008/11/26 Vogtle RAI for SER - Request for Additional Information Letter No. 017 Related to SRP Section 02.04.13 for the Vogtle 3 and 4 Combined License Application 6 Page(s)	E-Mail Request for Additional Information (RAI)	NRC/NRO	NRC/NRO/DNRL/NWE1	05200025 05200026
11/26/2008	ML083310358	2008/11/26 Vogtle COL Review - RAI Letter No. 017 Related SRP Section 2.4.13 for Vogtle Units 3 and 4 8 Page(s)	E-Mail	NRC/NRO	NRC/NRO/DNRL/NWE1	05200025 05200026
12/02/2008	ML083370521	2008/12/02-Establishment of Atomic Safety and Licensing Board for Vogtle, Units 3 and 4	Legal-Order	NRC/ASLBP		05200025 05200026

Chronology of the Combined License Application for Vogtle Electric Generating Plant Units 3 and 4

This appendix contains a chronological listing of routine licensing correspondence between the staff of the U.S. Nuclear Regulatory Commission (NRC) regarding the review of the Vogtle Electric Generating Plant, Units 3 and 4 plant design under Docket Nos. 052-000025 and 052-000026

Document Date	Accession Number	Title & Estimated Page Count	Document Type	Author Affiliation	Addressee Affiliation	Docket Number
		Proceedings. 5 Page(s)				
12/02/2008	ML083370699	2008/12/02 Vogtle COL Review - No Conference Call Required for Electronic RAI # 1486 2 Page(s)	E-Mail	NRC/NRO	NRC/NRO/DNRL/NWE1	05200025 05200026
12/02/2008	ML083390120	Vogtle, Units 3 and 4, Combined License Application, Response to Request for Additional Information Letter No. 009. 6 Page(s)	Letter	Southern Nuclear Operating Co, Inc	NRC/Document Control Desk NRC/NRO	05200025 05200026
12/05/2008	ML083400446	2008/12/05-Notice of Appearance of Kathryn M. Sutton on Behalf of Southern Nuclear Operating Co. Regarding Vogtle, Units 3 and 4. 3 Page(s)	Legal-Notice of Appearance Legal-Pleading	Morgan, Lewis & Bockius, LLP Southern Nuclear Operating Co, Inc	NRC/ASLBP	05200025 05200026
12/11/2008	ML092151139	2008/12/11 Vogtle COL Review - DRAFT - RAI 1721 - SRP section 13.1.1 - Vogtle Units 3 and 4 Combined License Application 3 Page(s)	E-Mail	NRC/NRO	NRC/NRO/DNRL/NWE1	05200025 05200026
12/11/2008	ML083510076	Vogtle, Units 3 & 4, Combined License Application, Response to Request for Additional Information Letter No. 010 Regarding Liquid Waste Management Systems. 8 Page(s)	Letter	Southern Nuclear Operating Co, Inc	NRC/Document Control Desk NRC/NRO	05200025 05200026
12/11/2008	ML083510077	Vogtle Electric Generating Plant Units 3 and 4 Combined License	Letter	Southern Nuclear Operating Co, Inc	NRC/Document Control Desk	05200025 05200026

Appendix B

Chronology of the Combined License Application for Vogtle Electric Generating Plant Units 3 and 4

This appendix contains a chronological listing of routine licensing correspondence between the staff of the U.S. Nuclear Regulatory Commission (NRC) regarding the review of the Vogtle Electric Generating Plant, Units 3 and 4 plant design under Docket Nos. 052-000025 and 052-000026

Document Date	Accession Number	Title & Estimated Page Count	Document Type	Author Affiliation	Addressee Affiliation	Docket Number
		Application - Response to Request for Additional Information Letter No. 012 Concerning Control Room Habitability System Information Need. 7 Page(s)			NRC/NRO	
12/11/2008	ML083510078	Vogtle, Units 3 & 4, Combined License Application, Response to Request for Additional Information Letter No. 011 Regarding Gaseous Waste Management System. 8 Page(s)	Letter	Southern Nuclear Operating Co, Inc	NRC/Document Control Desk NRC/NRO	05200025 05200026
12/11/2008	ML092151140	2008/12/11 Vogtle COL Review - DRAFT - RAI 1722 - SRP Sections 13.1.2 & 13.1.3 - Vogtle Units 3 and 4 Combined License Application 3 Page(s)	E-Mail	NRC/NRO	NRC/NRO/DNRL/NWE1	05200025 05200026
12/11/2008	ML092151141	2008/12/11 Vogtle COL Review - Draft RAIs 1748, 1751, and 1752 related to Chapter 12 for Vogtle 7 Page(s)	E-Mail	NRC/NRO	NRC/NRO/DNRL/NWE1	05200025 05200026
12/12/2008	ML083470212	2008/12/12 Vogtle COL Review - Draft RAI 1596 Related SRP Section 8.2 for Vogtle Units 3 and 4 5 Page(s)	E-Mail	NRC/NRO	NRC/NRO/DNRL/NWE1	05200025 05200026
12/11/2008	ML083530555	Vogtle, Unit 3 and 4 Combined License Application, Response to Request for Additional Information Letter No. 013.	Letter	Southern Nuclear Operating Co, Inc	NRC/Document Control Desk NRC/NRO	05200025 05200026

Appendix B

Chronology of the Combined License Application for Vogtle Electric Generating Plant Units 3 and 4

This appendix contains a chronological listing of routine licensing correspondence between the staff of the U.S. Nuclear Regulatory Commission (NRC) regarding the review of the Vogtle Electric Generating Plant, Units 3 and 4 plant design under Docket Nos. 052-000025 and 052-000026

Document Date	Accession Number	Title & Estimated Page Count	Document Type	Author Affiliation	Addressee Affiliation	Docket Number
12/12/2008	ML083470948	2008/12/12 Vogtle RAI for SER - Request for Additional Information Letter No. 019 Related to SRP Section 02.02.03 for the Vogtle 3 and 4 Combined License Application 6 Page(s)	E-Mail Request for Additional Information (RAI)	NRC/NRO	NRC/NRO/DNRL/NWE1	05200025 05200026
12/12/2008	ML083470951	2008/12/12 Vogtle COL Review - RAI Letter No. 019 Related SRP Section 2.2.3 for Vogtle Units 3 and 4 8 Page(s)	E-Mail	NRC/NRO	NRC/NRO/DNRL/NWE1	05200025 05200026
12/15/2008	ML083500265	2008/12/15 Vogtle RAI for SER - RAI LETTER NO. 018 Related to SRP Section 03.07.02 for the Vogtle 3 and 4 Combined License Applciation 7 Page(s)	E-Mail Request for Additional Information (RAI)	NRC/NRO	NRC/NRO/DNRL/NWE1	05200025 05200026
12/16/2008	ML083540149	Vogtle, Units 3 and 4 - Combined License Application Contract for Disposal of High-Level Radioactive Waste. 4 Page(s)	Letter	Southern Nuclear Operating Co, Inc	NRC/Document Control Desk NRC/NRO	05200025 05200026

Chronology of the Combined License Application for Vogtle Electric Generating Plant Units 3 and 4

This appendix contains a chronological listing of routine licensing correspondence between the staff of the U.S. Nuclear Regulatory Commission (NRC) regarding the review of the Vogtle Electric Generating Plant, Units 3 and 4 plant design under Docket Nos. 052-000025 and 052-000026

Document Date	Accession Number	Title & Estimated Page Count	Document Type	Author Affiliation	Addressee Affiliation	Docket Number
12/17/2008	ML083520028	2008/12/17 Vogtle RAI for SER – Request for Additional Information Letter No. 020 Related to SRP Section 19 fro the Vogtle Electric Generating Plant Untis 3 and 4 Combined License Application 6 Page(s)	E-Mail Request for Additional Information (RAI)	NRC/NRO	NRC/NRO/DNRL/NWE1	05200025 05200026
12/17/2008	ML083520373	Vogtle Electric Generating Plant Units 3 and 4 Combined Operating License, Revision 0, Phase 2 Inputs for Final Safety Evaluation Report Section 10.4.6. 7 Page(s)	Memoranda	NRC/NRO/DE/CIB1	NRC/NRO/DNRL/NWE1	05200025 05200026
12/17/2008	ML083520693	2008/12/17 Vogtle COL Review - Draft RAI 1854 Related SRP Section 3.8.4 for Vogtle Units 3 and 4 3 Page(s)	E-Mail	NRC/NRO	NRC/NRO/DNRL/NWE1	05200025 05200026
12/17/2008	ML083570397	Vogtle Electric Generating Plant Units 3 & 4 Combined License Application Response to Request for Additional Information Letter No. 015. 6 Page(s)	Letter	Southern Nuclear Operating Co, Inc	NRC/Document Control Desk NRC/NRO	05200025 05200026
12/17/2008	ML083570588	Vogtle, Units 3 & 4, Combined License Application Response to Request for Additional Information Letter No. 016 Concerning Fire Protection Engineer Qualification	Letter	Southern Nuclear Operating Co, Inc	NRC/Document Control Desk NRC/NRO	05200025 05200026

Appendix B

Chronology of the Combined License Application for Vogtle Electric Generating Plant Units 3 and 4

This appendix contains a chronological listing of routine licensing correspondence between the staff of the U.S. Nuclear Regulatory Commission (NRC) regarding the review of the Vogtle Electric Generating Plant, Units 3 and 4 plant design under Docket Nos. 052-000025 and 052-000026

Document Date	Accession Number	Title & Estimated Page Count	Document Type	Author Affiliation	Addressee Affiliation	Docket Number
12/17/2008	ML083570590	Requirements. 6 Page(s) Vogtle, Units 3 and 4 - Combined License Application - Endorsement of Bellefonte R-COLA Standard Content Requests for Additional Information. 19 Page(s)	Letter	Southern Nuclear Operating Co, Inc	NRC/Document Control Desk NRC/NRO	05200025 05200026
12/17/2008	ML083570594	Vogtle, Units 3 and 4, Combined License Application, Response to Request for Additional Information Letter No. 014. 9 Page(s)	Letter	Southern Nuclear Operating Co, Inc	NRC/Document Control Desk NRC/NRO	05200025 05200026
12/17/2008	ML092151142	2008/12/17 Vogtle COL Review - RAI letter number 020 related to Section 19 for Vogtle units 3 and 4 8 Page(s)	E-Mail	NRC/NRO	NRC/NRO/DNRL/NWE1	05200025 05200026
12/17/2008	ML092600709	2008/12/17 Vogtle COL Review - FW: RAI for Vogtle Roof Design Load.doc 3 Page(s)	E-Mail	NRC/NRO	NRC/NRO/DNRL/NWE1	05200025 05200026
12/18/2008	ML083530816	2008/12/18 Vogtle RAI for SER - RAI Section: 03.08.04 for the Vogtle COL Application 6 Page(s)	E-Mail Request for Additional Information (RAI)	NRC/NRO	NRC/NRO/DNRL/NWE1	05200025 05200026
12/18/2008	ML092151143	2008/12/18 Vogtle COL Review - Draft RAI No 568 related to SRP Section: 14.02 - Initial Plant Test Program for Vogtle Electric	E-Mail	NRC/NRO	NRC/NRO/DNRL/NWE1	05200025 05200026

Appendix B

Chronology of the Combined License Application for Vogtle Electric Generating Plant Units 3 and 4

This appendix contains a chronological listing of routine licensing correspondence between the staff of the U.S. Nuclear Regulatory Commission (NRC) regarding the review of the Vogtle Electric Generating Plant, Units 3 and 4 plant design under Docket Nos. 052-000025 and 052-000026

Document Date	Accession Number	Title & Estimated Page Count	Document Type	Author Affiliation	Addressee Affiliation	Docket Number
		Generating Plant Units 3 and 4 3 Page(s)				
12/19/2008	ML083540133	2008/12/19 Vogtle RAI for SER – Request for Additional Information Letter No. 023 Related to SRP Section 13.1.1 For the Vogtle Electric Generating Plant 6 Page(s)	E-Mail Request for Additional Information (RAI)	NRC/NRO	NRC/NRO/DNRL/NWE1	05200025 05200026
12/19/2008	ML083540135	2008/12/19 Vogtle RAI for SER – Request for Additional Information Letter No. 024 Related to SRP Section 13.1.2-13.1.3 for the Vogtle Electric Generating Plant 6 Page(s)	E-Mail Request for Additional Information (RAI)	NRC/NRO	NRC/NRO/DNRL/NWE1	05200025 05200026
12/19/2008	ML083540273	2008/12/19 Vogtle COL Review - Vogtle SCOL Teleconference on 12/18 for Chapter 3 2 Page(s)	E-Mail	NRC/NRO	NRC/NRO/DNRL/NWE1	05200025 05200026
12/19/2008	ML083540805	2008/12/19 Vogtle RAI for SER – Request for Additional Information Letter No. 021 Related to SRP Section 12.3-12.4 for the Vogtle Electric Generating Plant Units 3 and 4 Combined License Application 8 Page(s)	E-Mail Request for Additional Information (RAI)	NRC/NRO	NRC/NRO/DNRL/NWE1	05200025 05200026

Chronology of the Combined License Application for Vogtle Electric Generating Plant Units 3 and 4

This appendix contains a chronological listing of routine licensing correspondence between the staff of the U.S. Nuclear Regulatory Commission (NRC) regarding the review of the Vogtle Electric Generating Plant, Units 3 and 4 plant design under Docket Nos. 052-000025 and 052-000026

Document Date	Accession Number	Title & Estimated Page Count	Document Type	Author Affiliation	Addressee Affiliation	Docket Number
12/19/2008	ML083540807	2008/12/19 Vogtle RAI for SER - Request for Additional Information Letter No. 025 Related to SRP Section 08.02 for the Vogtle 3 and 4 Combined License Application 8 Page(s)	E-Mail Request for Additional Information (RAI)	NRC/NSIR	NRC/NRO/DNRL/NWE1	05200025 05200026
12/19/2008	ML083540809	2008/12/19 Vogtle COL Review - Conference Call Summary – December 18, 2008 – Southern Nuclear Vogtle Units 3 and 4 COLA - RAIs # 1721, 1722 - Draft RAIs related to SRP Section 13.1 2 Page(s)	E-Mail	NRC/NRO	NRC/NRO/DNRL/NWE1	05200025 05200026
12/19/2008	ML083540810	2008/12/19 Vogtle COL Review - Teleconference with SNC on RAIs for Vogtle SCOL Chapter 8 2 Page(s)	E-Mail	NRC/NRO	NRC/NRO/DNRL/NWE1	05200025 05200026
12/19/2008	ML092151144	2008/12/19 Vogtle COL Review – Request for Additional Information Letter No. 023 Related to SRP Section 13.1.1 for the Vogtle Units 3 and 4 Combined License Application 8 Page(s)	E-Mail	NRC/NRO	NRC/NRO/DNRL/NWE1	05200025 05200026
12/19/2008	ML092151145	2008/12/19 Vogtle COL Review – Request for Additional Letter No. 024 Related to SRP Section 13.1.2-13.1.3 for the Vogtle Units 3 and 4 Ccombined License Application	E-Mail	NRC/NRO	NRC/NRO/DNRL/NWE1	05200025 05200026

Appendix B

Chronology of the Combined License Application for Vogtle Electric Generating Plant Units 3 and 4

This appendix contains a chronological listing of routine licensing correspondence between the staff of the U.S. Nuclear Regulatory Commission (NRC) regarding the review of the Vogtle Electric Generating Plant, Units 3 and 4 plant design under Docket Nos. 052-000025 and 052-000026

Document Date	Accession Number	Title & Estimated Page Count	Document Type	Author Affiliation	Addressee Affiliation	Docket Number
12/19/2008	ML092151146	2008/12/19 Vogtle COL Review - RAI Letter No. 021 Related to SRP Section 12.03-12.04 for Vogtle 8 Page(s)	E-Mail	NRC/NRO	NRC/NRO/DNRL/NWE1	05200025 05200026
12/19/2008	ML083540823	2008/12/19 Vogtle COL Review - RAI Letter No. 025 Related SRP Section 8.2 for Vogtle Units 3 and 4 10 Page(s)	E-Mail	NRC/NRO	NRC/NRO/DNRL/NWE1	05200025 05200026
12/23/2008	ML083640476	Vogtle, Units 3 and 4 Combined License Application, Supplemental Response to Request for Additional Information Letter No. 005. 11 Page(s)	Letter	Southern Nuclear Operating Co, Inc	NRC/Document Control Desk NRC/NRO	05200025 05200026
12/23/2008	ML083640477	Vogtle, Units 3 and 4 Combined License Application, Response to Request for Additional Information Letter No. 017. 7 Page(s)	Letter	Southern Nuclear Operating Co, Inc	NRC/Document Control Desk NRC/NRO	05200025 05200026
12/23/2008	ML083640478	Vogtle, Units 3 and 4 Combined License Application, Update to Part 1 of the COLA. 3 Page(s)	Letter	Southern Nuclear Operating Co, Inc	NRC/Document Control Desk NRC/NRO	05200025 05200026
12/23/2008	ML083640479	Vogtle, Units 3 and 4 Combined License Application Future Incorporation of Vogtle Early Site Permit Application (ESPA) Revision 5. 3 Page(s)	Letter	Southern Nuclear Operating Co, Inc	NRC/Document Control Desk NRC/NRO	05200025 05200026

Appendix B

Chronology of the Combined License Application for Vogtle Electric Generating Plant Units 3 and 4

This appendix contains a chronological listing of routine licensing correspondence between the staff of the U.S. Nuclear Regulatory Commission (NRC) regarding the review of the Vogtle Electric Generating Plant, Units 3 and 4 plant design under Docket Nos. 052-000025 and 052-000026

Document Date	Accession Number	Title & Estimated Page Count	Document Type	Author Affiliation	Addressee Affiliation	Docket Number
12/30/2008	ML091130217	Vogtle COL SER/OI Section 2.3.5: XOQDOQ Atmospheric Dispersion Model Runs for Site Boundary and Special Receptors. 21 Page(s)	Graphics incl Charts and Tables	NRC/NRO		05200025 05200026
12/23/2008	ML091050318	2008/12/23 Vogtle COL Review - ND-08-1929 VEGP Update to Part 1 of COLA 5 Page(s)	E-Mail	- No Known Affiliation	NRC/NRO/DNRL/NWE1	05200025 05200026
12/23/2008	ML091050319	2008/12/23 Vogtle COL Review - ND-08-1930 Future Inc. VEGP of ESPA Rev.5 5 Page(s)	E-Mail	- No Known Affiliation	NRC/NRO/DNRL/NWE1	05200025 05200026
01/09/2009	ML090140355	Vogtle, Units 3 & 4 Combined License Application, Response to Request for Additional Information Letter No. 019. 6 Page(s)	Letter	Southern Nuclear Operating Co, Inc	NRC/Document Control Desk NRC/NRO	05200025 05200026
01/14/2009	ML090150432	Vogtle, Units 3 and 4 Combined License Application - Response to Request for Additional Information Letter No. 018. 6 Page(s)	Letter	Southern Nuclear Operating Co, Inc	NRC/Document Control Desk NRC/NRO	05200025 05200026
01/15/2009	ML090150449	2009/01/15 Vogtle COL Review - Draft RAIs 1931 Related SRP Section 9.2.1 for Vogtle Units 3 and 4 5 Page(s)	E-Mail	NRC/NRO	NRC/NRO/DNRL/NWE1	05200025 05200026
01/16/2009	ML090160119	2009/01/16 Vogtle COL Review -	E-Mail	NRC/NRO	NRC/NRO/DNRL/NWE1	05200025

Appendix B

Chronology of the Combined License Application for Vogtle Electric Generating Plant Units 3 and 4

This appendix contains a chronological listing of routine licensing correspondence between the staff of the U.S. Nuclear Regulatory Commission (NRC) regarding the review of the Vogtle Electric Generating Plant, Units 3 and 4 plant design under Docket Nos. 052-000025 and 052-000026

Document Date	Accession Number	Title & Estimated Page Count	Document Type	Author Affiliation	Addressee Affiliation	Docket Number
01/16/2009		Telcon Summary – Telcon with Vogtle 11/24/2008 2 Page(s)				05200026
01/16/2009	ML090220179	Vogtle, Units 3 and 4, Combined License Application, Response to Request for Additional Information Letter No. 024. 18 Page(s)	Letter	Southern Nuclear Operating Co, Inc	NRC/Document Control Desk NRC/NRO	05200025 05200026
01/16/2009	ML090220180	Vogtle, Units 3 and 4 - Response to Request for Additional Information Letter No. 025. 14 Page(s)	Letter	Southern Nuclear Operating Co, Inc	NRC/Document Control Desk NRC/NRO	05200025 05200026
01/16/2009	ML090220267	Vogtle, Units 3 and 4 Combined License Application, Response to Request for Additional Information Letter No. 021 Regarding Radiation Protection. 9 Page(s)	Letter	Southern Nuclear Operating Co, Inc	NRC/Document Control Desk NRC/NRO	05200025 05200026
01/16/2009	ML090220268	Vogtle, Units 3 and 4 Combined License Application, Response to Request for Additional Information Letter No. 022 Regarding Roof Loading. 6 Page(s)	Letter	Southern Nuclear Operating Co, Inc	NRC/Document Control Desk NRC/NRO	05200025 05200026
01/16/2009	ML090220269	Vogtle, Units 3 and 4 Combined License Application - Response to Request for Additional Information Letter No. 023 Regarding Management Organizational	Letter	Southern Nuclear Operating Co, Inc	NRC/Document Control Desk NRC/NRO	05200025 05200026

Appendix B

Chronology of the Combined License Application for Vogtle Electric Generating Plant Units 3 and 4

This appendix contains a chronological listing of routine licensing correspondence between the staff of the U.S. Nuclear Regulatory Commission (NRC) regarding the review of the Vogtle Electric Generating Plant, Units 3 and 4 plant design under Docket Nos. 052-000025 and 052-000026

Document Date	Accession Number	Title & Estimated Page Count	Document Type	Author Affiliation	Addressee Affiliation	Docket Number
01/22/2009	ML090220539	Press Release-09-015: NRC Licensing Board to Discuss Vogtle Nuclear Plant Combined License, Early Site Permit Applications on Jan. 28. 1 Page(s)	Press Release	NRC/OPA		05200011 05200025 05200026
01/26/2009	ML090260496	2009/01/26 Vogtle RAI for SER - Request for Additional Information Letter No. 026 Related to SRP Section 09.02.01 for the Vogtle Units 3 and 4 Combined License Application 8 Page(s)	E-Mail Request for Additional Information (RAI)	NRC/NRO	NRC/NRO/DNRL/NWE1	05200025 05200026
01/26/2009	ML090260529	2009/01/26 Vogtle COL Review - RAI Letter No. 026 Related to SRP Section 9.2.1 for Vogtle Units 3 and 4 10 Page(s)	E-Mail	NRC/NRO	NRC/NRO/DNRL/NWE1	05200025 05200026
01/28/2009	ML090350072	2009/01/28-Transcript of Southern Nuclear Operating Company Vogtle Units 3 & 4 Hearing, January 28, 2009, Pages 1-113. 115 Page(s)	Legal-Hearing Transcript	NRC/ASLBP		05200025 05200026
02/04/2009	ML091350305	(1) Letter to FEMA, and (2) Enclosed Vogtle ESP Application Final Safety Evaluation Report (FSER), Section 13.3, Emergency	- No Document Type Applies	- No Known Affiliation	NRC/NSIR	05200025 05200026

Chronology of the Combined License Application for Vogtle Electric Generating Plant Units 3 and 4

This appendix contains a chronological listing of routine licensing correspondence between the staff of the U.S. Nuclear Regulatory Commission (NRC) regarding the review of the Vogtle Electric Generating Plant, Units 3 and 4 plant design under Docket Nos. 052-000025 and 052-000026

Document Date	Accession Number	Title & Estimated Page Count	Document Type	Author Affiliation	Addressee Affiliation	Docket Number
		Planning. 148 Page(s)				
02/06/2009	ML090371065	2009/02/06 Vogtle COL Review - Draft RAIs 2057 and 2091 Related SRP Sections 1.0, 2.4.2, 2.4.13 for Vogtle Units 3 and 4 5 Page(s)	E-Mail	NRC/NRO	NRC/NRO/DNRL/NWE1	05200025 05200026
02/06/2009	ML090400995	Tables Showing Cumulative Water Withdrawals of Vogtle Units 3 & 4. 1 Page(s)	Graphics incl Charts and Tables	NRC/NRO		05200025 05200026
02/08/2009	ML092151147	2009/02/08 Vogtle COL Review - FW: Participant list 1/16/09 2 Page(s)	E-Mail	NRC/NRO	NRC/NRO/DNRL/NWE1	05200025 05200026
02/08/2009	ML092151148	2009/02/08 Vogtle COL Review - FW: 2/12/09 AP1000 DCWG Agenda topics 2 Page(s)	E-Mail	NRC/NRO	NRC/NRO/DNRL/NWE1	05200025 05200026
02/08/2009	ML092250106	2009/02/08 Vogtle COL Review - FW: NRO Directors Issue Resolution Process 4 Page(s)	E-Mail	NRC/NRO	NRC/NRO/DNRL/NWE1	05200025 05200026
02/09/2009	ML090340581	Draft Safety Evaluation Report (SER) with Open Items for FSAR Chapter 11 of the Vogtle Units 3 and 4 Combined License Application (COLA). 34 Page(s)	Memoranda Safety Evaluation Report	NRC/NRO/DCIP/CHPB	NRC/NRO/DNRL/NWE1	05200025 05200026
02/10/2009	ML090490095	Vogtle, Units 3 & 4 Combined License Application, Response to	Letter	Southern Nuclear Operating Co, Inc	NRC/Document Control Desk	05200025 05200026

Appendix B

Chronology of the Combined License Application for Vogtle Electric Generating Plant Units 3 and 4

This appendix contains a chronological listing of routine licensing correspondence between the staff of the U.S. Nuclear Regulatory Commission (NRC) regarding the review of the Vogtle Electric Generating Plant, Units 3 and 4 plant design under Docket Nos. 052-000025 and 052-000026

Document Date	Accession Number	Title & Estimated Page Count	Document Type	Author Affiliation	Addressee Affiliation	Docket Number
		Request for Additional Information Letter No. 020. 13 Page(s)			NRC/NRO	
02/13/2009	ML090440272	Press Release-09-028: Licensing Board To Hold Hearing In Georgia March 16-25, Take Public Comments On Vogtle Site Permit Application. 2 Page(s)	Press Release	NRC/OPA		05200011 05200025 05200026
02/18/2009	ML090490044	Inspection Plan for the Quality Assurance Program Implementation Inspection of Vogtle Units 3 and 4 Combined License Application. 4 Page(s)	Inspection Plan Memoranda	NRC/NRO/DCIP/CQVP	NRC/NRO/DCIP/CQVP	05200025 05200026
02/18/2009	ML090490104	Inspection of Southern Nuclear Operating Company Quality Assurance Program Implementation for Vogtle Units 3 & 4 Combined License Application. 2 Page(s)	Letter	NRC/NRO/DCIP/CQVP	Southern Nuclear Operating Co, Inc	05200025 05200026
02/23/2009	ML090700518	2009/02/23 Vogtle COL Review - DRAFT - RAI 2087 - SRP section 13.3 - Vogtle Units 3 and 4 Combined License Application 7 Page(s)	E-Mail	NRC/NRO	NRC/NRO/DNRL/NWE1	05200025 05200026

Appendix B

Chronology of the Combined License Application for Vogtle Electric Generating Plant Units 3 and 4

This appendix contains a chronological listing of routine licensing correspondence between the staff of the U.S. Nuclear Regulatory Commission (NRC) regarding the review of the Vogtle Electric Generating Plant, Units 3 and 4 plant design under Docket Nos. 052-000025 and 052-000026

Document Date	Accession Number	Title & Estimated Page Count	Document Type	Author Affiliation	Addressee Affiliation	Docket Number
02/25/2009	ML090560655	2009/02/25 Vogtle RAI for SER - Request for Additional Information Letter No. 027 Related to SRP Sections 1 and 02.04.13 for the Vogtle 3 and 4 Combined License Application 6 Page(s)	E-Mail Request for Additional Information (RAI)	NRC/NRO	NRC/NRO/DNRL/NWE1	05200025 05200026
02/25/2009	ML090560717	2009/02/25 Vogtle COL Review - RAI Letter No. 027 Related SRP Sections 1 and 2.4.13 for Vogtle Units 3 and 4 8 Page(s)	E-Mail	NRC/NRO	NRC/NRO/DNRL/NWE1	05200025 05200026
02/25/2009	ML090560817	2009/02/25 Vogtle RAI for SER - Request for Additional Information Letter No. 028 Related to SRP Section 02.04.02 for the Vogtle 3 and 4 Combined License Application 7 Page(s)	E-Mail Request for Additional Information (RAI)	NRC/NRO	NRC/NRO/DNRL/NWE1	05200025 05200026
02/25/2009	ML090560818	2009/02/25 Vogtle COL Review - RAI Letter No. 028 Related SRP Section 2.4.2 for Vogtle Units 3 and 4 9 Page(s)	E-Mail	NRC/NRO	NRC/NRO/DNRL/NWE1	05200025 05200026
02/27/2009	ML090580630	2009/02/27 Vogtle COL Review - Draft RAI 2171 Related SRP Section 2.5.4 for Vogtle Units 3 and 4 4 Page(s)	E-Mail	NRC/NRO	NRC/NRO/DNRL/NWE1	05200025 05200026

Appendix B

Chronology of the Combined License Application for Vogtle Electric Generating Plant Units 3 and 4

This appendix contains a chronological listing of routine licensing correspondence between the staff of the U.S. Nuclear Regulatory Commission (NRC) regarding the review of the Vogtle Electric Generating Plant, Units 3 and 4 plant design under Docket Nos. 052-000025 and 052-000026

Document Date	Accession Number	Title & Estimated Page Count	Document Type	Author Affiliation	Addressee Affiliation	Docket Number
03/02/2009	ML090700628	Vogtle, Units 3 and 4 - Combined License Application, Supplemental Response to Request for Additional Information Letter No. 18. 43 Page(s)	Letter	Southern Nuclear Operating Co, Inc	NRC/Document Control Desk NRC/NRO	05200025 05200026
03/04/2009	ML090970032	2009/03/04 Vogtle COL Review - RE: Draft RAI 2171 Related SRP Section 2.5.4 for Vogtle Units 3 and 4 2 Page(s)	E-Mail	NRC/NRO	NRC/NRO/DNRL/NWE1	05200025 05200026
03/06/2009	ML090650536	2009/03/06 Vogtle RAI for SER – Request for Information Letter No. 029 Related to SRP Section 13.3 for the Vogtle Electric Generating Plant 9 Page(s)	E-Mail Request for Additional Information (RAI)	NRC/NRO	NRC/NRO/DNRL/NWE1	05200025 05200026
03/06/2009	ML090650710	2009/03/06 Vogtle COL Review - Conference Call Summary - March 3, 2009 - Southern Nuclear Vogtle Units 3 and 4 COLA - RAI # 2087 - Draft RAI related to SRP Section 13.3 3 Page(s)	E-Mail	NRC/NRO	NRC/NRO/DNRL/NWE1	05200025 05200026
03/06/2009	ML090680091	2009/03/06, Letter from Kathryn Winsberg to Michael R. Johnson Re: Atomic Safety and Licensing Board Memorandum and Order in Southern Nuclear Operating Co. (Vogtle Electric Generating Plant,	Memoranda	NRC/OGC	NRC/NRO	05200025 05200026

Appendix B

Chronology of the Combined License Application for Vogtle Electric Generating Plant Units 3 and 4

This appendix contains a chronological listing of routine licensing correspondence between the staff of the U.S. Nuclear Regulatory Commission (NRC) regarding the review of the Vogtle Electric Generating Plant, Units 3 and 4 plant design under Docket Nos. 052-000025 and 052-000026

Document Date	Accession Number	Title & Estimated Page Count	Document Type	Author Affiliation	Addressee Affiliation	Docket Number
03/06/2009	ML090690215	2009/03/06 Vogtle COL Review – Request for Additional Information Letter No. 029 Related to SRP Section 13.3 for the Vogtle Untis 3 and 4 Combined License Application 11 Page(s)	E-Mail	NRC/NRO	NRC/NRO/DNRL/NWE1	05200025 05200026
03/09/2009	ML090700541	Vogtle, Units 3 and 4 - Combined License Application, Submittal of Departure Report Update. 3 Page(s)	Letter	Southern Nuclear Operating Co, Inc	NRC/Document Control Desk NRC/NRO	05200025 05200026
03/10/2009	ML090690604	2009/03/10 Vogtle COL Review - Telecon Summary– Telcon with Vogtle COL 3/10/2009 2 Page(s)	E-Mail	NRC/NRO	NRC/NRO/DNRL/NWE1	05200025 05200026
03/10/2009	ML090690661	2009/03/10 Vogtle COL Review - Draft RAI 2300 Related SRP Section 2.4.2 for Vogtle Units 3 and 4 3 Page(s)	E-Mail	NRC/NRO	NRC/NRO/DNRL/NWE1	05200025 05200026
03/11/2009	ML090700044	2009/03/11 Vogtle RAI for SER - RAI-LTR-030 RELATED TO SRP Section 02-05-04 for the Vogtle Nuclear Site Units 3 and 4 COLA 7 Page(s)	E-Mail Request for Additional Information (RAI)	NRC/NRO	NRC/NRO/DNRL/NWE1	05200025 05200026
03/11/2009	ML090700201	2009/03/11 Vogtle COL Review - FW: RAI-LTR-030 Related to SRP	E-Mail	NRC/NRO	NRC/NRO/DNRL/NWE1	05200025 05200026

Units 3 and 4) (LBP-09-03). 1 Page(s)

Appendix B

Chronology of the Combined License Application for Vogtle Electric Generating Plant Units 3 and 4

This appendix contains a chronological listing of routine licensing correspondence between the staff of the U.S. Nuclear Regulatory Commission (NRC) regarding the review of the Vogtle Electric Generating Plant, Units 3 and 4 plant design under Docket Nos. 052-000025 and 052-000026

Document Date	Accession Number	Title & Estimated Page Count	Document Type	Author Affiliation	Addressee Affiliation	Docket Number
		Section 02-05-04 for the Vogtle Nuclear Site Units 3 and 4 COLA 9 Page(s)				05200025 05200026
03/11/2009	ML090700478	2009/03/11 Vogtle COL Review - FW: Attendees on Conference Call - Draft RAI 2171 2 Page(s)	E-Mail	NRC/NRO	NRC/NRO/DNRL/NWE1	05200025 05200026
03/11/2009	ML090710003	2009/03/11 Vogtle COL Review - RAI-LTR-030 Related to SRP Section 02-05-04 for the Vogtle Nuclear Site Units 3 and 4 COLA 9 Page(s)	E-Mail	NRC/NRO	NRC/NRO/DNRL/NWE1	05200025 05200026
03/12/2009	ML090720809	2009/03/12-Limited Appearance Statement of Unknown Author, re: New New Reactors. (Vogtle) 5 Page(s)	Legal-Limited Appearance Statement	- No Known Affiliation	NRC/SECY/RAS	05200025 05200026
03/12/2009	ML090760819	Vogtle, Units 3 & 4 Combined License Application, Response to Request for Additional Information Letter No. 026. 28 Page(s)	Letter	Southern Nuclear Operating Co, Inc	NRC/Document Control Desk NRC/NRO	05200025 05200026
03/13/2009	ML090760608	2009/03/13-Limited Appearance Statement of Jennifer V. Posey On Behalf of Himself regarding Expansion of Vogtle. 2 Page(s)	Legal-Limited Appearance Statement	Burke County, GA	NRC/SECY/RAS	05200025 05200026
03/13/2009	ML090760609	2009/03/13-Limited Appearance Statement of Lisa Kishoni Regarding Proposed Expansion at	Legal-Limited Appearance Statement	- No Known Affiliation	NRC/ASLBP	05200025 05200026

Appendix B

Chronology of the Combined License Application for Vogtle Electric Generating Plant Units 3 and 4

This appendix contains a chronological listing of routine licensing correspondence between the staff of the U.S. Nuclear Regulatory Commission (NRC) regarding the review of the Vogtle Electric Generating Plant, Units 3 and 4 plant design under Docket Nos. 052-000025 and 052-000026

Document Date	Accession Number	Title & Estimated Page Count	Document Type	Author Affiliation	Addressee Affiliation	Docket Number
03/13/2009	ML090760818	Vogtle Plant. 4 Page(s) Vogtle, Units 3 & 4 COL, Response to Regulatory Issue Summary 2009-03. 4 Page(s)	Letter	Southern Nuclear Operating Co, Inc	NRC/Document Control Desk NRC/NRO	05200025 05200026
03/14/2009	ML092161034	2009/03/14 Vogtle COL Review - FW: ND-09-0294-Departure Report Updates 5 Page(s)	E-Mail	NRC/NRO	NRC/NRO/DNRL/NWE1	05200025 05200026
03/17/2009	ML090570651	Revision to Schedule for the Draft Supplemental Environmental Impact Statement for the Combined Operating License Application for the Plant Vogtle Site. 16 Page(s)	Letter	NRC/NRO/DSER/RAP1	Southern Nuclear Operating Co, Inc	05200025 05200026
03/17/2009	ML090770813	2009/03/17-Limited Appearance Statement of Jo Claire Hickson, Opposing Expansion of Vogtle. 2 Page(s)	Legal-Limited Appearance Statement	- No Known Affiliation	NRC/SECY/RAS	05200025 05200026
03/18/2009	ML090770029	2009/03/18 Vogtle COL Review - call to discuss draft RAI 2300 2 Page(s)	E-Mail	NRC/NRO	NRC/NRO/DNRL/NWE1	05200025 05200026
03/18/2009	ML090770112	2009/03/18 Vogtle COL Review - FW: call to discuss draft RAI 2300 2 Page(s)	E-Mail	NRC/NRO	NRC/NRO/DNRL/NWE1	05200025 05200026
03/19/2009	ML090780346	2009/03/19-Limited Appearance Statement of Mal McKibben, Executive Director Emeritus, CNTA,	Legal-Limited Appearance Statement	NRC/ASLBP	NRC/SECY/RAS	05200025 05200026

Appendix B

Chronology of the Combined License Application for Vogtle Electric Generating Plant Units 3 and 4

This appendix contains a chronological listing of routine licensing correspondence between the staff of the U.S. Nuclear Regulatory Commission (NRC) regarding the review of the Vogtle Electric Generating Plant, Units 3 and 4 plant design under Docket Nos. 052-000025 and 052-000026

Document Date	Accession Number	Title & Estimated Page Count	Document Type	Author Affiliation	Addressee Affiliation	Docket Number
		regarding New Nuclear Reactors, Testimony in the Vogtle ESP/COL to ASLB . 3 Page(s)				
03/19/2009	ML090830186	2009/03/19-Limited Appearance Statement of Jerry L. McCollum and Jim Manley on Behalf of Georgia Wildlife Federation Re: Vogtle Electric Generating Plant, Unit 3 and 4. 7 Page(s)	Legal-Limited Appearance Statement	State of GA, Wildlife Federation	NRC/SECY/RAS	05200025 05200026
03/22/2009	ML090930412	2009/03/22-Limited Appearance Statement of Curtis Barton on Behalf of Lake Hartwell Association Opposing Proposed Vogtle Plants. 4 Page(s)	Legal-Limited Appearance Statement	Lake Hartwell Association, Inc	NRC/SECY/RAS	05200025 05200026
03/22/2009	ML091120489	2009/03/22-Limited Appearance Statement of Tom Clements Regarding Vogtle ESP Proceeding. 22 Page(s)	Legal-Limited Appearance Statement	Friends of the Earth	NRC/SECY/RAS	05200025 05200026
03/23/2009	ML090830015	2009/03/23 Vogtle COL Review - Re: last Thursday attendees 2 Page(s)	E-Mail	- No Known Affiliation	NRC/NRO/DNRL/NWE1	05200025 05200026
03/24/2009	ML090830475	2009/03/24 Vogtle COL Review - review of revised RAI 2300 for Vogtle COL 2 Page(s)	E-Mail	NRC/NRO	NRC/NRO/DNRL/NWE1	05200025 05200026
03/24/2009	ML090840012	2009/03/24 Vogtle COL Review - Revision 1 to the Vogtle Units 3 and	E-Mail	- No Known Affiliation	NRC/NRO/DNRL/NWE1	05200025 05200026

Appendix B

Chronology of the Combined License Application for Vogtle Electric Generating Plant Units 3 and 4

This appendix contains a chronological listing of routine licensing correspondence between the staff of the U.S. Nuclear Regulatory Commission (NRC) regarding the review of the Vogtle Electric Generating Plant, Units 3 and 4 plant design under Docket Nos. 052-000025 and 052-000026

Document Date	Accession Number	Title & Estimated Page Count	Document Type	Author Affiliation	Addressee Affiliation	Docket Number
03/24/2009	ML090840427	4 COLA 2 Page(s) 2009/03/24-Limited Appearance Statement of Clint Wolfe on Supporting the Two New Units at Plant Vogtle. 2 Page(s)	Legal-Limited Appearance Statement	Citizens for Nuclear Technology Awareness (CNTA)	NRC/SECY/RAS	05200025 05200026
03/27/2009	ML090900179	Vogtle, Units 3 and 4 Combined License Application, Response to Request for Additional Information Letter No. 027 Involving Chelating Agents in Radioactive Waste Liquids. 5 Page(s)	Letter	Southern Nuclear Operating Co, Inc	NRC/Document Control Desk NRC/NRO	05200025 05200026
03/27/2009	ML090920416	Vogtle, Units 3 and 4 - Combined License Application, Response to Request for Additional Information Letter No. 028. 10 Page(s)	Letter	Southern Nuclear Operating Co, Inc	NRC/Document Control Desk NRC/NRO	05200025 05200026
03/27/2009	ML090920495	Vogtle, Units 3 and 4 - FSAR Figure 2.4-201, "Site Plan with PMP Drainage Boundaries and Flow Paths." 1 Page(s)	Drawing Final Safety Analysis Report (FSAR)	Southern Nuclear Operating Co, Inc	NRC/NRO	05200025 05200026
03/27/2009	ML090920496	Vogtle, Units 3 and 4 - FSAR Figure 2.4-201a, "Cross-Section Location Map for HEC-RAS Model of Local PMF for Units 3 and 4." 1 Page(s)	Final Safety Analysis Report (FSAR) Map	Southern Nuclear Operating Co, Inc	NRC/NRO	05200025 05200026

Chronology of the Combined License Application for Vogtle Electric Generating Plant Units 3 and 4

This appendix contains a chronological listing of routine licensing correspondence between the staff of the U.S. Nuclear Regulatory Commission (NRC) regarding the review of the Vogtle Electric Generating Plant, Units 3 and 4 plant design under Docket Nos. 052-000025 and 052-000026

Document Date	Accession Number	Title & Estimated Page Count	Document Type	Author Affiliation	Addressee Affiliation	Docket Number
03/28/2009	ML091040766	2009/03/28-Limited Appearance of Claude C. Howard Regarding Vogtle, Units 3 and 4 Early Site Permit and Combined License Proceedings. 7 Page(s)	Legal-Limited Appearance Statement	- No Known Affiliation	NRC/SECY	05200025 05200026
03/30/2009	ML090890314	2009/03/30 Vogtle RAI for SER - Request for Additional Information Letter No. 031 Related to SRP Section 02.04.02 for The Vogtle Electric Generating Plant Units 3 and 4 Combined License Application 6 Page(s)	E-Mail Request for Additional Information (RAI)	NRC/NRO	NRC/NRO/DNRL/NWE1	05200025 05200026
03/30/2009	ML090890375	2009/03/30 Vogtle COL Review - RAI Letter No. 031 Related SRP Section 2.4.2 for Vogtle Units 3 and 4 8 Page(s)	E-Mail	NRC/NRO	NRC/NRO/DNRL/NWE1	05200025 05200026
03/30/2009	ML090890582	FSAR Section 2.3 Safety Evaluation Report Input for Vogtle Units 3 and 4 Combined License Application. 2 Page(s)	Memoranda	NRC/NRO/DSER	NRC/NRO/DNRL	05200025 05200026
03/30/2009	ML090890590	Vogtle SCOL Safety Evaluation Report Section 3.5 PSER. 18 Page(s)	Safety Evaluation Report	NRC/NRO/DSER		05200025 05200026
04/01/2009	ML090960350	Vogtle, Units 3 and 4 - Combined License Application - Fitness For	Letter License-Fitness for	Southern Nuclear Operating Co, Inc	NRC/Document Control Desk	05200025 05200026

Appendix B

Chronology of the Combined License Application for Vogtle Electric Generating Plant Units 3 and 4

This appendix contains a chronological listing of routine licensing correspondence between the staff of the U.S. Nuclear Regulatory Commission (NRC) regarding the review of the Vogtle Electric Generating Plant, Units 3 and 4 plant design under Docket Nos. 052-000025 and 052-000026

Document Date	Accession Number	Title & Estimated Page Count	Document Type	Author Affiliation	Addressee Affiliation	Docket Number
		Duty Program, Physical Security During Construction, Physical Security License Condition, and Physical Security Inspections, Tests, Analyses, and Acceptance Criteria Changes. 14 Page(s)	Duty (FFD) Performance Report		NRC/NRO	
04/01/2009	ML090970164	Vogtle Electric Generating Plant, Units 3 & 4 - Combined License Application Proposed Revision to Physical Security Plan. 5 Page(s)	Letter	Southern Nuclear Operating Co, Inc	NRC/Document Control Desk NRC/NRR	05200025 05200026
04/01/2009	ML091050315	2009/04/01 Vogtle COL Review - SNC Letter ND-09-0480 transmitting VEGP Units 3 & 4 COLA FFD, Physical Security During Construction, Physical Security License Condition and Physical Security Information 16 Page(s)	E-Mail	- No Known Affiliation	NRC/NRO/DNRL/NWE1	05200025 05200026
04/01/2009	ML091050317	2009/04/01 Vogtle COL Review - SNC Letter ND-09-0494 transmitting VEGP Units 3 and 4 COLA Proposed Revision to Physical Security Plan 7 Page(s)	E-Mail	- No Known Affiliation	NRC/NRO/DNRL/NWE1	05200025 05200026
04/03/2009	ML090990453	Vogtle Units 3 & 4 Combined License Application, Response to Request for Additional Information	Letter	Southern Nuclear Operating Co, Inc	NRC/Document Control Desk NRC/NRO	05200025 05200026

Appendix B

Chronology of the Combined License Application for Vogtle Electric Generating Plant Units 3 and 4

This appendix contains a chronological listing of routine licensing correspondence between the staff of the U.S. Nuclear Regulatory Commission (NRC) regarding the review of the Vogtle Electric Generating Plant, Units 3 and 4 plant design under Docket Nos. 052-000025 and 052-000026

Document Date	Accession Number	Title & Estimated Page Count	Document Type	Author Affiliation	Addressee Affiliation	Docket Number
04/03/2009	ML092161033	Letter No. 029. 14 Page(s)				
04/03/2009	ML092161033	2009/04/03 Vogtle COL Review - Draft RAI 2306 and 2476 related to SRP section 19 for Vogtle units 3 and 4 7 Page(s)	E-Mail	NRC/NRO	NRC/NRO/DNRL/NWE1	05200025 05200026
04/08/2009	ML092161032	2009/04/08 Vogtle COL Review - RE: Draft RAI 2306 and 2476 related to SRP Section 19 for Vogtle Units 3 and 4 2 Page(s)	E-Mail	NRC/NRO	NRC/NRO/DNRL/NWE1	05200025 05200026
04/09/2009	ML091030217	Vogtle, Units 3 & 4 Combined License Application, Response to Request for Additional Information Letter No. 030, Involving Lateral Earth Pressures and Hydrostatic Pressures. 13 Page(s)	Letter	Southern Nuclear Operating Co, Inc	NRC/Document Control Desk NRC/NRO	05200025 05200026
04/09/2009	ML092161031	2009/04/09 Vogtle COL Review - DRAFT - RAI 2312 - SRP Section 13.3 - Vogtle Units 3 and 4 Combined License Application 6 Page(s)	E-Mail	NRC/NRO	NRC/NRO/DNRL/NWE1	05200025 05200026
04/13/2009	ML092161030	2009/04/13 Vogtle COL Review - FW: Draft PRA RAI phone call 2 Page(s)	E-Mail	NRC/NRO	NRC/NRO/DNRL/NWE1	05200025 05200026
04/14/2009	ML091040017	2009/04/14 Vogtle COL Review - Draft RAIs 2339 Related SRP	E-Mail	NRC/NRO	NRC/NRO/DNRL/NWE1	05200025 05200026

Appendix B

Chronology of the Combined License Application for Vogtle Electric Generating Plant Units 3 and 4

This appendix contains a chronological listing of routine licensing correspondence between the staff of the U.S. Nuclear Regulatory Commission (NRC) regarding the review of the Vogtle Electric Generating Plant, Units 3 and 4 plant design under Docket Nos. 052-000025 and 052-000026

Document Date	Accession Number	Title & Estimated Page Count	Document Type	Author Affiliation	Addressee Affiliation	Docket Number
04/15/2009	ML091050099	Section 8.2 for Vogtle Units 3 and 4 3 Page(s)	Meeting Agenda Meeting Notice	NRC/NRO/DNRL/NWE1	NRC/NRO/DNRL/NWE1	05200025 05200026
04/15/2009	ML091050328	04/29/2009 Notice of Meeting with Southern Nuclear Operating Company to Discuss Future Plans to Supplement the Vogtle Electric Generating Plant Units 3 and 4 Combined License Application with a Limited Work Authorization Request. 7 Page(s)				
04/15/2009	ML091050328	2009/04/15 Vogtle COL Review - Public Hearing File Emails 2 Page(s)	E-Mail	NRC/NRO	NRC/NRO/DNRL/NWE1	C5200025 C5200026
04/17/2009	ML091070271	2009/04/17 Vogtle RAI for SER – Request for Additional Information letter No. 032 Related to SRP Section 13.3 for the Vogtle Electric Generating Plant 9 Page(s)	E-Mail Request for Additional Information (RAI)	NRC/NRO	NRC/NRO/DNRL/NWE1	05200025 05200026
04/17/2009	ML091110311	Vogtle Electric Generating Plant, Units 3 & 4 - Combined License Application 10 CFR 50.46 Annual Report. 2 Page(s)	Letter	Southern Nuclear Operating Co, Inc	NRC/Document Control Desk NRC/NRO	05200025 05200026
04/17/2009	ML092161029	2009/04/17 Vogtle COL Review – Request for Additional Information Letter No. 032 Related to SRP Section for the Vogtle Units 3 and 4	E-Mail	NRC/NRO	NRC/NRO/DNRL/NWE1	05200025 05200026

Chronology of the Combined License Application for Vogtle Electric Generating Plant Units 3 and 4

This appendix contains a chronological listing of routine licensing correspondence between the staff of the U.S. Nuclear Regulatory Commission (NRC) regarding the review of the Vogtle Electric Generating Plant, Units 3 and 4 plant design under Docket Nos. 052-000025 and 052-000026

Document Date	Accession Number	Title & Estimated Page Count	Document Type	Author Affiliation	Addressee Affiliation	Docket Number
		Combined License Application 11 Page(s)				
04/22/2009	ML091120011	2009/04/22 Vogtle RAI for SER – Request for Additional Information Letter No. 033 Related to SRP Section 19 for the Vogtle Electric Generating Plant Units 3 and 4 Combined License Application 7 Page(s)	Request for Additional Information (RAI)	NRC/NRO	NRC/NRO/DNRL/NWE1	05200025 05200026
04/22/2009	ML091120284	2009/04/22 Vogtle RAI for SER – Request for Additional Information Letter No. 034 Related to SRP Section 08.02 for the Vogtle Units 3 and 4 Combined License Application 6 Page(s)	E-Mail Request for Additional Information (RAI)	NRC/NRO	NRC/NRO/DNRL/NWE1	05200025 05200026
04/22/2009	ML091120305	2009/04/22 Vogtle COL Review - RAI Letter No. 034 Related to SRP Section 8.2 for Vogtle Units 3 and 4 8 Page(s)	E-Mail	NRC/NRO	NRC/NRO/DNRL/NWE1	05200025 05200026
04/22/2009	ML092161027	2009/04/22 Vogtle COL Review - RE: RAI Letter Number 033 Related to SRP Section 19 for Vogtle units 3 and 4 2 Page(s)	E-Mail	NRC/NRO	NRC/NRO/DNRL/NWE1	05200025 05200026
04/22/2009	ML092161028	2009/04/22 Vogtle COL Review - RAI letter number 033 related to SRP section 19 for Vogtle Units 3 and 4	E-Mail	NRC/NRO	NRC/NRO/DNRL/NWE1	05200025 05200026

Appendix B

Chronology of the Combined License Application for Vogtle Electric Generating Plant Units 3 and 4

This appendix contains a chronological listing of routine licensing correspondence between the staff of the U.S. Nuclear Regulatory Commission (NRC) regarding the review of the Vogtle Electric Generating Plant, Units 3 and 4 plant design under Docket Nos. 052-000025 and 052-000026

Document Date	Accession Number	Title & Estimated Page Count	Document Type	Author Affiliation	Addressee Affiliation	Docket Number
04/23/2009	ML091130004	2009/04/23 Vogtle COL Review - Telcon Summary - Telcon with Vogtle 4/22/09 2 Page(s) 9 Page(s)	E-Mail	NRC/NRO	NRC/NRO/DNRL/NWE1	05200025 05200026
04/24/2009	ML091180673	Vogtle Plants - Management Organization Changes. 4 Page(s)	Letter	Southern Nuclear Operating Co, Inc	NRC/Document Control Desk NRC/NRO	05200011 05200025 05200026
04/27/2009	ML092161026	2009/04/27 Vogtle COL Review - Status Call 4-28-09 Data 16 Page(s)	E-Mail	NRC/NRO	NRC/NRO/DNRL/NWE1	05200025 05200026
04/29/2009	ML091200579	Vogtle, Units 3 & 4 Combined License Application Response to Request for Additional Information Letter No. 031, Involving Local Intense Precipitation Flooding. 6 Page(s)	Letter	Southern Nuclear Operating Co, Inc	NRC/Document Control Desk NRC/NRO	05200025 05200026
04/29/2009	ML091480175	04/29/09 Agenda for Meeting with Southern Nuclear Operating Company to Discuss an Additional Limited Work Authorization Request Submittal for The Vogtle Electric Generating Plant Proposed Units 3 and 4. 1 Page(s)	Meeting Agenda	NRC/NRO/DNRL/NWE1	Southern Nuclear Operating Co, Inc	05200025 05200026
04/29/2009	ML091480204	04/29/2009, List of Attendees, Meeting To Discuss An Additional Limited Work Authorization Request For The Vogtle Electric Generating	Meeting Summary	NRC/NRO/DNRL/NWE1		05200025 05200026

Appendix B

Chronology of the Combined License Application for Vogtle Electric Generating Plant Units 3 and 4

This appendix contains a chronological listing of routine licensing correspondence between the staff of the U.S. Nuclear Regulatory Commission (NRC) regarding the review of the Vogtle Electric Generating Plant, Units 3 and 4 plant design under Docket Nos. 052-000025 and 052-000026

Document Date	Accession Number	Title & Estimated Page Count	Document Type	Author Affiliation	Addressee Affiliation	Docket Number
04/29/2009	ML091530097	Plant Proposed Units 3 and 4. 1 Page(s)	- No Document Type Applies	Southern Nuclear Operating Co, Inc	NRC/NRO	05200025 05200026
04/29/2009	ML092151036	Vogtle, Units 3 & 4, Loss of Large Areas of the Plant Due to Explosions or Fire Mitigative Strategies Description and Plans. 29 Page(s)	E-Mail	NRC/NRO	NRC/NRO/DNRL/NWE1	05200025 05200026
04/29/2009	ML092151037	2009/04/29 Vogtle COL Review - FW: RAI Communications 2 Page(s)	E-Mail	NRC/NRO	NRC/NRO/DNRL/NWE1	05200025 05200026
04/29/2009	ML092161021	2009/04/29 Vogtle COL Review - FW: Potential agenda additions for April 9 meeting 2 Page(s)	E-Mail	NRC/NRO	NRC/NRO/DNRL/NWE1	05200025 05200026
04/29/2009	ML092161024	2009/04/29 Vogtle COL Review - FW: conf line # 2 Page(s)	E-Mail	NRC/NRO	NRC/NRO/DNRL/NWE1	05200025 05200026
04/29/2009	ML092161025	2009/04/29 Vogtle COL Review - FW: ACTION - draft ACRS slides 16 Page(s)	E-Mail	NRC/NRO	NRC/NRO/DNRL/NWE1	05200025 05200026
04/30/2009	ML092151033	2009/04/29 Vogtle COL Review - FW: Vogtle LWA Meeting 2 Page(s)	E-Mail	NRC/NRO	NRC/NRO/DNRL/NWE1	05200025 05200026
05/04/2009	ML090370696	2009/04/30 Vogtle COL Review - FW: ACTION - Information of Lead and Chapter PM assignments 3 Page(s) The Vogtle Electric Generating Plant Request for Withholding	Letter Proprietary	NRC/NRO/DNRL/NWE1	Southern Nuclear Operating Co, Inc	05200025 05200026

Appendix B

Chronology of the Combined License Application for Vogtle Electric Generating Plant Units 3 and 4

This appendix contains a chronological listing of routine licensing correspondence between the staff of the U.S. Nuclear Regulatory Commission (NRC) regarding the review of the Vogtle Electric Generating Plant, Units 3 and 4 plant design under Docket Nos. 052-000025 and 052-000026

Document Date	Accession Number	Title & Estimated Page Count	Document Type	Author Affiliation	Addressee Affiliation	Docket Number
		Information from Public Disclosure for Units 3 and 4 NRC Project Number 755 6 Page(s)	Information Review			PROJ0755
05/04/2009	ML091890652	2009/05/04 Vogtle COL Review - Data for the status call May 5, 2009. 14 Page(s)	E-Mail	NRC/NRO	NRC/NRO/DNRL/NWE1	C5200025 C5200026
05/05/2009	ML091890650	2009/05/05 Vogtle COL Review - DRAFT - RAI 2616 - SRP section 13.3 - Vogtle Units 3 and 4 Combined License Application 3 Page(s)	E-Mail	NRC/NRO	NRC/NRO/DNRL/NWE1	05200025 05200026
05/06/2009	ML091890648	2009/05/06 Vogtle COL Review - RE: This week 3 Page(s)	E-Mail	NRC/NRO	NRC/NRO/DNRL/NWE1	05200025 05200026
05/08/2009	ML091890646	2009/05/08 Vogtle COL Review - Vogtle data for the status call – Tuesday May 12, 2009 15 Page(s)	E-Mail	NRC/NRO	NRC/NRO/DNRL/NWE1	05200025 05200026
05/08/2009	ML091890647	2009/05/08 Vogtle COL Review - FW: ACRS slides 20 Page(s)	E-Mail	NRC/NRO	NRC/NRO/DNRL/NWE1	05200025 05200026
05/12/2009	ML091890644	2009/05/12 Vogtle COL Review - Draft RAI 2653 3 Page(s)	E-Mail	NRC/NRO	NRC/NRO/DNRL/NWE1	05200025 05200026
05/13/2009	ML091330816	2009/05/13 Vogtle RAI for SER – Request for Additional Information Letter No. 035 Related to SRP	E-Mail Request for Additional	NRC/NRO	NRC/NRO/DNRL/NWE1	05200025 05200026

Appendix B

Chronology of the Combined License Application for Vogtle Electric Generating Plant Units 3 and 4

This appendix contains a chronological listing of routine licensing correspondence between the staff of the U.S. Nuclear Regulatory Commission (NRC) regarding the review of the Vogtle Electric Generating Plant, Units 3 and 4 plant design under Docket Nos. 052-000025 and 052-000026

Document Date	Accession Number	Title & Estimated Page Count	Document Type	Author Affiliation	Addressee Affiliation	Docket Number
		Section 13.3 for the Vogtle Electric Generating Plant 6 Page(s)	Information (RAI)			
05/13/2009	ML091890598	2009/05/13 Vogtle COL Review – Request for Additional Information Letter No. 035 Related to SRP Section 13.3 for the Vogtle Untis 3 and 4 Combined License Application 8 Page(s)	E-Mail	NRC/NRO	NRC/NRO/DNRL/NWE1	05200025 05200026
05/13/2009	ML091890599	2009/05/13 Vogtle COL Review - FW: Vogtle channel question 2 Page(s)	E-Mail	NRC/NRO	NRC/NRO/DNRL/NWE1	05200025 05200026
05/15/2009	ML091390050	Vogtle Electric Generating Plant Units 3 and 4 Combined License Application Response to Request for Additional Information Letter No. 032. 9 Page(s)	Letter	Southern Nuclear Operating Co, Inc	NRC/Document Control Desk NRC/NRO	05200025 05200026
05/15/2009	ML091390051	Vogtle Electric Generating Plant Units 3 and 4 Combined License Application, Reply to a Notice of Violation. 14 Page(s)	Letter Licensee Response to Notice of Violation	Southern Nuclear Operating Co, Inc	NRC/Document Control Desk NRC/NRO	05200025 05200026
05/15/2009	ML091390566	Southern Nuclear Operating Co., Review of Bellefonte Response to Request for Additional Information Number 08.01-02 for Applicability to Vogtle Electric Generating Plant	Letter	Southern Nuclear Operating Co, Inc	NRC/Document Control Desk NRC/NRO	05200014 05200015 05200025 05200026

Appendix B

Chronology of the Combined License Application for Vogtle Electric Generating Plant Units 3 and 4

This appendix contains a chronological listing of routine licensing correspondence between the staff of the U.S. Nuclear Regulatory Commission (NRC) regarding the review of the Vogtle Electric Generating Plant, Units 3 and 4 plant design under Docket Nos. 052-000025 and 052-000026

Document Date	Accession Number	Title & Estimated Page Count	Document Type	Author Affiliation	Addressee Affiliation	Docket Number
		Units 3 and 4 Combined License Application. 4 Page(s)				
05/15/2009	ML091390567	Southern Nuclear Operating Co., Endorsement of Bellefonte R-COLA Standard Content Requests for Additional Information for Vogtle Electric Generating Plant Units 3 and 4 Combined License Application. 9 Page(s)	Letter	Southern Nuclear Operating Co, Inc	NRC/Document Control Desk NRC/NRO	05200014 05200015 05200025 05200026
05/18/2009	ML091890522	2009/05/18 Vogtle COL Review - Vogtle Data for the status call (Tuesday, May 19, 2009) 14 Page(s)	E-Mail	NRC/NRO	NRC/NRO/DNRL/NWE1	05200025 05200026
05/20/2009	ML091330853	Vogtle Early Site Permit Application Final Safety Evaluation Report - Section 13.3, Emergency Planning. 2 Page(s)	Letter	NRC/NSIR/DPR/EPD	US Federal Emergency Mgmt Agency (FEMA)	05200025 05200026
05/20/2009	ML091890518	2009/05/20 Vogtle COL Review - DRAFT - RAI 2863 - SRP section 13.3 - Vogtle Units 3 and 4 Combined License Application 3 Page(s)	E-Mail	NRC/NRO	NRC/NRO/DNRL/NWE1	05200025 05200026
05/20/2009	ML091890519	2009/05/20 Vogtle COL Review - Summary of conference call with applicant (Vogtle)--May 19, 2009 -- Financial information 2 Page(s)	E-Mail	NRC/NRO	NRC/NRO/DNRL/NWE1	05200025 05200026

Appendix B

Chronology of the Combined License Application for Vogtle Electric Generating Plant
Units 3 and 4

This appendix contains a chronological listing of routine licensing correspondence between the staff of the U.S. Nuclear Regulatory Commission (NRC) regarding the review of the Vogtle Electric Generating Plant, Units 3 and 4 plant design under Docket Nos. 052-000025 and 052-000026

Document Date	Accession Number	Title & Estimated Page Count	Document Type	Author Affiliation	Addressee Affiliation	Docket Number
05/21/2009	ML091410637	Southern Nuclear Operating Company (SNC) Vogtle Electric Generating Plant Units 3 and 4 Combined License Application's Response to U.S. Nuclear Regulatory Commission Inspection Reports 05200025-09-201 and 05200026-09-201, Notice of Violation. 5 Page(s)	Inspection Report Letter Notice of Violation	NRC/NRO/DCIP/CQVP	Southern Nuclear Operating Co, Inc	05200025 05200026
05/22/2009	ML091470574	Vogtle, Units 3 and 4 - Combined License Application, Response to Request for Additional Information Letter No. 033. 24 Page(s)	Letter	Southern Nuclear Operating Co, Inc	NRC/Document Control Desk NRC/NRO	05200025 05200026
05/22/2009	ML091470575	Vogtle, Units 3 and 4 - Combined License Application Response to Request for Additional Information Letter No. 034. 7 Page(s)	Letter	Southern Nuclear Operating Co, Inc	NRC/Document Control Desk NRC/NRO	05200025 05200026
05/22/2009	ML091480383	Vogtle, Units 3 and 4 - Region II AP1000 Module Status Meeting Presentation 10 CFR 2.390 Affidavit. 11 Page(s)	Letter	Southern Nuclear Operating Co, Inc	NRC/Document Control Desk NRC/NRO	05200025 05200026

Appendix B

Chronology of the Combined License Application for Vogtle Electric Generating Plant Units 3 and 4

This appendix contains a chronological listing of routine licensing correspondence between the staff of the U.S. Nuclear Regulatory Commission (NRC) regarding the review of the Vogtle Electric Generating Plant, Units 3 and 4 plant design under Docket Nos. 052-000025 and 052-000026

Document Date	Accession Number	Title & Estimated Page Count	Document Type	Author Affiliation	Addressee Affiliation	Docket Number
05/22/2009	ML091480384	Vogtle, Units 3 and 4, Enclosure 3 to ND-09-0831, AP1000 Module Status Presentation Slides. 53 Page(s)	Meeting Briefing Package/Handouts Slides and Viewgraphs	Southern Nuclear Operating Co, Inc	NRC/NRO	05200025 05200026
05/22/2009	ML091480392	Vogtle, Units 3 and 4, Enclosure 2 to ND-09-0831, AP1000 Module Status Presentation Slides. 54 Page(s)	Meeting Briefing Package/Handouts Slides and Viewgraphs	Southern Nuclear Operating Co, Inc	NRC/NRO	05200025 05200026
05/22/2009	ML091630226	Vogtle, Units 3 and 4, Combined License Application, Revision 1 to the Application. 8 Page(s)	Letter License-Combined License (COL)	Southern Nuclear Operating Co, Inc	NRC/Document Control Desk NRC/NRO	05200025 05200026
05/22/2009	ML091890515	2009/05/22 Vogtle COL Review - Vogtle data for status call 5-26-09 14 Page(s)	E-Mail	NRC/NRO	NRC/NRO/DNRL/NWE1	05200025 05200026
05/22/2009	ML091890516	2009/05/22 Vogtle COL Review - RE: DRAFT - RAI 2863 - SRP section 13.3 - Vogtle Units 3 and 4 Combined License Application 2 Page(s)	E-Mail	- No Known Affiliation	NRC/NRO/DNRL/NWE1	05200025 05200026
05/29/2009	ML091490253	2009/05/29 Vogtle COL Review - Conference Call Summary - May 28, 2009 - Southern Nuclear Vogtle Units 3 and 4 COLA - RAI # 2863 - Draft RAI related to SRP Section 13.3 3 Page(s)	E-Mail	NRC/NRO	NRC/NRO/DNRL/NWE1	05200025 05200026
05/29/2009	ML091490633	2009/05/29 Vogtle COL Review - Conference Call Summary - May	E-Mail	NRC/NRO	NRC/NRO/DNRL/NWE1	05200025 05200026

Appendix B

Chronology of the Combined License Application for Vogtle Electric Generating Plant Units 3 and 4

This appendix contains a chronological listing of routine licensing correspondence between the staff of the U.S. Nuclear Regulatory Commission (NRC) regarding the review of the Vogtle Electric Generating Plant, Units 3 and 4 plant design under Docket Nos. 052-000025 and 052-000026

Document Date	Accession Number	Title & Estimated Page Count	Document Type	Author Affiliation	Addressee Affiliation	Docket Number
		13, 2009 - Southern Nuclear Vogtle Units 3 and 4 COLA - RAI # 2616 - Draft RAI related to SRP Section 13.3 2 Page(s)				
05/29/2009	ML091490779	2009/05/29 Vogtle RAI for SER – Request for Additional Information Letter No. SRP Section 3.7.2 for the Vogtle Electric Generating Plant Units 3 and 4 Combined License Application 6 Page(s)	E-Mail Request for Additional Information (RAI)	NRC/NRO	NRC/NRO/DNRL/NWE1	05200025 05200026
05/29/2009	ML09530095	Vogtle, Units 3 & 4 Combined License Application, Response to RAIs Regarding Loss of Large Areas of the Plant Due to Explosions or Fire - Mitigative Strategies Description and Plans. 12 Page(s)	Letter	Southern Nuclear Operating Co, Inc	NRC/Document Control Desk NRC/NRO	05200025 05200026
05/29/2009	ML091890512	2009/05/29 Vogtle COL Review - FW: Draft RAI 2653 2 Page(s)	E-Mail	NRC/NRO	NRC/NRO/DNRL/NWE1	05200025 05200026
05/31/2009	ML091890511	2009/05/31 Vogtle COL Review - RAI Letter 036 for Vogtle 3 and 4 COLA 8 Page(s)	E-Mail	NRC/NRO	NRC/NRO/DNRL/NWE1	05200025 05200026
06/01/2009	ML091890510	2009/06/01 Vogtle COL Review - RAI data for June 2 status call 13 Page(s)	E-Mail	NRC/NRO	NRC/NRO/DNRL/NWE1	05200025 05200026

Appendix B

Chronology of the Combined License Application for Vogtle Electric Generating Plant Units 3 and 4

This appendix contains a chronological listing of routine licensing correspondence between the staff of the U.S. Nuclear Regulatory Commission (NRC) regarding the review of the Vogtle Electric Generating Plant, Units 3 and 4 plant design under Docket Nos. 052-000025 and 052-000026

Document Date	Accession Number	Title & Estimated Page Count	Document Type	Author Affiliation	Addressee Affiliation	Docket Number
06/05/2009	ML091600149	Vogtle, Units 3 and 4 - Combined License Application, Mitigative Strategies Description and Plans - Reviewer's Aid. 6 Page(s)	Letter	Southern Nuclear Operating Co, Inc	NRC/Document Control Desk NRC/NRO	05200025 05200026
06/05/2009	ML091600150	Vogtle, Units 3 and 4 - Combined License Application, Attachment A, Loss of Large Areas of the Plant Due to Explosions or Fire, Mitigative Strategies Description and Plans - Reviewer's Aid. 29 Page(s)	- No Document Type Applies	Southern Nuclear Operating Co, Inc	NRC/NRO	05200025 05200026
06/082009	ML091480155	04/29/2009-Summary of Public Meeting To Discuss An Additional Limited Work Authorization Request For The Vogtle Electric Generating Plant Proposed Units 3 And 4. 3 Page(s)	Meeting Summary	NRC/NRO/DNRL/NWE1		05200025 05200026
06/08/2009	ML091890506	2009/06/08 Vogtle COL Review - Vogtle RAI data for the Tuesday status call 14 Page(s)	E-Mail	NRC/NRO	NRC/NRO/DNRL/NWE1	05200025 05200026
06/10/2009	ML091890504	2009/06/10 Vogtle COL Review - FW: Bellefonte chapter 4 SER with open items 18 Page(s)	E-Mail	NRC/NRO	NRC/NRO/DNRL/NWE1	05200025 05200026
06/11/2009	ML091890368	2009/06/11 Vogtle COL Review - FW: SNC Letter ND-09-0927 transmitting VEGP Units 3&4 COLA	E-Mail	NRC/NRO	NRC/NRO/DNRL/NWE1	05200025 05200026

Appendix B

Chronology of the Combined License Application for Vogtle Electric Generating Plant Units 3 and 4

This appendix contains a chronological listing of routine licensing correspondence between the staff of the U.S. Nuclear Regulatory Commission (NRC) regarding the review of the Vogtle Electric Generating Plant, Units 3 and 4 plant design under Docket Nos. 052-000025 and 052-000026

Document Date	Accession Number	Title & Estimated Page Count	Document Type	Author Affiliation	Addressee Affiliation	Docket Number
		Mitigative Strategies Description and Plans - Reviewer's Aid 37 Page(s)				
06/11/2009	ML091890500	2009/06/11 Vogtle COL Review - FW: Bellefonte chapter 4 SER with open items 18 Page(s)	E-Mail	NRC/NRO	NRC/NRO/DNRL/NWE1	05200025 05200026
06/16/2009	ML091670295	2009/06/16 Vogtle COL Review - Telcon Summary – Telcon with Vogtle 6/16/09 2 Page(s)	E-Mail	NRC/NRO	NRC/NRO/DNRL/NWE1	05200025 05200026
06/17/2009	ML091680204	2009/06/17 Vogtle RAI for SER - RAI Letter No. 037 Related to SRP Section 01 for the Vogtle Electric Generating Plant Units 3 and 4 COLA 6 Page(s)	E-Mail Request for Additional Information (RAI)	NRC/NRO	NRC/NRO/DNRL/NWE1	05200025 05200026
06/17/2009	ML091890332	2009/06/17 Vogtle COL Review - RE: RAI Letter Number 37 related to SRP Section 1 for Vogtle Units 3 and 4 2 Page(s)	E-Mail	NRC/NRO	NRC/NRO/DNRL/NWE1	05200025 05200026
06/17/2009	ML091890346	2009/06/17 Vogtle COL Review - RAI Letter Number 37 related to SRP Section 1 for Vogtle Units 3 and 4 8 Page(s)	E-Mail	NRC/NRO	NRC/NRO/DNRL/NWE1	05200025 05200026
06/17/2009	ML091890359	2009/06/17 Vogtle COL Review - 8 Page(s)	E-Mail	NRC/NRO	NRC/NRO/DNRL/NWE1	05200025 05200026

Appendix B

Chronology of the Combined License Application for Vogtle Electric Generating Plant Units 3 and 4

This appendix contains a chronological listing of routine licensing correspondence between the staff of the U.S. Nuclear Regulatory Commission (NRC) regarding the review of the Vogtle Electric Generating Plant, Units 3 and 4 plant design under Docket Nos. 052-000025 and 052-000026

Document Date	Accession Number	Title & Estimated Page Count	Document Type	Author Affiliation	Addressee Affiliation	Docket Number
06/18/2009	ML091750106	Vogtle Electric Plant Units 3 & 4 - Combined License Application Supplemental Response to Request for Additional Information Letter No. 029. 14 Page(s)	Letter	Southern Nuclear Operating Co, Inc	NRC/Document Control Desk NRC/NRO	05200025 05200026
06/18/2009	ML091890144	2009/06/18 Vogtle COL Review - Cover letter and Chapter 14 SER with OI for Bellefonte 50 Page(s)	E-Mail	NRC/NRO	NRC/NRO/DNRL/NWE1	05200025 05200026
06/19/2009	ML091730366	Vogtle, Units 3 and 4, Combined License Application, Additional Financial Information. 4 Page(s)	Letter	Southern Nuclear Operating Co, Inc	NRC/Document Control Desk NRC/NRO	05200025 05200026
06/19/2009	ML091890309	2009/06/19 Vogtle COL Review - FW: Bellefonte chapter 19 SER with open items 35 Page(s)	E-Mail	NRC/NRO	NRC/NRO/DNRL/NWE1	05200025 05200026
06/22/2009	ML091890265	2009/06/22 Vogtle COL Review - FW: SNC Letter ND-09-1013 transmitting VEGP 3&4 COLA Additional Financial Information 6 Page(s)	E-Mail	NRC/NRO	NRC/NRO/DNRL/NWE1	05200025 05200026
06/22/2009	ML091890294	2009/06/22 Vogtle COL Review - RAI Data for Tuesday's call 6-23-09 14 Page(s)	E-Mail	NRC/NRO	NRC/NRO/DNRL/NWE1	05200025 05200026
06/22/2009	ML092151032	2009/06/22 Vogtle COL Review - RE: Vogtle LWA? 3 Page(s)	E-Mail	NRC/NRO	NRC/NRO/DNRL/NWE1	05200025 05200026

Chronology of the Combined License Application for Vogtle Electric Generating Plant Units 3 and 4

This appendix contains a chronological listing of routine licensing correspondence between the staff of the U.S. Nuclear Regulatory Commission (NRC) regarding the review of the Vogtle Electric Generating Plant, Units 3 and 4 plant design under Docket Nos. 052-000025 and 052-000026

Document Date	Accession Number	Title & Estimated Page Count	Document Type	Author Affiliation	Addressee Affiliation	Docket Number
06/23/2009	ML091890150	2009/06/23 Vogtle COL Review - FW: Tracking #s for COLA Rev1 3 Page(s)	E-Mail	NRC/NRO	NRC/NRO/DNRL/NWE1	05200025 05200026
06/23/2009	ML091890238	2009/06/23 Vogtle COL Review - FW: SNC Letter ND-09-0786 Transmitting COLA Revision 1 to the Application 10 Page(s)	E-Mail	NRC/NRO	NRC/NRO/DNRL/NWE1	05200025 05200026
06/23/2009	ML091890242	2009/06/23 Vogtle COL Review - RE: Bellefonte chapter 12 SER with open items 37 Page(s)	E-Mail	NRC/NRO	NRC/NRO/DNRL/NWE1	05200025 05200026
06/23/2009	ML091890255	2009/06/23 Vogtle COL Review - RE: Bellefonte chapter 5 SER with open items 58 Page(s)	E-Mail	NRC/NRO	NRC/NRO/DNRL/NWE1	05200025 05200026
06/24/2009	ML091890149	2009/06/24 Vogtle COL Review - RE Bellefonte Chapter 11 SER with Open Items 46 Page(s)	E-Mail	NRC/NRO	NRC/NRO/DNRL/NWE1	05200025 05200026
06/25/2009	ML091760702	M090625A - Affirmation Session: I. SECY-09-0063 Crow Butte License Amendment for the North Trend Expansion Area, Appeals of LBP-08-6; II. SECY-09-0084 - Southern Nuclear (Vogtle Units 3 and 4), LBP-09-3 (Ruling on Standing & Contention Admissibility). 4 Page(s)	Commission Meeting Transcript/Exhibit	NRC/OCM		04008943 05200025 05200026

Appendix B

Chronology of the Combined License Application for Vogtle Electric Generating Plant Units 3 and 4

This appendix contains a chronological listing of routine licensing correspondence between the staff of the U.S. Nuclear Regulatory Commission (NRC) regarding the review of the Vogtle Electric Generating Plant, Units 3 and 4 plant design under Docket Nos. 052-000025 and 052-000026

Document Date	Accession Number	Title & Estimated Page Count	Document Type	Author Affiliation	Addressee Affiliation	Docket Number
06/25/2009	MI091760728	Srm-m090625A – Affirmation Session: I. SECY-09-0063 Crow Butte License Amendment for the North Trend Expansion Airea, Appeals of LBP-08-6, II. SECY-09-0084 – Southern Nuclear (Vogtle Untis 3 and 4), lbp-09-3 Ruling on Standing & Contention Admissibility. 1 Page(s)	Commission Staff Requirements Memo (SRM)	NRC/SECY	NRC/OCAA	04008943 05200025 05200026
06/26/2009	ML091810095	Vogtle, Units 3 & 4 Combined License Application, Response to Request for Additional Information Letter No. 035. 11 Page(s)	Letter	Southern Nuclear Operating Co, Inc	NRC/Document Control Desk NRC/NRO	05200025 05200026
06/29/2009	ML091890148	2009/06/29 Vogtle COL Review - RE: RAI Data for Tuesday's call 6-30-09 14 Page(s)	E-Mail	NRC/NRO	NRC/NRO/DNRL/NWE1	05200025 05200026
06/30/2009	ML091600577	Vogtle, Units 3 And 4 Combined License Application - Revised Safety Review Schedule. 7 Page(s)	Graphics incl Charts and Tables Letter	NRC/NRO/DNRL/NWE1	Southern Nuclear Operating Co, Inc	05200025 05200026
06/30/2009	ML091890139	2009/06/30 Vogtle COL Review - FW: SNC Letter ND-09-0806 transmitting VEGP Units 3 and 4 COLA Supplemental Response to RAI Letter No. 029 17 Page(s)	E-Mail	NRC/NRO	NRC/NRO/DNRL/NWE1	05200025 05200026

Appendix B

Chronology of the Combined License Application for Vogtle Electric Generating Plant Units 3 and 4

This appendix contains a chronological listing of routine licensing correspondence between the staff of the U.S. Nuclear Regulatory Commission (NRC) regarding the review of the Vogtle Electric Generating Plant, Units 3 and 4 plant design under Docket Nos. 052-000025 and 052-000026

Document Date	Accession Number	Title & Estimated Page Count	Document Type	Author Affiliation	Addressee Affiliation	Docket Number
06/30/2009	ML091890140	2009/06/30 Vogtle COL Review - FW: SNC Response Letter to NRC RAI Letter No. 035 on the VEGP Units 3 and 4 COL Application 13 Page(s)	E-Mail	NRC/NRO	NRC/NRO/DNRL/NWE1	05200025 05200026
06/30/2009	ML091890142	2009/06/30 Vogtle COL Review - FW: Cover letter and Chapter 14 SER with OI for Bellefonte 50 Page(s)	E-Mail	NRC/NRO	NRC/NRO/DNRL/NWE1	05200025 05200026
06/30/2009	ML091890145	2009/06/30 Vogtle COL Review - FW: Cover letter and Chapter 14 SER with OI for Bellefonte 2 Page(s)	E-Mail	NRC/NRO	NRC/NRO/DNRL/NWE1	05200025 05200026
07/01/2009	ML091890133	2009/07/01 Vogtle COL Review - FW: SNC Letter ND-09-1040 transmitting VEGP Units 3&4 COLA Response to RAI Letter No. 036 25 Page(s)	E-Mail	NRC/NRO	NRC/NRO/DNRL/NWE1	05200025 05200026
07/01/2009	ML091890137	2009/07/01 Vogtle COL Review - Vogtle Electric Generating Plants Units 3 and 4 COL Application-Revised Review Schedule 22 Page(s)	E-Mail	NRC/NRO	NRC/NRO/DNRL/NWE1	05200025 05200026
07/01/2009	ML092080390	Vogtle Electric Generating Plant Units 3 and 4 Combined License Application, Response to Request for Additional Information Letter No. 036. 23 Page(s)	Letter	Southern Nuclear Operating Co, Inc	NRC/Document Control Desk NRC/NRO	05200025 05200026

Appendix B

Chronology of the Combined License Application for Vogtle Electric Generating Plant Units 3 and 4

This appendix contains a chronological listing of routine licensing correspondence between the staff of the U.S. Nuclear Regulatory Commission (NRC) regarding the review of the Vogtle Electric Generating Plant, Units 3 and 4 plant design under Docket Nos. 052-000025 and 052-000026

Document Date	Accession Number	Title & Estimated Page Count	Document Type	Author Affiliation	Addressee Affiliation	Docket Number
07/02/2009	ML092120402	Vogtle Site-Specific Expert Panel Review. 1 Page(s)	Memoranda	NRC/RGN-II/DCP/CPB2	NRC/RGN-II/DCP/CPB2	05200025 05200026
07/06/2009	ML091810775	Cover Letter to Buzz Miller (Vogtle) - Bellefonte Units 3 and 4 Safety Evaluation Report for Chapter 1, "Introduction and General Description of Plant". 12 Page(s)	Letter	NRC/NRO/DNRL/NWE1	Southern Nuclear Operating Co, Inc	05200025 05200026
07/06/2009	ML091890129	2009/07/06 Vogtle COL Review - Vogtle RAI data for the July 7 2009 status call 15 Page(s)	E-Mail	NRC/NRO	NRC/NRO/DNRL/NWE1	05200025 05200026
07/06/2009	ML091890131	2009/07/06 Vogtle COL Review - FW: SNC Letter ND-09-1040 transmitting VEGP Units 3&4 COLA Response to RAI Letter No. 036 25 Page(s)	E-Mail	NRC/NRO	NRC/NRO/DNRL/NWE1	05200025 05200026
07/07/2009	ML091890327	2009/07/07 Vogtle COL Review - RE: Bellefonte Chapter 1 SER with Open Items 47 Page(s)	E-Mail	NRC/NRO	NRC/NRO/DNRL/NWE1	05200025 05200026
07/09/2009	ML092161020	2009/07/09 Vogtle COL Review - FW: ACRS Presenters 2 Page(s)	E-Mail	NRC/NRO	NRC/NRO/DNRL/NWE1	05200025 05200026
07/13/2009	ML092151031	2009/07/13 Vogtle COL Review - Vogtle RAI data for July 14 status call 14 Page(s)	E-Mail	NRC/NRO	NRC/NRO/DNRL/NWE1	05200025 05200026

Appendix B

Chronology of the Combined License Application for Vogtle Electric Generating Plant Units 3 and 4

This appendix contains a chronological listing of routine licensing correspondence between the staff of the U.S. Nuclear Regulatory Commission (NRC) regarding the review of the Vogtle Electric Generating Plant, Units 3 and 4 plant design under Docket Nos. 052-000025 and 052-000026

Document Date	Accession Number	Title & Estimated Page Count	Document Type	Author Affiliation	Addressee Affiliation	Docket Number
07/14/2009	ML09212040B	Vogtle Site-specific Expert Panel Results. 4 Page(s)	Memoranda	NRC/RGN-II/DCP/CPB2	NRC/RGN-II/DCP/CPB2	05200025 05200026
07/16/2009	ML092010089	Vogtle, Units 3 & 4 Combined License Application, Response to Bellefonte Units 3 & 4 Safety Evaluation Report Open Items for Chapter 12. 8 Page(s)	Letter	Southern Nuclear Operating Co, Inc	NRC/Document Control Desk NRC/NRO	05200014 05200015 05200025 05200026
07/16/2009	ML092010090	Vogtle, Units 3 and 4 Combined License Application, Response to Bellefonte Units 3 and 4 Safety Evaluation Report Open Items for Chapter 14. 7 Page(s)	Letter	Southern Nuclear Operating Co, Inc	NRC/Document Control Desk NRC/NRO	05200014 05200015 05200025 05200026
07/16/2009	ML092010091	Vogtle, Units 3 & 4 Combined License Application Response to Bellefonte Units 3 & 4 Safety Evaluation Report Open Items for Chapter 10. 7 Page(s)	Letter	Southern Nuclear Operating Co, Inc	NRC/Document Control Desk NRC/NRO	05200014 05200015 05200025 05200026
07/16/2009	ML092010092	Vogtle, Units 3 and 4 Combined License Application, Response to Bellefonte, Units 3 and 4 Safety Evaluation Report Open Items for Chapter 19. 6 Page(s)	Letter	Southern Nuclear Operating Co, Inc	NRC/Document Control Desk NRC/NRO	05200014 05200015 05200025 05200026
07/16/2009	ML092010093	Vogtle, Units 3 and 4, Combined License Application - Response to	Letter	Southern Nuclear Operating Co, Inc	NRC/Document Control Desk	05200025 05200026

B-72

Appendix B

Chronology of the Combined License Application for Vogtle Electric Generating Plant Units 3 and 4

This appendix contains a chronological listing of routine licensing correspondence between the staff of the U.S. Nuclear Regulatory Commission (NRC) regarding the review of the Vogtle Electric Generating Plant, Units 3 and 4 plant design under Docket Nos. 052-000025 and 052-000026

Document Date	Accession Number	Title & Estimated Page Count	Document Type	Author Affiliation	Addressee Affiliation	Docket Number
		Request for Additional Information Letter No. 037 Involving Interfaces with Standard AP1000 Design. 13 Page(s)			NRC/NRO	
07/17/2009	ML092030409	Vogtle, Units 3 and 4, Combined License Application - Response to Bellefonte Units 3 and 4 Safety Evaluation Report Open Items for Chapter 16. 6 Page(s)	Letter	Southern Nuclear Operating Co, Inc	NRC/Document Control Desk NRC/NRO	05200025 05200026
07/17/2009	ML092151028	2009/07/17 Vogtle COL Review - RE: OI responses submitted 7/16/09 2 Page(s)	E-Mail	NRC/NRO	NRC/NRO/DNRL/NWE1	05200025 05200026
07/17/2009	ML092151029	2009/07/17 Vogtle COL Review - FW: OI responses submitted 7/16/09 30 Page(s)	E-Mail	NRC/NRO	NRC/NRO/DNRL/NWE1	05200025 05200026
07/20/2009	ML092150969	2009/07/20 Vogtle COL Review - Vogtle RAI data for Tuesday's status call--7-21-09 16 Page(s)	E-Mail	NRC/NRO	NRC/NRO/DNRL/NWE1	05200025 05200026
07/21/2009	ML092150948	2009/07/21 Vogtle COL Review - FW: SNC Letter ND-09-1090 transmitting VEGP Units 3 & 4 COLA Response to Bellefonte Units 3 & 4 Safety Evaluation Report Open Items for Chapter 12 10 Page(s)	E-Mail	NRC/NRO	NRC/NRO/DNRL/NWE1	05200025 05200026

Appendix B

Chronology of the Combined License Application for Vogtle Electric Generating Plant Units 3 and 4

This appendix contains a chronological listing of routine licensing correspondence between the staff of the U.S. Nuclear Regulatory Commission (NRC) regarding the review of the Vogtle Electric Generating Plant, Units 3 and 4 plant design under Docket Nos. 052-000025 and 052-000026

Document Date	Accession Number	Title & Estimated Page Count	Document Type	Author Affiliation	Addressee Affiliation	Docket Number
07/21/2009	ML092150950	2009/07/21 Vogtle COL Review - FW: SNC Letter ND-09-1092 transmitting VEGP Units 3 & 4 COLA Response to Bellefonte Units 3 & 4 Safety Evaluation Report Open Items for Chapter 14 9 Page(s)	E-Mail	NRC/NRO	NRC/NRO/DNRL/NWE1	05200025 05200026
07/21/2009	ML092150951	2009/07/21 Vogtle COL Review - FW: SNC Letter ND-09-1089 transmitting VEGP Units 3 & 4 COLA Response to Bellefonte Units 3 & 4 Safety Evaluation Report Open Items for Chapter 10 9 Page(s)	E-Mail	NRC/NRO	NRC/NRO/DNRL/NWE1	05200025 05200026
07/21/2009	ML092150952	2009/07/21 Vogtle COL Review - FW: SNC Letter ND-09-1113 transmitting VEGP Units 3 & 4 COLA Response to Bellefonte Units 3 & 4 Safety Evaluation Report Open Items for Chapter 19 8 Page(s)	E-Mail	NRC/NRO	NRC/NRO/DNRL/NWE1	05200025 05200026
07/21/2009	ML092150953	2009/07/21 Vogtle COL Review - FW: SNC Letter ND-09-1114 transmitting VEGP Units 3 & 4 COLA Response to Request for Additional Information Letter No. 037 15 Page(s)	E-Mail	NRC/NRO	NRC/NRO/DNRL/NWE1	05200025 05200026
07/21/2009	ML092150955	2009/07/21 Vogtle COL Review -	E-Mail	NRC/NRO	NRC/NRO/DNRL/NWE1	05200025

Appendix B

Chronology of the Combined License Application for Vogtle Electric Generating Plant Units 3 and 4

This appendix contains a chronological listing of routine licensing correspondence between the staff of the U.S. Nuclear Regulatory Commission (NRC) regarding the review of the Vogtle Electric Generating Plant, Units 3 and 4 plant design under Docket Nos. 052-000025 and 052-000026

Document Date	Accession Number	Title & Estimated Page Count	Document Type	Author Affiliation	Addressee Affiliation	Docket Number
		FW: SNC Letter ND-09-1090 transmitting VEGP Units 3 & 4 COLA Response to Bellefonte Units 3 & 4 Safety Evaluation Report Open Items for Chapter 12 10 Page(s)				05200026
07/21/2009	ML092150956	2009/07/21 Vogtle COL Review - FW: SNC Letter ND-09-1140 transmitting VEGP 3 & 4 COLA Response to Bellefonte Units 3 & 4 Safety Evaluation Report Open Items for Chatper 16 8 Page(s)	E-Mail	NRC/NRO	NRC/NRO/DNRL/NWE1	05200025 05200026
07/23/2009	ML092040481	2009/07/23 Vogtle COL Review - Draft RAIs 3340 Related SRP Section 8.1 for Vogtle Units 3 and 4 3 Page(s)	E-Mail	NRC/NRO	NRC/NRO/DNRL/NWE1	05200025 05200026
07/24/2009	ML092050155	2009/07/24 Vogtle COL Review - Draft RAIs 3335 Related SRP Section 8.2 for Vogtle Units 3 and 4 3 Page(s)	E-Mail	NRC/NRO	NRC/NRO/DNRL/NWE1	05200025 05200026
07/24/2009	ML092080427	Vogtle, Units 3 and 4 Combined License Application Proposed Cyber Security Plan and Proposed Implementation Schedule. 8 Page(s)	Letter	Southern Nuclear Operating Co, Inc	NRC/Document Control Desk NRC/NRO	05200025 05200026
07/27/2009	ML092150947	2009/07/27 Vogtle COL Review - Vogtle RAI data for July 28, 2009 call	E-Mail	NRC/NRO	NRC/NRO/DNRL/NWE1	05200025 05200026

Chronology of the Combined License Application for Vogtle Electric Generating Plant Units 3 and 4

This appendix contains a chronological listing of routine licensing correspondence between the staff of the U.S. Nuclear Regulatory Commission (NRC) regarding the review of the Vogtle Electric Generating Plant, Units 3 and 4 plant design under Docket Nos. 052-000025 and 052-000026

Document Date	Accession Number	Title & Estimated Page Count	Document Type	Author Affiliation	Addressee Affiliation	Docket Number
		18 Page(s)				
07/28/2009	ML092150943	2009/07/28 Vogtle COL Review - RE: Points of Contact--Chapter PM assignments for the Vogtle COLA 4 Page(s)	E-Mail	NRC/NRO	NRC/NRO/DNRL/NWE1	05200025 05200026
07/29/2009	ML092120063	Vogtle, Units 3 and 4 Combined License Application, Response to Bellefonte Units 3 and 4 Safety Evaluation Report Open Items for Chapter 17. 12 Page(s)	Letter	Southern Nuclear Operating Co, Inc	NRC/Document Control Desk NRC/NRO	05200014 05200015 05200025 05200026
07/29/2009	ML092120064	Vogtle, Units 3 and 4 Combined License Application, Response to Bellefonte Safety Evaluation Report Open Items for Chapter 01. 14 Page(s)	Letter	Southern Nuclear Operating Co, Inc	NRC/Document Control Desk NRC/NRO	05200014 05200015 05200025 05200026
07/30/2009	ML092150942	2009/07/30 Vogtle COL Review - FW: Addition to Vogtle Points of Contact 2 Page(s)	E-Mail	NRC/NRO	NRC/NRO/DNRL/NWE1	05200025 05200026
07/31/2009	ML092120476	M090731 - Affirmation Session: I. SECY-09-0076 - Southern Nuclear Operating Co. (Vogtle Electric Generating Plant, Units 3 and 4) 4 Page(s)	Commission Meeting Transcript/Exhibit	NRC/OCM		05200025 05200026

Appendix B

Chronology of the Combined License Application for Vogtle Electric Generating Plant Units 3 and 4

This appendix contains a chronological listing of routine licensing correspondence between the staff of the U.S. Nuclear Regulatory Commission (NRC) regarding the review of the Vogtle Electric Generating Plant, Units 3 and 4 plant design under Docket Nos. 052-000025 and 052-000026

Document Date	Accession Number	Title & Estimated Page Count	Document Type	Author Affiliation	Addressee Affiliation	Docket Number
07/31/2009	ML092120620	SRM-M090731 - Affirmation: I. SECY-09-0076 - Southern Nuclear Operating Co. (Vogtle Electric Generating Plant, Units 3 and 4), LBP-09-3 (Ruling on Standing and Contention Admissibility). 1 Page(s)	Commission Staff Requirements Memo (SRM)	NRC/SECY	NRC/OCAA	05200025 05200026
07/31/2009	ML092080428	Vogtle, Units 3 and 4, Proposed Cyber Security Plan. 89 Page(s)	Security Plan	Southern Nuclear Operating Co, Inc	NRC/NRO	05200025 05200026
07/31/2009	ML092170636	Vogtle, Units 3 and 4 Combined License Application - Attachment A - Proposed Cyber Security Plan, Reviewer's Aid. 86 Page(s)	Security Plan	Southern Nuclear Operating Co, Inc	NRC/NRO	05200025 05200026
08/03/2009	ML092151039	2009/08/03 Vogtle COL Review - RE: Vogtle RAI Data -8-3 17 Page(s)	E-Mail	NRC/NRO	NRC/NRO/DNRL/NWE1	05200025 05200026
08/03/2009	ML092170635	Vogtle, Units 3 and 4, Combined License Application - Submittal of Proposed Cyber Security Plan, Reviewer's Aid. 5 Page(s)	Letter	Southern Nuclear Operating Co, Inc	NRC/Document Control Desk NRC/NRO	05200025 05200026
08/04/2009	ML092160007	2009/08/04 Vogtle RAI for SER – Requestion for Additional Information Letter No. 038 Related to SRP Section 8.2 for the Vogtle Electric Generating Plant Units 3 and 4 Combined License	E-Mail Request for Additional Information (RAI)	NRC/NRO	NRC/NRO/DNRL/NWE1	05200025 05200026

Appendix B

Chronology of the Combined License Application for Vogtle Electric Generating Plant Units 3 and 4

This appendix contains a chronological listing of routine licensing correspondence between the staff of the U.S. Nuclear Regulatory Commission (NRC) regarding the review of the Vogtle Electric Generating Plant, Units 3 and 4 plant design under Docket Nos. 052-000025 and 052-000026

Document Date	Accession Number	Title & Estimated Page Count	Document Type	Author Affiliation	Addressee Affiliation	Docket Number
		Application 6 Page(s)				
08/04/2009	ML092160310	2009/08/04 Vogtle COL Review - RAI letter 38 related to SRP Section 8.2 for Vogtle units 3 and 4 8 Page(s)	E-Mail	NRC/NRO	NRC/NRO/DNRL/NWE1	05200025 05200026
08/04/2009	ML092190400	Vogtle, Units 3 & 4, Quarterly and Ongoing Construction Schedule Information and Request for Withholding. 20 Page(s)	Letter	Southern Nuclear Operating Co, Inc	NRC/Document Control Desk NRC/NRO	05200025 05200026
08/04/2009	ML092190401	Enclosures 1 & 2, to ND-09-1155 - Vogtle, Units 3 and 4, Overall Project Status and Electronic Schedule Export Information. 8 Page(s)	Letter	Southern Nuclear Operating Co, Inc	NRC/NRO	05200025 05200026
08/05/2009	ML092160518	08/21/2009 Meeting Notice, Closed Meeting with the Southern Nuclear Operating Company on the Vogtle Cyber Security Plan Submittal and the Draft Regulatory Guide 5.71 Appendix, "Cyber Security Plan Template". 7 Page(s)	Meeting Agenda Meeting Notice Memoranda	NRC/NRO/DNRL/NGE2	NRC/NRO/DNRL/NWE1	05200025 05200026
08/05/2009	ML092170304	2009/08/05 Vogtle COL Review - FW: Plant Vogtle Combined Operating License activities 2 Page(s)	E-Mail	NRC/NRO	NRC/NRO/DNRL/NWE1	05200025 05200026
08/05/2009	ML092170305	2009/08/05 Vogtle COL Review -	E-Mail	NRC/NRO	NRC/NRO/DNRL/NWE1	05200025

Appendix B

Chronology of the Combined License Application for Vogtle Electric Generating Plant Units 3 and 4

This appendix contains a chronological listing of routine licensing correspondence between the staff of the U.S. Nuclear Regulatory Commission (NRC) regarding the review of the Vogtle Electric Generating Plant, Units 3 and 4 plant design under Docket Nos. 052-000025 and 052-000026

Document Date	Accession Number	Title & Estimated Page Count	Document Type	Author Affiliation	Addressee Affiliation	Docket Number
		FW: ADAMS number for Trip Report 2 Page(s)				05200026
08/05/2009	ML092170520	2009/08/05 Vogtle COL Review - FW: BAs for Vogtle COL 2 Page(s)	E-Mail	NRC/NRO	NRC/NRO/DNRL/NWE1	05200025 05200026
08/05/2009	ML092180964	2009/08/05 Vogtle COL Review - RE: Hydrology supplemental response 2 Page(s)	E-Mail	NRC/NRO	NRC/NRO/DNRL/NWE1	05200025 05200026
08/05/2009	ML092180965	2009/08/05 Vogtle COL Review - RE: Necessary Excavation Handout for phone call 2 Page(s)	E-Mail	NRC/NRO	NRC/NRO/DNRL/NWE1	05200025 05200026
08/05/2009	ML092230107	2009/08/05 Vogtle COL Review - SNC Letter ND-09-1155 Transmitting VEGP 3&4 Quarterly and Ongoing Construction Schedule Information and Request for Withholding_Cover 7 Page(s)	E-Mail	- No Known Affiliation	NRC/NRO/DNRL/NWE1	05200025 05200026
08/05/2009	ML092230147	Vogtle, Units 3 and 4, Combined License Application Supplemental Response to Request for Additional Information Number 02.04.02-1. 13 Page(s)	Letter	Southern Nuclear Operating Co, Inc	NRC/Document Control Desk NRC/NRO	05200025 05200026
08/06/2009	ML092180962	2009/08/06 Vogtle COL Review - Draft RAI 3380 3 Page(s)	E-Mail	NRC/NRO	NRC/NRO/DNRL/NWE1	05200025 05200026

Appendix B

Chronology of the Combined License Application for Vogtle Electric Generating Plant Units 3 and 4

This appendix contains a chronological listing of routine licensing correspondence between the staff of the U.S. Nuclear Regulatory Commission (NRC) regarding the review of the Vogtle Electric Generating Plant, Units 3 and 4 plant design under Docket Nos. 052-000025 and 052-000026

Document Date	Accession Number	Title & Estimated Page Count	Document Type	Author Affiliation	Addressee Affiliation	Docket Number
08/06/2009	ML092180963	2009/08/06 Vogtle COL Review - Draft RAI 3381 related to SRP section 2.5.4 for Vogtle Units 3 and 4 3 Page(s)	E-Mail	NRC/NRO	NRC/NRO/DNRL/NWE1	05200025 05200026
08/06/2009	ML092220514	Vogtle, Units 3 & 4, Remote Creation and Storage of Safeguards Information at Shaw Stone & Webster Offices. 3 Page(s)	Letter	Southern Nuclear Operating Co, Inc	NRC/Document Control Desk NRC/NRO	05200025 05200026
08/07/2009	ML092190198	2009/08/07-Commission Memorandum and Order in Southern Nuclear Operating Co., (Vogtle Electric Generating Plant, Units 3 and 4) (CLI-09-16). 1 Page(s)	Legal-Memorandum and Order	NRC/OGC	NRC/NRO	05200025 05200026
08/10/2009	ML092230154	2009/08/10 Vogtle COL Review - Vogtle RAI data for 8-10-09 17 Page(s)	E-Mail	NRC/NRO	NRC/NRO/DNRL/NWE1	05200025 05200026
08/11/2009	ML092250093	2009/08/11 Vogtle COL Review - FW: Hydrology supplemental response 15 Page(s)	E-Mail	NRC/NRO	NRC/NRO/DNRL/NWE1	05200025 05200026
08/11/2009	ML092250096	2009/08/11 Vogtle COL Review - Vogtle e-RAI public report--includes standard SER Open Items from Bellefonte SER 9 Page(s)	E-Mail	NRC/NRO	NRC/NRO/DNRL/NWE1	05200025 05200026
08/11/2009	ML092250099	2009/08/11 Vogtle COL Review -	E-Mail	NRC/NRO	NRC/NRO/DNRL/NWE1	05200025

Appendix B

Chronology of the Combined License Application for Vogtle Electric Generating Plant Units 3 and 4

This appendix contains a chronological listing of routine licensing correspondence between the staff of the U.S. Nuclear Regulatory Commission (NRC) regarding the review of the Vogtle Electric Generating Plant, Units 3 and 4 plant design under Docket Nos. 052-000025 and 052-000026

Document Date	Accession Number	Title & Estimated Page Count	Document Type	Author Affiliation	Addressee Affiliation	Docket Number
		FW: SNC Letter ND-09-1171 transmitting Response to Bellefonte Safety Evaluation Report Open Items for Chapter 01 16 Page(s)				05200026
08/11/2009	ML092250100	2009/08/11 Vogtle COL Review - FW: SNC Letter ND-09-1115 transmitting Response to Bellefonte Units 3 and 4 Safety Evaluation Report Open Items for Chapter 17 14 Page(s)	E-Mail	NRC/NRO	NRC/NRO/DNRL/NWE1	05200025 05200026
08/12/2009	ML092250102	2009/08/12 Vogtle COL Review - FW: BLN SER Chapter 5 confirmatory items 3 Page(s)	E-Mail	NRC/NRO	NRC/NRO/DNRL/NWE1	05200025 05200026
08/13/2009	ML092250399	2009/08/13-Projected Schedule for Completion of Safety and Environmental Evaluations is Designated as "To be Determined", After a Final Decision is Reached Regarding the Vogtle Early Site Permit Application. 4 Page(s)	Legal-Correspondence/Miscellaneous	NRC/OGC	NRC/ASLBP	05200025 05200026
08/14/2009	ML092290010	Edwin I. Hatch, Units 1 and 2, Joseph M. Farley, Units 1 and 2, Vogtle Electric, Units 1, 2, 3 and 4, Response to Regulatory Issue Summary 2009-11 Preparation and Scheduling of Operator Licensing	Letter Schedule and Calendars	Southern Nuclear Operating Co, Inc	NRC/Document Control Desk NRC/NRO NRC/NRR	05000321 05000348 05000364 05000366 05000424 05000425

Appendix B

Chronology of the Combined License Application for Vogtle Electric Generating Plant Units 3 and 4

This appendix contains a chronological listing of routine licensing correspondence between the staff of the U.S. Nuclear Regulatory Commission (NRC) regarding the review of the Vogtle Electric Generating Plant, Units 3 and 4 plant design under Docket Nos. 052-000025 and 052-000026

Document Date	Accession Number	Title & Estimated Page Count	Document Type	Author Affiliation	Addressee Affiliation	Docket Number
		Examinations. 10 Page(s)				05200025 05200026
08/14/2009	ML092310485	Vogtle Electric Generating Plant Units 3 & 4 Combined License Application Revision 1 Roadmap. 61 Page(s)	Letter	Southern Nuclear Operating Co, Inc	NRC/Document Control Desk NRC/NRO	05200025 05200026
08/17/2009	ML092290664	2009/08/17 Vogtle COL Review - RAI data for August 18, 2009 call 19 Page(s)	E-Mail	NRC/NRO	NRC/NRO/DNRL/NWE1	05200025 05200026
08/18/2009	ML092300521	Transmittal of Vogtle SER Section 9.5.2 (E-mail). 1 Page(s)	E-Mail	NRC/NRO/DNRL/NWE1	NRC/NRO/DNRL/NWE1	05200011 05200025 05200026
08/18/2009	ML092300527	Transmittal of Vogtle SER Section 9.52 (Attachment). 7 Page(s)	- No Document Type Applies	NRC/NRO		05200011 05200025 05200026
08/18/2009	ML092300546	Transmittal of Vogtle SER section 9.3.3 (E-mail). 1 Page(s)	E-Mail	NRC/NRO/DNRL/NWE1	NRC/NRO/DNRL/NWE1	05200025 05200026
08/18/2009	ML092300553	Transmittal of Vogtle SER Section 9.3.3 (attachment). 2 Page(s)	Safety Evaluation Report	NRC/NRO/DNRL/NWE1		05200025 05200026
08/18/2009	ML092300570	Transmittal of Vogtle SER Section 9.2.8 and 9.2.11 (E-mail and attachment). 2 Page(s)	E-Mail	NRC/NRO/DNRL	NRC/NRO/DNRL	05200025 05200026
08/18/2009	ML092300582	Vogtle - Preliminary SER Section 9.2.11, "Raw Water System". 5 Page(s)	Safety Evaluation Report	NRC/NRO/DNRL		05200025 05200026
08/18/2009	ML092300584	Vogtle - Preliminary SER 9.28,	Safety Evaluation	NRC/NRO/DNRL		05200025

Appendix B

Chronology of the Combined License Application for Vogtle Electric Generating Plant Units 3 and 4

This appendix contains a chronological listing of routine licensing correspondence between the staff of the U.S. Nuclear Regulatory Commission (NRC) regarding the review of the Vogtle Electric Generating Plant, Units 3 and 4 plant design under Docket Nos. 052-000025 and 052-000026

Document Date	Accession Number	Title & Estimated Page Count	Document Type	Author Affiliation	Addressee Affiliation	Docket Number
08/18/2009		"Turbine Building Closed Cooling Water System". 3 Page(s)	Report			05200026
08/18/2009	ML092310055	Transmittal of Vogtle SER Section 9.2.1 (e-mail). 2 Page(s)	E-Mail	NRC/NRO/DNRL	NRC/NRO/DNRL/NWE1	05200025 05200026
08/18/2009	ML092310063	Transmittal of Vogtle SER Section 9.2.1 (attachment). 2 Page(s)	Safety Evaluation Report	NRC/NRO/DNRL		05200025 05200026
08/18/2009	ML092310069	Transmittal of Vogtle SER Section 10.4.2 (e-mail). 2 Page(s)	E-Mail	NRC/NRO/DNRL	NRC/NRO/DNRL/NWE1	05200025 05200026
08/18/2009	ML092310075	Transmittal of Vogtle SER Section 10.4.2 (attachment). 3 Page(s)	Safety Evaluation Report	NRC/NRO/DNRL		05200025 05200026
08/18/2009	ML092600675	2009/08/18 Vogtle COL Review - FW: NL-1312 - HNP, FNP,VEGP Response to Regulatory Issue Summary 2009-11 Preparation and Scheduling of Operator Licensing Examinations 2 Page(s)	E-Mail	NRC/NRO	NRC/NRO/DNRL/NWE1	05200025 05200026
08/18/2009	ML092600681	2009/08/18 Vogtle COL Review - FW: Vogtle Unit 3 and 4 Combined License Application Revision 1 Roadmap 63 Page(s)	E-Mail	NRC/NRO	NRC/NRO/DNRL/NWE1	05200025 05200026
08/18/2009	ML092600671	2009/08/18 Vogtle COL Review - FW: Necessary Excavation.doc	E-Mail	NRC/NRO	NRC/NRO/DNRL/NWE1	05200025 05200026

Appendix B

Chronology of the Combined License Application for Vogtle Electric Generating Plant Units 3 and 4

This appendix contains a chronological listing of routine licensing correspondence between the staff of the U.S. Nuclear Regulatory Commission (NRC) regarding the review of the Vogtle Electric Generating Plant, Units 3 and 4 plant design under Docket Nos. 052-000025 and 052-000026

Document Date	Accession Number	Title & Estimated Page Count	Document Type	Author Affiliation	Addressee Affiliation	Docket Number
08/18/2009	ML092600673	2009/08/18 Vogtle COL Review - FW: Necessary Excavation.doc 4 Page(s)	E-Mail	NRC/NRO	NRC/NRO/DNRL/NWE1	05200025 05200026
08/18/2009	ML092600686	2009/08/18 Vogtle COL Review - FW: VEGP COLA Rev1 Roadmap 63 Page(s)	E-Mail	NRC/NRO	NRC/NRO/DNRL/NWE1	05200025 05200026
08/18/2009	ML092600694	2009/08/18 Vogtle COL Review - Draft RAIs 3463 and 3464 for Vogtle Units 3 and 4 4 Page(s)	E-Mail	NRC/NRO	NRC/NRO/DNRL/NWE1	05200025 05200026
08/21/2009	ML092600697	2009/08/21 Vogtle COL Review - Summary of Telephone call with Applicant (Vogtle) -- August 21, 2009 2 Page(s)	E-Mail	NRC/NRO	NRC/NRO/DNRL/NWE1	05200025 05200026
08/24/2009	ML092360579	2009/08/24 Vogtle RAI for SER -- Request for Additional Information Letter No. 039 Related to SRP Section 11.04 for the Vogtle Electric Generating Plant, Units 3 and 4 Combined License Application 6 Page(s)	E-Mail	NRC/NRO	NRC/NRO/DNRL/NWE1	05200025 05200026
08/24/2009	ML092390080	Vogtle, Units 3 and 4, Combined License Application Response to Bellefonte Units 3 and 4 Safety Evaluation Report With Open Items for Chapter 05. 7 Page(s)	Letter	Southern Nuclear Operating Co, Inc	NRC/Document Control Desk NRC/NRO	05200025 05200026

Appendix B

Chronology of the Combined License Application for Vogtle Electric Generating Plant Units 3 and 4

This appendix contains a chronological listing of routine licensing correspondence between the staff of the U.S. Nuclear Regulatory Commission (NRC) regarding the review of the Vogtle Electric Generating Plant, Units 3 and 4 plant design under Docket Nos. 052-000025 and 052-000026

Document Date	Accession Number	Title & Estimated Page Count	Document Type	Author Affiliation	Addressee Affiliation	Docket Number
08/24/2009	ML092600698	2009/08/24 Vogtle COL Review - RAI Letter 039 for Vogtle Units 3 and 4 8 Page(s)	E-Mail	NRC/NRO	NRC/NRO/DNRL/NWE1	05200025 05200026
08/24/2009	ML092600707	2009/08/24 Vogtle COL Review - DRAFT - RAI 3391 - SRP section 13.6 - Vogtle Units 3 and 4 Combined License Application 3 Page(s)	E-Mail	NRC/NRO	NRC/NRO/DNRL/NWE1	05200025 05200026
08/24/2009	ML092600708	2009/08/24 Vogtle COL Review - DRAFT - RAI 3403 - SRP section 13.6 - Vogtle Units 3 and 4 Combined License Application 3 Page(s)	E-Mail	NRC/NRO	NRC/NRO/DNRL/NWE1	05200025 05200026
08/24/2009	ML092600873	2009/08/24 Vogtle COL Review - 4 DRAFT RAIs - SRP section 13.6 - Vogtle Units 3 and 4 Combined License Application 9 Page(s)	E-Mail	NRC/NRO	NRC/NRO/DNRL/NWE1	05200025 05200026
08/25/2009	ML092600878	2009/08/25 Vogtle COL Review - DRAFT - RAI 3384 - SRP section 13.6 - Vogtle Units 3 and 4 Combined License Application 5 Page(s)	E-Mail	NRC/NRO	NRC/NRO/DNRL/NWE1	05200025 05200026
08/25/2009	ML092600882	2009/08/25 Vogtle COL Review - FW: Telephone Service Outage 2 Page(s)	E-Mail	NRC/NRO	NRC/NRO/DNRL/NWE1	05200025 05200026
08/26/2009	ML092380589	Press Release-09-141: NRC Issues Early Site Permit, Work	Press Release	NRC/OPA		05200025 05200026

Appendix B

Chronology of the Combined License Application for Vogtle Electric Generating Plant Units 3 and 4

This appendix contains a chronological listing of routine licensing correspondence between the staff of the U.S. Nuclear Regulatory Commission (NRC) regarding the review of the Vogtle Electric Generating Plant, Units 3 and 4 plant design under Docket Nos. 052-000025 and 052-000026

Document Date	Accession Number	Title & Estimated Page Count	Document Type	Author Affiliation	Addressee Affiliation	Docket Number
		Authorization for Vogtle Site in Georgia. 1 Page(s)				
08/31/2009	ML092430096	2009/08/31-NRC Staff Response to Memorandum and Order Dated August 17, 2009 Directing Staff to Provide by August 31 its Projected Schedule for Completion of the Environmental Portion of its Review Regarding Vogtle, Units 3 and 4. 4 Page(s)	Legal-Correspondence/Miscellaneous	NRC/OGC	NRC/ASLBP	05200025 05200026
08/31/2009	ML092450485	Vogtle, Units 3 and 4 - Combined License Application - Response to Request for Additional Information Letter No. 038. 6 Page(s)	Letter	Southern Nuclear Operating Co, Inc	NRC/Document Control Desk NRC/NRO	05200025 05200026
08/31/2009	ML092540040	2009/08/31-Initial Mandatory Disclosures Pursuant to 10 CFR § 2.336 for Contention SAFETY-1; Southern Nuclear Operating Co., (COL for Vogtle Units 3 and 4). 18 Page(s)	Legal-Hearing File	Balch & Bingham, LLP	Emory Univ School of Law NRC/OGC	05200025 05200026
08/31/2009	ML092600973	2009/08/31 Vogtle COL Review - Vogtle data for 9/1 call 19 Page(s)	E-Mail	NRC/NRO	NRC/NRO/DNRL/NWE1	05200025 05200026

Appendix B

Chronology of the Combined License Application for Vogtle Electric Generating Plant Units 3 and 4

This appendix contains a chronological listing of routine licensing correspondence between the staff of the U.S. Nuclear Regulatory Commission (NRC) regarding the review of the Vogtle Electric Generating Plant, Units 3 and 4 plant design under Docket Nos. 052-000025 and 052-000026

Document Date	Accession Number	Title & Estimated Page Count	Document Type	Author Affiliation	Addressee Affiliation	Docket Number
08/31/2009	ML092960300	2009/08/31 Vogtle COL Review - SNC Letter ND-09-1329 transmitting VEGP Units 3 & 4 COLA Response to RAI Letter No. 038 8 Page(s)	E-Mail	- No Known Affiliation	NRC/NRO/DNRL/NWE1	05200025 05200026
09/01/2009	ML092600974	2009/09/01 Vogtle COL Review - RE: Necessary Excavation.doc 2 Page(s)	E-Mail	NRC/NRO	NRC/NRO/DNRL/NWE1	05200025 05200026
09/03/2009	ML092440831	09/16-17/2009 - Meeting Notice, Closed Meeting with the Southern Nuclear Operating Company on the Vogtle Cyber Security Plan Submittal and the Draft Regulatory Guide 5.71 Appendix, "Cyber Security Plan Template". 6 Page(s)	Meeting Agenda Meeting Notice Memoranda	NRC/NRO/DNRL/NGE2	NRC/NRO/DNRL/NWE1	05200025 05200026
09/08/2009	ML092600975	2009/09/08 Vogtle COL Review - FW: DRAFT RAI - Vogtle Units 3 and 4 Combined License Application 2 Page(s)	E-Mail	NRC/NRO	NRC/NRO/DNRL/NWE1	05200025 05200026
09/09/2009	ML092600976	2009/09/09 Vogtle COL Review - FW: SNC Letter ND-09-1415 transmitting VEGP Units 3 & 4 Periodic Update of Construction Schedule Information 6 Page(s)	E-Mail	NRC/NRO	NRC/NRO/DNRL/NWE1	05200025 05200026
09/10/2009	ML092600977	2009/09/10 Vogtle COL Review -	E-Mail	NRC/NRO	NRC/NRO/DNRL/NWE1	05200025

Appendix B

Chronology of the Combined License Application for Vogtle Electric Generating Plant Units 3 and 4

This appendix contains a chronological listing of routine licensing correspondence between the staff of the U.S. Nuclear Regulatory Commission (NRC) regarding the review of the Vogtle Electric Generating Plant, Units 3 and 4 plant design under Docket Nos. 052-000025 and 052-000026

Document Date	Accession Number	Title & Estimated Page Count	Document Type	Author Affiliation	Addressee Affiliation	Docket Number
		FW: SNC Letter ND-09-1447 transmitting VEGP Units 3 & 4 COLA Response to BLN Units 3 & 4 SER OI for Chapter 19 6 Page(s)				05200026
09/10/2009	ML092570239	Vogtle, Units 3 & 4 Combined License Application, Response to Bellefonte, Units 3 & 4, Safety Evaluation Report Open Items for Chapter 19. 4 Page(s)	Letter	Southern Nuclear Operating Co, Inc	NRC/Document Control Desk NRC/NRO	05200014 05200015 05200025 05200026
09/11/2009	ML092570539	2009/09/11-Revised Initial Mandatory Disclosures for Contention SAFETY-1, Southern Nuclear Operating Co. (COL for Vogtle Units 3 and 4). 16 Page(s)	Legal-Hearing File (For Informal Hearings)	Balch & Bingham, LLP	Emory Univ School of Law NRC/OGC	05200025 05200026
09/14/2009	ML092600667	2009/09/14 Vogtle COL Review - NMFS BA for Vogtle COL 2 Page(s)	E-Mail	NRC/NRO	NRC/NRO/DNRL/NWE1	05200025 05200026
09/14/2009	ML092600978	2009/09/14 Vogtle COL Review - Re: Vogtle RAI data for Tuesday's call 21 Page(s)	E-Mail	NRC/NRO	NRC/NRO/DNRL/NWE1	05200025 05200026
09/15/2009	ML092580294	2009/09/15 Vogtle COL Review - Conference Call Summary - August 31, 2009 - Southern Nuclear Vogtle Units 3 and 4 COLA - Draft RAIs related to SRP Section 13.6	E-Mail	NRC/NRO	NRC/NRO/DNRL/NWE1	05200025 05200026

Appendix B

Chronology of the Combined License Application for Vogtle Electric Generating Plant Units 3 and 4

This appendix contains a chronological listing of routine licensing correspondence between the staff of the U.S. Nuclear Regulatory Commission (NRC) regarding the review of the Vogtle Electric Generating Plant, Units 3 and 4 plant design under Docket Nos. 052-000025 and 052-000026

Document Date	Accession Number	Title & Estimated Page Count	Document Type	Author Affiliation	Addressee Affiliation	Docket Number
		3 Page(s)				
09/17/2009	ML092600660	2009/09/17 Vogtle RAI for SER – Request for Additional Information Letter No. 041 Related to SRP Section 13.6 for the Vogtle Electric Generating Plant Units 3 and 4 Combined License Application 11 Page(s)	E-Mail	NRC/NRO	NRC/NRO/DNRL/NWE1	05200025 05200026
09/17/2009	ML092720806	2009/09/17 Vogtle COL Review – Request for Additional Information Letter No. 041 Related to SRP Section 13.6 for the Vogtle Units 3 and 4 Combined License Application 13 Page(s)	E-Mail	NRC/NRO	NRC/NRO/DNRL/NWE1	05200025 05200026
09/18/2009	ML092330845	Notice of Intent to Prepare a Supplemental Environmental Impact Statement for Plant Vogtle Units 3 and 4.	Federal Register Notice Letter	NRC/NRO/DSER	Southern Nuclear Operating Co, Inc	05200025 05200026
09/18/2009	ML092650823	FRN- General Federal Notice of Intent to Prepare a Supplemental Environmental Impact Statement for Vogtle Units 3 and 4. 3 Page(s)	Federal Register Notice	NRC/NRO/DSER/RAP1		05200025 05200026
09/21/2009	ML092600338	Vogtle Audit Execution Plan for New and Significant Information. 19 Page(s)	Memoranda Project Plans and Schedules	NRC/NRO/DSER/RAP1	NRC/NRO/DSER	05200025 05200026
09/21/2009	ML092660091	Vogtle Electric Generating Plant	Letter	Southern Nuclear Operating	NRC/Document Control	05200025

Chronology of the Combined License Application for Vogtle Electric Generating Plant Units 3 and 4

This appendix contains a chronological listing of routine licensing correspondence between the staff of the U.S. Nuclear Regulatory Commission (NRC) regarding the review of the Vogtle Electric Generating Plant, Units 3 and 4 plant design under Docket Nos. 052-000025 and 052-000026

Document Date	Accession Number	Title & Estimated Page Count	Document Type	Author Affiliation	Addressee Affiliation	Docket Number
		Units 3 & 4, Combined License Application Response to Bellefonte Safety Evaluation Report Open Items for Chapter 01.		Co, Inc	Desk NRC/NRO	05200026
09/21/2009	ML092960026	2009/09/21 Vogtle COL Review - RE: Re: Vogtle RAI data for Tuesday's call-9-22-09 20 Page(s)	E-Mail	NRC/NRO	NRC/NRO/DNRL/NWE1	05200025 05200026
09/21/2009	ML092960296	2009/09/21 Vogtle COL Review - SNC Letter ND-09-1457 transmitting VEGP Units 3 & 4 COLA Response to BLN Safety Evaluation OIs for Chapter 01 10 Page(s)	E-Mail	- No Known Affiliation	NRC/NRO/DNRL/NWE1	05200025 05200026
09/23/2009	ML092680023	Vogtle, Units 3 and 4, Combined License Application, Response to Request for Additional Information Letter No. 039 Involving the Solid Waste Management System. 11 Page(s)	Letter	Southern Nuclear Operating Co, Inc	NRC/Document Control Desk NRC/NRO	05200025 05200026
09/23/2009	ML092740386	Vogtle, Units 3 and 4 Combined License Application, Revision 1 to the Environmental Report. 6 Page(s)	Letter	Southern Nuclear Operating Co, Inc	NRC/Document Control Desk NRC/NRO	05200025 05200026
09/23/2009	ML092740400	Vogtle, Units 3 & 4 COL Application Part 3, Environmental Report, Revision 1. 78 Page(s)	Environmental Report	Southern Nuclear Operating Co, Inc	NRC/NRO	05200025 05200026

Appendix B

Chronology of the Combined License Application for Vogtle Electric Generating Plant Units 3 and 4

This appendix contains a chronological listing of routine licensing correspondence between the staff of the U.S. Nuclear Regulatory Commission (NRC) regarding the review of the Vogtle Electric Generating Plant, Units 3 and 4 plant design under Docket Nos. 052-000025 and 052-000026

Document Date	Accession Number	Title & Estimated Page Count	Document Type	Author Affiliation	Addressee Affiliation	Docket Number
09/23/2009	ML092960171	2009/09/23 Vogtle COL Review - SNC Letter ND-09-1493 transmitting periodic VEGP Units 3 & 4 Construction Schedule Information to NRC Region II 6 Page(s)	E-Mail	- No Known Affiliation	NRC/NRO/DNRL/NWE1	05200025 05200026
09/24/2009	ML092670260	2009/09/24 Vogtle COL Review - FW: Plant Vogtle New and Significant Information Audit 2 Page(s)	E-Mail	NRC/NRO	NRC/NRO/DNRL/NWE1	05200025 05200026
09/24/2009	ML092670575	2009/09/24 Vogtle COL Review - RE: Vogtle N&S Audit & ER 3 Page(s)	E-Mail	NRC/NRO	NRC/NRO/DNRL/NWE1	05200025 05200026
09/24/2009	ML092960167	2009/09/24 Vogtle COL Review - SNC Letter ND-09-1540 - SNC VEGP Units 3&4 Combined License Application Response to Request for Additional Information Letter No. 039 14 Page(s)	E-Mail	- No Known Affiliation	NRC/NRO/DNRL/NWE1	05200025 05200026
09/24/2009	ML092960308	2009/09/24 Vogtle COL Review - SNC Letter ND-09-1501 - SNC VEGP Units 3&4 Combined License Application Revision 1 to the Environmental Report 9 Page(s)	E-Mail	- No Known Affiliation	NRC/NRO/DNRL/NWE1	05200025 05200026
09/28/2009	ML092960160	2009/09/28 Vogtle COL Review - RE: Re: Vogtle RAI data for Tuesday's call-9-29-09	E-Mail	NRC/NRO	NRC/NRO/DNRL/NWE1	05200025 05200026

Appendix B

Chronology of the Combined License Application for Vogtle Electric Generating Plant Units 3 and 4

This appendix contains a chronological listing of routine licensing correspondence between the staff of the U.S. Nuclear Regulatory Commission (NRC) regarding the review of the Vogtle Electric Generating Plant, Units 3 and 4 plant design under Docket Nos. 052-000025 and 052-000026

Document Date	Accession Number	Title & Estimated Page Count	Document Type	Author Affiliation	Addressee Affiliation	Docket Number
09/29/2009	ML092710546	Safety Evaluation Report Financial Qualifications Review Input for the Vogtle Electric Generating Plant, Units 3 and 4 Combined License. 20 Page(s)	Memoranda	NRC/NRR/DPR/PFPB	NRC/NRO/DNRL	05200025 05200026
09/29/2009	ML092720037	Safety Evaluation by the Office of Nuclear Reactor Regulation on Behalf of the Office of New Reactors Combined License Application Vogtle Electric Generating Plant, Units 3 and 4- Proprietary. 2 Page(s)	Safety Evaluation	NRC/NRR/DPR/PFPB	NRC/NRO/DNRL	05200025 05200026
09/29/2009	ML092720147	Safety Evaluation by the Office of Nuclear Reactor Regulation on Behalf of the Office of New Reactors Combined License Application Vogtle Electric Generating Plant, Units 3 and 4 - Non-Proprietary. 16 Page(s)	Safety Evaluation	NRC/NRR/DPR/PFPB	NRC/NRO/DNRL	05200025 05200026
09/29/2009	ML092960170	2009/09/29 Vogtle COL Review - FW: SNC Letter ND-09-1540 - SNC VEGP Units 3&4 Combined License Application Response to Request for Additional Information Letter No. 039 15 Page(s)	E-Mail	NRC/NRO	NRC/NRO/DNRL/NWE1	05200025 05200026

Appendix B

Chronology of the Combined License Application for Vogtle Electric Generating Plant Units 3 and 4

This appendix contains a chronological listing of routine licensing correspondence between the staff of the U.S. Nuclear Regulatory Commission (NRC) regarding the review of the Vogtle Electric Generating Plant, Units 3 and 4 plant design under Docket Nos. 052-000025 and 052-000026

Document Date	Accession Number	Title & Estimated Page Count	Document Type	Author Affiliation	Addressee Affiliation	Docket Number
09/29/2009	ML092960172	2009/09/29 Vogtle COL Review - FW: SNC Letter ND-09-1493 transmitting periodic VEGP Units 3 & 4 Construction Schedule Information to NRC Region II 2 Page(s)	E-Mail	NRC/NRO	NRC/NRO/DNRL/NWE1	05200025 05200026
09/30/2009	ML092780040	Vogtle Electric Generating Plant, Units 3 & 4 – Combined License Application Supplemental Change to Address Standard Content. 6 Page(s)	Letter	Southern Nuclear Operating Co, Inc	NRC/Document Control Desk NRC/NRO	05200025 05200026
09/30/2009	ML092730351	2009/09/30 Vogtle RAI for SER – Request for Additional Information Letter No. 040 Related to SRP Section 19.0 0 for the Vogtle Electric Generating Plant Unit 3 and 4 Combined License Application 6 Page(s)	E-Mail Request for Additional Information (RAI)	NRC/NRO	NRC/NRO/DNRL/NWE1	05200025 05200026
09/30/2009	ML092730618	2009/09/30– NRC Staff Submittal of Vogtle COL Hearing File Index, Update 1 – September 30, 2009. 10 Page(s)	Legal-Hearing File	NRC/OGC	NRC/ASLBP	05200025 05200026
09/30/2009	ML092790182	2009/09/30–First Supplemental Disclosures for Contention Safety-1; Southern Nuclear Operating Co. (COL for Plant Vogtle Units 3 and 4), Docket Nos. 52-025-COL and 52-026-COL. 6 Page(s)	Legal-Correspondence/Miscellaneous Letter	Emory Univ School of Law	Balch & Bingham, LLP NRC/OGC NRC/SECY/RAS	05200025 05200026

Appendix B

Chronology of the Combined License Application for Vogtle Electric Generating Plant Units 3 and 4

This appendix contains a chronological listing of routine licensing correspondence between the staff of the U.S. Nuclear Regulatory Commission (NRC) regarding the review of the Vogtle Electric Generating Plant, Units 3 and 4 plant design under Docket Nos. 052-000025 and 052-000026

Document Date	Accession Number	Title & Estimated Page Count	Document Type	Author Affiliation	Addressee Affiliation	Docket Number
09/30/2009	ML092790183	2009/09/30-Vogtle, Units 3 & 4, First Supplemental Disclosures Pursuant to 10 CFR 2.336 for Contention SAFETY-1. 7 Page(s)	Legal-Correspondence	Balch & Bingham, LLP Southern Nuclear Operating Co, Inc	NRC/OGC Turner Environmental Law Clinic	05200025 05200026
09/30/2009	ML092960173	2009/09/30 Vogtle COL Review - Request for Additional Information Letter No. 040 Related to SRP Section 19.0 0 for the Vogtle Electric Generating Plant Units 3 and 4 Combined License Application 8 Page(s)	E-Mail	NRC/NRO	NRC/NRO/DNRL/NWE1	05200025 05200026
09/30/2009	ML092960175	2009/09/30 Vogtle COL Review - FW: Bellefonte SER with OI 41 Page(s)	E-Mail	NRC/NRO	NRC/NRO/DNRL/NWE1	05200025 05200026
09/30/2009	ML092960177	2009/09/30 Vogtle COL Review - FW: Cover letter and Chapter 14 SER with OI for Bellefonte 50 Page(s)	E-Mail	NRC/NRO	NRC/NRO/DNRL/NWE1	05200025 05200026
09/30/2009	ML092960179	2009/09/30 Vogtle COL Review - FW: Draft RAI No. 3512 related to SRP Section: 19 - Probabilistic Risk Assessment and Severe Accident Evaluation for Vogtle Units 3 and 4 3 Page(s)	E-Mail	NRC/NRO	NRC/NRO/DNRL/NWE1	05200025 05200026
09/30/2009	ML092960180	2009/09/30 Vogtle COL Review - FW: Confirmation of agreement on LOLA Teleconference	E-Mail	NRC/NRO	NRC/NRO/DNRL/NWE1	05200025 05200026

Appendix B

Chronology of the Combined License Application for Vogtle Electric Generating Plant Units 3 and 4

This appendix contains a chronological listing of routine licensing correspondence between the staff of the U.S. Nuclear Regulatory Commission (NRC) regarding the review of the Vogtle Electric Generating Plant, Units 3 and 4 plant design under Docket Nos. 052-000025 and 052-000026

Document Date	Accession Number	Title & Estimated Page Count	Document Type	Author Affiliation	Addressee Affiliation	Docket Number
09/30/2009	ML092960182	2009/09/30 Vogtle COL Review - FW: LOLA Draft RAI information 8 Page(s)	E-Mail	NRC/NRO	NRC/NRO/DNRL/NWE1	05200025 05200026
10/01/2009	ML092100687	Vogtle COL Review - Draft RAI 3409 Related SRP Section 19.0 for Vogtle Units 3 and 4 - Loss of Large Areas (Security-Related Information). 11 Page(s)	Request for Additional Information (RAI)	NRC/NRO/DNRL/NWE1	Southern Nuclear Operating Co, Inc	05200025 05200026
10/01/2009	ML092750349	Letter, Request for Additional Information Letter No. 042 Related to SRP Section 19.0 for the Vogtle Electric Generating Plant Units 3 and 4 Combined License Application. 3 Page(s)	Letter	NRC/NRO/DNRL/NWE1	Southern Nuclear Operating Co, Inc	05200025 05200026
10/01/2009	ML092960184	2009/10/01 Vogtle COL Review - Phone Call Summary Vogtle RCOL 093009.doc 3 Page(s)	E-Mail	NRC/NRO	NRC/NRO/DNRL/NWE1	05200025 05200026
10/01/2009	ML092960187	2009/10/01 Vogtle COL Review - SNC Letter ND-09-1589 transmitting VEGP Units 3 & 4 COLA Supplemental Information for RAI # 05.02.01.01-01 9 Page(s)	E-Mail	- No Known Affiliation	NRC/NRO/DNRL/NWE1	05200025 05200026
10/02/2009	ML092960295	2009/10/02 Vogtle COL Review - SNC Letter ND-09-1568 - SNC	E-Mail	- No Known Affiliation	NRC/NRO/DNRL/NWE1	05200025 05200026

Appendix B

Chronology of the Combined License Application for Vogtle Electric Generating Plant Units 3 and 4

This appendix contains a chronological listing of routine licensing correspondence between the staff of the U.S. Nuclear Regulatory Commission (NRC) regarding the review of the Vogtle Electric Generating Plant, Units 3 and 4 plant design under Docket Nos. 052-000025 and 052-000026

Document Date	Accession Number	Title & Estimated Page Count	Document Type	Author Affiliation	Addressee Affiliation	Docket Number
		VEGP Units 3&4 Combined License Application Revision 1 to Part 6, LWA Request 9 Page(s)				
10/02/2009	ML092960549	Vogtle, Units 3 & 4, Combined License Application, Revision 1 to Part 6, LWA Request. 6 Page(s)	Letter	Southern Nuclear Operating Co, Inc	NRC/Document Control Desk NRC/NRO	05200025 05200026
10/05/2009	ML092780338	2009/10/05 Vogtle COL Review - FW: N&S Key Inputs 240 Page(s)	E-Mail	NRC/NRO	NRC/NRO/DNRL/NWE1	05200025 05200026
10/05/2009	ML092960273	2009/10/05 Vogtle COL Review - Telecon to discuss SER open items for chapter 17 for Vogtle 3 Page(s)	E-Mail	NRC/NRO	NRC/NRO/DNRL/NWE1	05200025 05200026
10/05/2009	ML092960275	2009/10/05 Vogtle COL Review - RE: Telecon to discuss SER open items for chapter 17 for Vogtle 2 Page(s)	E-Mail	- No Known Affiliation	NRC/NRO/DNRL/NWE1	05200025 05200026
10/05/2009	ML092960288	2009/10/05 Vogtle COL Review - RE: Vogtle e-RAI data for Tuesday 10-06-09 call 24 Page(s)	E-Mail	NRC/NRO	NRC/NRO/DNRL/NWE1	05200025 05200026
10/07/2009	ML092960301	2009/10/07 Vogtle COL Review - RE: Hydrology Call 2 Page(s)	E-Mail	NRC/NRO	NRC/NRO/DNRL/NWE1	05200025 05200026
10/08/2009	ML092960302	2009/10/08 Vogtle COL Review - FW: FA- Cyber Security Conference Call NEW Bridge Line	E-Mail	NRC/NRO	NRC/NRO/DNRL/NWE1	05200025 05200026

Appendix B

Chronology of the Combined License Application for Vogtle Electric Generating Plant Units 3 and 4

This appendix contains a chronological listing of routine licensing correspondence between the staff of the U.S. Nuclear Regulatory Commission (NRC) regarding the review of the Vogtle Electric Generating Plant, Units 3 and 4 plant design under Docket Nos. 052-000025 and 052-000026

Document Date	Accession Number	Title & Estimated Page Count	Document Type	Author Affiliation	Addressee Affiliation	Docket Number
		Number 2 Page(s)				
10/09/2009	ML092960306	2009/10/09 Vogtle COL Review - Vogtle RAI data for Tuesday's Call-10-13-09 25 Page(s)	E-Mail	NRC/NRO	NRC/NRO/DNRL/NWE1	05200025 05200026
10/15/2009	ML092920173	Vogtle Electric Generating Plant Units 3 and 4 - Response to Questions on the Waterproof Membrane. 20 Page(s)	Letter	Southern Nuclear Operating Co, Inc	NRC/Document Control Desk NRC/NRO	05200025 05200026
10/15/2009	ML092881301	2009/10/15 Vogtle COL Review - RE: New & Significant Evaluation for New Info 2 Page(s)	E-Mail	NRC/NRO	NRC/NRO/DNRL/NWE1	05200025 05200026
10/15/2009	ML093200704	2009/10/15 Vogtle COL Review - SNC Letter ND-09-1578, Response to NRC Waterproof Membrane Questions 23 Page(s)	E-Mail	- No Known Affiliation	NRC/NRO/DNRL/NWE1	05200025 05200026
10/15/2009	ML100630201	Vogtle, Units 3 & 4, Combined License Application, Post New and Significant Audit Supporting Information. 4 Page(s)	Letter	Southern Nuclear Operating Co, Inc	NRC/Document Control Desk NRC/NRO	05200025 05200026
10/16/2009	ML092930117	Vogtle, Units 3 and 4 - Combined License Application - Response to Request for Additional Information Letter No. 041.	Letter	Southern Nuclear Operating Co, Inc	NRC/Document Control Desk NRC/NRO	05200025 05200026

Appendix B

Chronology of the Combined License Application for Vogtle Electric Generating Plant Units 3 and 4

This appendix contains a chronological listing of routine licensing correspondence between the staff of the U.S. Nuclear Regulatory Commission (NRC) regarding the review of the Vogtle Electric Generating Plant, Units 3 and 4 plant design under Docket Nos. 052-000025 and 052-000026

Document Date	Accession Number	Title & Estimated Page Count	Document Type	Author Affiliation	Addressee Affiliation	Docket Number
10/16/2009	ML092930120	Vogtle Electric Generating Plant, Units 3 & 4 Combined License Application Response to Bellefonte Units 3 and 4 Safety Evaluation Report confirmatory item for Chapter 12. 26 Page(s)	Letter	Southern Nuclear Operating Co, Inc	NRC/Document Control Desk NRC/NRO	05200025 05200026
10/16/2009	ML092960312	2009/10/16 Vogtle COL Review - FW: SNC Letter ND-09-1673, Environmental Report Post New and Significant Audit Supporting Information 19 Page(s)	E-Mail	NRC/NRO	NRC/NRO/DNRL/NWE1	05200025 05200026
10/16/2009	ML092960391	2009/10/16 Vogtle COL Review - FW: SNC Letter ND-09-1578, Response to NRC Waterproof Membrane Questions 6 Page(s)	E-Mail	NRC/NRO	NRC/NRO/DNRL/NWE1	05200025 05200026
10/16/2009	ML092960392	2009/10/16 Vogtle COL Review - FW: Update on NEI Template 07-08 6 Page(s)	E-Mail	NRC/NRO	NRC/NRO/DNRL/NWE1	05200025 05200026
10/16/2009	ML092960393	2009/10/16 Vogtle COL Review - FW: Update on NEI Template 07-08 2 Page(s)	E-Mail	NRC/NRO	NRC/NRO/DNRL/NWE1	05200025 05200026
10/19/2009	ML092960396	2009/10/19 Vogtle COL Review - RAI data for Tuesday 10-20 call 2 Page(s)	E-Mail	NRC/NRO	NRC/NRO/DNRL/NWE1	05200025 05200026

Appendix B

Chronology of the Combined License Application for Vogtle Electric Generating Plant Units 3 and 4

This appendix contains a chronological listing of routine licensing correspondence between the staff of the U.S. Nuclear Regulatory Commission (NRC) regarding the review of the Vogtle Electric Generating Plant, Units 3 and 4 plant design under Docket Nos. 052-000025 and 052-000026

Document Date	Accession Number	Title & Estimated Page Count	Document Type	Author Affiliation	Addressee Affiliation	Docket Number
10/19/2009	ML092960397	2009/10/19 Vogtle COL Review - FW: COLA - Southern Co ND 09 1568 Vogtle Electric Generating Plant Units 3 and 4 Combined License Application. 4 Page(s)	E-Mail	NRC/NRO	NRC/NRO/DNRL/NWE1	05200025 05200026
10/19/2009	ML092960398	2009/10/19 Vogtle COL Review - RE: COLA - Southern Co ND 09 1568 Vogtle Electric Generating Plant Units 3 and 4 Combined License Application. 3 Page(s)	E-Mail	- No Known Affiliation	NRC/NRO/DNRL/NWE1	05200025 05200026
10/20/2009	ML092931714	2009/10/20 Vogtle COL Review – Service Accident N&S Evaluation Call Information 2 Page(s)	E-mail	NRC/NRO	NRC/NRO/DNRL/NWE1	05200025 05200026
10/20/2009	ML092960402	2009/10/20 Vogtle COL Review - Update on NEI Template 08-08 22 Page(s)	E-Mail	NRC/NRO	NRC/NRO/DNRL/NWE1	05200025 05200026
10/21/2009	ML092920183	Memo to K. Williams: Support to NSIR: Technical Support Center Habitability Input for SRP Section 13.3, "Emergency Planning" for Vogtle Units 3 and 4 Combined Operating License Application. 2 Page(s)	Memoranda	NRC/NRO/DSER/RSAC	NRC/NSIR/DPR/DDEP	05200025 05200026
10/21/2009	ML092960400	2009/10/21 Vogtle COL Review - Phone call on October 14, 2009 to discuss Loss of large area Fire with	E-Mail	NRC/NRO	NRC/NRO/DNRL/NWE1	05200025 05200026

Appendix B

Chronology of the Combined License Application for Vogtle Electric Generating Plant Units 3 and 4

This appendix contains a chronological listing of routine licensing correspondence between the staff of the U.S. Nuclear Regulatory Commission (NRC) regarding the review of the Vogtle Electric Generating Plant, Units 3 and 4 plant design under Docket Nos. 052-000025 and 052-000026

Document Date	Accession Number	Title & Estimated Page Count	Document Type	Author Affiliation	Addressee Affiliation	Docket Number
10/21/2009	ML092960401	Vogtle 2 Page(s) 2009/10/21 Vogtle COL Review - RE: Stephanie Coffin 2 Page(s)	E-Mail	NRC/NRO	NRC/NRO/DNRL/NWE1	05200025 05200026
10/21/2009	ML093000387	Vogtle, Units 3 & 4 Combined License Application, Request for Additional Information Letter No. 043 Related to SRP Section 13.6. 19 Page(s)	Letter Request for Additional Information (RAI)	NRC/NRO/DNRL/NWE1	Southern Nuclear Operating Co, Inc	05200025 05200026
10/22/2009	ML092930324	10/22/2009 Timothy Frye Memo re: Advance Safety Evaluation Report (SER) with No Open Items for FSAR Chapter 11 of the Vogtle Units 3 and 4 Combined License Application (COLA). 33 Page(s)	Memoranda Safety Evaluation Report	NRC/NRO/DCIP/CHPB	NRC/NRO/DNRL	05200025 05200026
10/22/2009	ML092950137	2009/10/22 Vogtle COL Review - FW: New & Significant Evaluation for New Info 2 Page(s)	E-Mail	NRC/NRO	NRC/NRO/DNRL/NWE1	05200025 05200026
10/22/2009	ML092960406	2009/10/22 Vogtle COL Review - Phone Call Summary Vogtle RCOL 102009 3 Page(s)	E-Mail	NRC/NRO	NRC/NRO/DNRL/NWE1	05200025 05200026
10/22/2009	ML092960614	2009/10/22-Petition for Review of Respondent Nuclear Regulatory Commission's Issuance of an Early	Legal-Correspondence/Miscellaneous	Blue Ridge Environmental Defense League Center for a Sustainable	NRC/ASLBP US Federal Judiciary, District Court for the Central	05200025 05200026

Appendix B

Chronology of the Combined License Application for Vogtle Electric Generating Plant Units 3 and 4

This appendix contains a chronological listing of routine licensing correspondence between the staff of the U.S. Nuclear Regulatory Commission (NRC) regarding the review of the Vogtle Electric Generating Plant, Units 3 and 4 plant design under Docket Nos. 052-000025 and 052-000026

Document Date	Accession Number	Title & Estimated Page Count	Document Type	Author Affiliation	Addressee Affiliation	Docket Number
		Site Permit and Limited Work Authorization to SNC for Units 3 and 4 of the Vogtle Nuclear Power Plant. 99 Page(s)		Coast Emory Univ School of Law Harmon, Curran, Spielberg & Eisenberg, LLP Savannah Riverkeeper Southern Alliance for Clean Energy Turner Environmental Law Clinic Women's Action for New Directions	District of Utah	
10/22/2009	ML093200702	2009/10/22 Vogtle COL Review - SNC Letter ND-09-1723 - SNC VEGP Units 3&4 Periodic Update of Construction Schedule Information 8 Page(s)	E-Mail	- No Known Affiliation	NRC/NRO/DNRL/NWE1	05200025 05200026
10/23/2009	ML093010571	Vogtle, Units 3 and 4 Combined License Application, Revised Response to Request for Additional Information Number 19-4. 7 Page(s)	Letter	Southern Nuclear Operating Co, Inc	NRC/Document Control Desk NRC/NRO	05200025 05200026
10/23/2009	ML093010572	Vogtle Electric, Units 3 and 4 Combined License Application, Revised Response to Request for Additional Information Letter No. 037. 14 Page(s)	Letter	Southern Nuclear Operating Co, Inc	NRC/Document Control Desk NRC/NRO	05200025 05200026
10/23/2009	ML093010573	Vogtle, Units 3 and 4 Combined License Application, Revised	Letter	Southern Nuclear Operating Co, Inc	NRC/Document Control Desk	05200025 05200026

Appendix B

Chronology of the Combined License Application for Vogtle Electric Generating Plant Units 3 and 4

This appendix contains a chronological listing of routine licensing correspondence between the staff of the U.S. Nuclear Regulatory Commission (NRC) regarding the review of the Vogtle Electric Generating Plant, Units 3 and 4 plant design under Docket Nos. 052-000025 and 052-000026

Document Date	Accession Number	Title & Estimated Page Count	Document Type	Author Affiliation	Addressee Affiliation	Docket Number
		Response to Request for Additional Information Letter No. 034. 8 Page(s)			NRC/NRO	
10/23/2009	ML093200694	2009/10/23 Vogtle COL Review - SNC Letter ND-09-1709 - SNC VEGP Units 3&4 Combined License Application Revised Response to RAI Letter No. 034 11 Page(s)	E-Mail	- No Known Affiliation	NRC/NRO/DNRL/NWE1	05200025 05200026
10/23/2009	ML093200695	2009/10/23 Vogtle COL Review - SNC Letter ND-09-1710 - SNC VEGP Units 3&4 Combined License Application Revised Response to RAI Letter No. 037 17 Page(s)	E-Mail	- No Known Affiliation	NRC/NRO/DNRL/NWE1	05200025 05200026
10/23/2009	ML093200698	2009/10/23 Vogtle COL Review - SNC Letter ND-09-1714 - SNC VEGP Units 3&4 Combined License Application Revised Response to RAI Number 19-4 10 Page(s)	E-Mail	- No Known Affiliation	NRC/NRO/DNRL/NWE1	05200025 05200026
10/23/2009	ML093200699	2009/10/23 Vogtle COL Review - SNC Letter ND-09-1708 - SNC VEGP Units 3&4 Combined License Application Revised Response to RAI Numbers 08.02-1, -5 and -8 13 Page(s)	E-Mail	- No Known Affiliation	NRC/NRO/DNRL/NWE1	05200025 05200026
10/23/2009	ML093240096	Vogtle, Units 3 and 4, Combined License Application Revised	Letter	Southern Nuclear Operating Co, Inc	NRC/Document Control Desk	05200025 05200026

Appendix B

Chronology of the Combined License Application for Vogtle Electric Generating Plant Units 3 and 4

This appendix contains a chronological listing of routine licensing correspondence between the staff of the U.S. Nuclear Regulatory Commission (NRC) regarding the review of the Vogtle Electric Generating Plant, Units 3 and 4 plant design under Docket Nos. 052-000025 and 052-000026

Document Date	Accession Number	Title & Estimated Page Count	Document Type	Author Affiliation	Addressee Affiliation	Docket Number
		Response to Request for Additional Information Numbers 08.02-1. -5 and -8. 10 Page(s)			NRC/NRO	
10/26/2009	ML093200692	2009/10/26 Vogtle COL Review - RE: RAI data for Tuesday 10-27 call 25 Page(s)	E-Mail	NRC/NRO	NRC/NRO/DNRL/NWE1	05200025 05200026
10/28/2009	ML093010343	2009/10/28 Vogtle COL Review - FW: Requested Reference 7 Page(s)	E-Mail	NRC/NRO	NRC/NRO/DNRL/NWE1	05200025 05200026
10/282009	ML093000052	Environmental Revision 1 by Southern Nuclear Operating Company for a Combined License for Units 3 & 4 at the Vogtle Electric Generating Plant. 5 Page(s)	Letter	NRC/NRO/DSER/RAP1	Burke County, GA	05200025 05200026 PROJ0755
10/29/2009	ML093070086	Vogtle, Units 3 and 4, Combined License Application, Response to Request for Additional Information Letter No. 042 Involving Loss of Large Areas of the Plant Mitigative Strategies Plan. 4 Page(s)	Letter	Southern Nuclear Operating Co, Inc	NRC/Document Control Desk NRC/NRO	05200025 05200026
10/30/2009	ML093070283	Southern Nuclear Operating Company (SNC) Vogtle Electric Generating Plant Units 3 and 4 Combined License Application, Supplemental Response to	Letter	Southern Nuclear Operating Co, Inc	NRC/Document Control Desk NRC/NRO	05200025 05200026

Appendix B

Chronology of the Combined License Application for Vogtle Electric Generating Plant Units 3 and 4

This appendix contains a chronological listing of routine licensing correspondence between the staff of the U.S. Nuclear Regulatory Commission (NRC) regarding the review of the Vogtle Electric Generating Plant, Units 3 and 4 plant design under Docket Nos. 052-000025 and 052-000026

Document Date	Accession Number	Title & Estimated Page Count	Document Type	Author Affiliation	Addressee Affiliation	Docket Number
		Bellefonte Units 3 and 4 Safety Evaluation Report Open Items for Chapter 12. 10 Page(s)				
10/30/2009	ML093070284	Southern Nuclear Operating Company Vogtle Electric Generating Plant Units 3 and 4 Combined License Application, Response to Request for Additional Information Letter No. 040. 10 Page(s)	Letter	Southern Nuclear Operating Co, Inc	NRC/Document Control Desk NRC/NRO	05200025 05200026
10/30/2009	ML093200724	2009/10/30 Vogtle COL Review - SNC Letter ND-09-1768 transmitting VEGP Units 3 & 4 COLA Response to RAI Letter No. 040 13 Page(s)	E-Mail	- No Known Affiliation	NRC/NRO/DNRL/NWE1	05200025 05200026
10/30/2009	ML093200725	2009/10/30 Vogtle COL Review - SNC Letter ND-09-1770 transmitting VEGP Units 3 & 4 COLA Supplemental Response to Bellefonte Units 3 & 4 SER OIs for Chapter 12 13 Page(s)	E-Mail	- No Known Affiliation	NRC/NRO/DNRL/NWE1	05200025 05200026
11/02/2009	ML093200705	2009/11/02 Vogtle COL Review - Vogtle RAI data for Tuesday's call 11-3-09 25 Page(s)	E-Mail	NRC/NRO	NRC/NRO/DNRL/NWE1	05200025 05200026
11/02/2009	ML093200706	2009/11/02 Vogtle COL Review -	E-Mail	- No Known Affiliation	NRC/NRO/DNRL/NWE1	05200025

Appendix B

Chronology of the Combined License Application for Vogtle Electric Generating Plant Units 3 and 4

This appendix contains a chronological listing of routine licensing correspondence between the staff of the U.S. Nuclear Regulatory Commission (NRC) regarding the review of the Vogtle Electric Generating Plant, Units 3 and 4 plant design under Docket Nos. 052-000025 and 052-000026

Document Date	Accession Number	Title & Estimated Page Count	Document Type	Author Affiliation	Addressee Affiliation	Docket Number
		FW: Address for Mike Price 2 Page(s)				05200026
11/03/2009	ML093200707	2009/11/03 Vogtle COL Review - chemical hazards review 2 Page(s)	E-Mail	- No Known Affiliation	NRC/NRO/DNRL/NWE1	05200025 05200026
11/05/2009	ML093200708	2009/11/05 Vogtle COL Review - Bellefonte chapter 15 ser open items	E-Mail	NRC/NRO	NRC/NRO/DNRL/NWE1	05200025 05200026
11/05/2009	ML093280281	Trip Report - Watts Bar Unit 2 and Vogtle Site - Enclosure 3. 13 Page(s)	Meeting Briefing Package/Handouts Slides and Viewgraphs	Southern Nuclear Operating Co, Inc	NRC/NSIR/DSP/ISCPB	05000390 05000391 05000424 05000425 05200025 05200026
11/09/2009	ML093200721	2009/11/09 Vogtle COL Review - Vogtle RAI data for Tuesday's call 11-10-09 26 Page(s)	E-Mail	NRC/NRO	NRC/NRO/DNRL/NWE1	05200025 05200026
11/09/2009	ML093200722	2009/11/09 Vogtle COL Review - RE: chemical hazards review 2 Page(s)	E-Mail	NRC/NRO	NRC/NRO/DNRL/NWE1	05200025 05200026
11/09/2010	ML092870782	VEGP SER Chapter 8 Vogtle Electric Generating Plant Units 3 And 4 Advanced Final Safety Evaluation Report With No Open Items For Chapter 8, "Electric Power." 39 Page(s)	NRO Safety Evaluation Report (SER)-Delayed Safety Evaluation Report	NRC/NRO/DNRL/NWE1		05200025 05200026

Appendix B

Chronology of the Combined License Application for Vogtle Electric Generating Plant Units 3 and 4

This appendix contains a chronological listing of routine licensing correspondence between the staff of the U.S. Nuclear Regulatory Commission (NRC) regarding the review of the Vogtle Electric Generating Plant, Units 3 and 4 plant design under Docket Nos. 052-000025 and 052-000026

Document Date	Accession Number	Title & Estimated Page Count	Document Type	Author Affiliation	Addressee Affiliation	Docket Number
11/10/2009	ML093200726	2009/11/10 Vogtle COL Review - FW: QAPD 2 Page(s)	E-Mail	NRC/NRO	NRC/NRO/DNRL/NWE1	05200025 05200026
11/13/2009	ML093200727	2009/11/13 Vogtle COL Review - RE: chemical hazards review 2 Page(s)	E-Mail	NRC/NRO	NRC/NRO/DNRL/NWE1	05200025 05200026
11/13/2009	ML093210475	Vogtle, Units 3 and 4, Combined License Application, Supplemental Response to Request for Additional Information Letter No. 042, Loss of Large Areas of the Plant due to Explosions or Fire. 4 Page(s)	Letter	Southern Nuclear Operating Co, Inc	NRC/Document Control Desk NRC/NRO	05200025 05200026
11/15/2009	ML093200728	2009/11/15 Vogtle COL Review - Conference call to discuss Loss of large area fire questions with COL applicants 2 Page(s)	E-Mail	NRC/NRO	NRC/NRO/DNRL/NWE1	05200025 05200026
11/16/2009	ML093441563	2009/11/16 Vogtle COL Review - RE: Vogtle RAI data for Tuesday's call 25 Page(s)	E-Mail	NRC/NRO	NRC/NRO/DNRL/NWE1	05200025 05200026
11/16/2009	ML093441567	2009/11/16 Vogtle COL Review - RE: chemical hazards review 3 Page(s)	E-Mail	NRC/NRO	NRC/NRO/DNRL/NWE1	05200025 05200026
11/16/2009	ML093441579	2009/11/16 Vogtle COL Review - Response to Request for Additional Information Letter No. 042 related to Loss of Large Areas of the Plant	E-Mail	Southern Nuclear Operating Co, Inc	NRC/NRO/DNRL/NWE1	05200025 05200026

Appendix B

Chronology of the Combined License Application for Vogtle Electric Generating Plant Units 3 and 4

This appendix contains a chronological listing of routine licensing correspondence between the staff of the U.S. Nuclear Regulatory Commission (NRC) regarding the review of the Vogtle Electric Generating Plant, Units 3 and 4 plant design under Docket Nos. 052-000025 and 052-000026

Document Date	Accession Number	Title & Estimated Page Count	Document Type	Author Affiliation	Addressee Affiliation	Docket Number
11/17/2009	ML093441571	due to Explosions or Fire 8 Page(s) 2009/11/17 Vogtle COL Review - RE: chemical hazards review 3 Page(s)	E-Mail	- No Known Affiliation	NRC/NRO/DNRL/NWE1	05200025 05200026
11/18/2009	ML093441584	2009/11/18 Vogtle COL Review - RE: Conference Call Participants 2 Page(s)	E-Mail	NRC/NRO	NRC/NRO/DNRL/NWE1	05200025 05200026
11/20/2009	MI093280746	Endorsement of Bellefonte R-COLA Standard Content Requests for Additional Information for Vogtle Electric Generating Plant Units 3 & 4 Combined License Application. 7 Page(s)	Letter	Southern Nuclear Operating Co, Inc.	NRC/Document Control Desk NRC/NRO	05200025 05200026
11/20/2009	ML093280749	Vogtle, Units 3 & 4 - Combined License Application Supplemental Response to Request for Additional Information Letter No. 021 Related to Radiation Protection. 9 Page(s)	Letter	Southern Nuclear Operating Co, Inc	NRC/Document Control Desk NRC/NRO	05200025 05200026
11/20/2009	ML093280750	Vogtle, Units 3 & 4 - Combined License Application Departure Report Update. 4 Page(s)	Letter	Southern Nuclear Operating Co, Inc	NRC/Document Control Desk NRC/NRO	05200025 05200026
11/20/2009	ML093280751	Vogtle, Units 3 & 4 - Combined License Application Summary Identification of Standard Content in Response to Safety Evaluation Report Open Items.	Letter	Southern Nuclear Operating Co, Inc	NRC/Document Control Desk NRC/NRO	05200025 05200026

Appendix B

Chronology of the Combined License Application for Vogtle Electric Generating Plant Units 3 and 4

This appendix contains a chronological listing of routine licensing correspondence between the staff of the U.S. Nuclear Regulatory Commission (NRC) regarding the review of the Vogtle Electric Generating Plant, Units 3 and 4 plant design under Docket Nos. 052-000025 and 052-000026

Document Date	Accession Number	Title & Estimated Page Count	Document Type	Author Affiliation	Addressee Affiliation	Docket Number
11/20/2009	ML093280901	Vogtle Electric Generating Plant, Units 3 and 4 - Combined License Application - Response to Request for Additional Information Letter No. 043, Cyber Security Plan. 6 Page(s)	Letter	Southern Nuclear Operating Co, Inc	NRC/Document Control Desk NRC/NRO	05200025 05200026
11/20/2009	ML093280903	Vogtle Electric Generating Plant, Units 3 and 4 - Combined License Application - Response to Request for Additional Information Letter No. 043, Cyber Security Plan 4 Page(s)	- No Document Type Applies	Southern Nuclear Operating Co, Inc	NRC/NRO	05200025 05200026
11/20/2009	ML093280906	Vogtle Electric Generating Plant, Units 3 and 4 - Combined License Application - Response to Request for Additional Information Letter No. 043. 153 Page(s)	- No Document Type Applies	Southern Nuclear Operating Co, Inc	NRC/NRO	05200025 05200026
11/20/2009	ML093441597	2009/11/20 Vogtle COL Review - RE: Vogtle RAI data for Tuesday's call (11-24-09) 148 Page(s)	E-Mail	NRC/NRO	NRC/NRO/DNRL/NWE1	05200025 05200026
11/20/2009	ML093441605	2009/11/20 Vogtle COL Review - SNC Letter ND-09-1837 transmitting VEGP 3 & 4 Construction Schedule Information 25 Page(s)	E-Mail	- No Known Affiliation	NRC/NRO/DNRL/NWE1	05200025 05200026
11/20/2009	ML093441611	2009/11/20 Vogtle COL Review - 8 Page(s)	E-Mail	- No Known Affiliation	NRC/NRO/DNRL/NWE1	05200025

Appendix B

Chronology of the Combined License Application for Vogtle Electric Generating Plant Units 3 and 4

This appendix contains a chronological listing of routine licensing correspondence between the staff of the U.S. Nuclear Regulatory Commission (NRC) regarding the review of the Vogtle Electric Generating Plant, Units 3 and 4 plant design under Docket Nos. 052-000025 and 052-000026

Document Date	Accession Number	Title & Estimated Page Count	Document Type	Author Affiliation	Addressee Affiliation	Docket Number
		SNC Letter ND-09-1869 transmitting VEGP 3 & 4 Periodic Update of Construction Schedule Information 7 Page(s)				05200026
11/20/2009	ML093441614	2009/11/20 Vogtle COL Review - SNC Letter ND-09-1867 transmitting VEGP 3 & 4 COLA Summary Identification of Standard Content in Response to SER OIs 9 Page(s)	E-Mail	- No Known Affiliation	NRC/NRO/DNRL/NWE1	05200025 05200026
11/20/2009	ML093441616	2009/11/20 Vogtle COL Review - SNC Letter ND-09-1834 Transmitting Supplemental Response to RAI Letter No. 021 Related to Radiation Protection 12 Page(s)	E-Mail	- No Known Affiliation	NRC/NRO/DNRL/NWE1	05200025 05200026
11/20/2009	ML093441618	2009/11/20 Vogtle COL Review - SNC Letter ND-09-1821 Transmitting VEGP Units 3 & 4 COLA Departure Report Update 8 Page(s)	E-Mail	- No Known Affiliation	NRC/NRO/DNRL/NWE1	05200025 05200026
11/20/2009	ML093441625	2009/11/20 Vogtle COL Review - Bellefonte Chapter 9 SER with Open Items 84 Page(s)	E-Mail	NRC/NRO	NRC/NRO/DNRL/NWE1	05200025 05200026
11/20/2009	ML093441627	2009/11/20 Vogtle COL Review - Bellefonte Chapter 9 SER with Open Items - RESEND	E-Mail	NRC/NRO	NRC/NRO/DNRL/NWE1	05200025 05200026

Appendix B

Chronology of the Combined License Application for Vogtle Electric Generating Plant Units 3 and 4

This appendix contains a chronological listing of routine licensing correspondence between the staff of the U.S. Nuclear Regulatory Commission (NRC) regarding the review of the Vogtle Electric Generating Plant, Units 3 and 4 plant design under Docket Nos. 052-000025 and 052-000026

Document Date	Accession Number	Title & Estimated Page Count	Document Type	Author Affiliation	Addressee Affiliation	Docket Number
11/20/2009	ML093441661	2009/11/20 Vogtle COL Review - SNC Letter ND-09-1873 Transmitting Response to RAI Letter No. 043 Related to Cyber Security Plan 8 Page(s) 84 Page(s)	E-Mail	- No Known Affiliation	NRC/NRO/DNRL/NWE1	05200025 05200026
11/20/2009	ML093441664	2009/11/20 Vogtle COL Review - SNC Letter ND-09-1852 transmitting Endorsement of Bellefonte R-COLA Standard Content RAI for VEGP Units 3 & 4 COLA 10 Page(s)	E-Mail	- No Known Affiliation	NRC/NRO/DNRL/NWE1	05200025 05200026
11/25/2009	ML093441662	2009/11/25 Vogtle COL Review - SNC Letter ND-09-1901 transmitting VEGP Units 3 & 4 Construction Schedule Information Involving Inspections, Tests, Analyses and Acceptance Criteria 7 Page(s)	E-Mail	- No Known Affiliation	NRC/NRO/DNRL/NWE1	05200025 05200026
11/30/2009	ML093340016	2009/11/30 Vogtle COL Review - Draft RAI 4067 Related SRP Section 8.2 for Vogtle Units 3 and 4 3 Page(s)	E-Mail	NRC/NRO	NRC/NRO/DNRL/NWE1	05200025 05200026
11/30/2009	ML100081560	2009/11/30 Vogtle COL Review - RE: Vogtle RAI data for Tuesday's call, 12/1/09 25 Page(s)	E-Mail	NRC/NRO	NRC/NRO/DNRL/NWE1	05200025 05200026

Appendix B

Chronology of the Combined License Application for Vogtle Electric Generating Plant Units 3 and 4

This appendix contains a chronological listing of routine licensing correspondence between the staff of the U.S. Nuclear Regulatory Commission (NRC) regarding the review of the Vogtle Electric Generating Plant, Units 3 and 4 plant design under Docket Nos. 052-000025 and 052-000026

Document Date	Accession Number	Title & Estimated Page Count	Document Type	Author Affiliation	Addressee Affiliation	Docket Number
12/03/2009	ML093280272	Trip Report - Site Visit to Watts Bar, Unit 2, and Vogtle Commercial Nuclear Power Stations to Meet Licensee Fitness-For-Duty Program Staff. 3 Page(s)	Memoranda	NRC/NSIR/DSP/ISCPB	NRC/NSIR/DSP/ISCPB	05000390 05000391 05000424 05000425 05200025 05200026
12/03/2009	ML093280276	Trip Report - Watts Bar Unit 2 and Vogtle Site - Enclosure 1, Staff Participants. 1 Page(s)	Trip Report	NRC/NSIR/DSP/ISCPB		05000390 05000391 05000424 05000425 05200025 05200026
12/03/2009	ML093280279	Trip Report - Watts Bar Unit 2 and Vogtle Site - Enclosure 2, Sign Sheet. 1 Page(s)	Trip Report	NRC/NSIR/DSP/ISCPB		05000390 05000391 05000424 05000425 05200025 05200026
12/03/2009	ML093441665	2009/12/03 Vogtle COL Review - draft rais associated control room habitability training 4 Page(s)	E-Mail	NRC/NRO	NRC/NRO/DNRL/NWE1	05200025 05200026
12/04/2009	ML093380131	12/16/2009 Meeting Notice With Southern Nuclear Operating Company, Inc. to Discuss the NRC's Construction Inspection Program for the Time Frame Covered by the Approved Early Site Permit for Vogtle Units 3 & 4.	Letter Meeting Agenda Meeting Notice	NRC/RGN-II/DCP	Southern Nuclear Operating Co, Inc	05200011 05200025 05200026

B-111

Appendix B

Chronology of the Combined License Application for Vogtle Electric Generating Plant Units 3 and 4

This appendix contains a chronological listing of routine licensing correspondence between the staff of the U.S. Nuclear Regulatory Commission (NRC) regarding the review of the Vogtle Electric Generating Plant, Units 3 and 4 plant design under Docket Nos. 052-000025 and 052-000026

Document Date	Accession Number	Title & Estimated Page Count	Document Type	Author Affiliation	Addressee Affiliation	Docket Number
12/04/2009	ML093441667	2009/12/04 Vogtle COL Review - SNC Letter ND-09-1774 transmitting VEGP LWA Notification of Commencement of Limited Construction Activities Involving Engineered Backfill 9 Page(s)	E-Mail	- No Known Affiliation	NRC/NRO/DNRL/NWE1	05200025 05200026
12/08/2009	ML093420690	2009/12/08 Vogtle RAI for SER - Request for Additional Information Letter No. 044 Related to SRP Section 08.02 for the Vogtle Units 3 and 4 Combined License Application 10 Page(s)	E-Mail	NRC/NRO	NRC/NRO/DNRL/NWE1	05200025 05200026
12/08/2009	ML093420721	2009/12/08 Vogtle COL Review - RAI Letter No. 044 Related to SRP Section 8.2 for Vogtle Units 3 and 4 6 Page(s)	E-Mail	NRC/NRO	NRC/NRO/DNRL/NWE1	05200025 05200026
12/08/2009	ML093420918	2009/12/08 Vogtle COL Review - Telcon Summary - Telcon with Vogtle 12/8/09 8 Page(s)	E-Mail	NRC/NRO	NRC/NRO/DNRL/NWE1	05200025 05200026
12/09/2009	ML092600744	Atlanta, GA (Vogtle Electric Generating Plant, Units 3 and 4 Combined License Application Review. 2 Page(s)	Letter	NRC/NRO/DSER/RAP1	State of GA, Dept of Natural Resources	05200025 05200026
12/09/2009	ML093491048	2009/12/09 Vogtle COL Review - 6 Page(s)	E-Mail	Battelle Memorial Institute,	Battelle Memorial Institute,	05200025

Appendix B

Chronology of the Combined License Application for Vogtle Electric Generating Plant Units 3 and 4

This appendix contains a chronological listing of routine licensing correspondence between the staff of the U.S. Nuclear Regulatory Commission (NRC) regarding the review of the Vogtle Electric Generating Plant, Units 3 and 4 plant design under Docket Nos. 052-000025 and 052-000026

Document Date	Accession Number	Title & Estimated Page Count	Document Type	Author Affiliation	Addressee Affiliation	Docket Number
		FW: Fwd: Plant Vogtle Comment Letter for Dan 3 Page(s)		Pacific Northwest National Lab	Pacific Northwest National Lab NRC/NRO	05200026
12/10/2009	ML092670288	Vogtle, Units 3 and 4, NRC's Supplemental Environmental	Letter	NRC/NRO/DSER/RAP1	Poarch Band of Creek Nation	05200025 05200026
12/10/2009	ML092730038	Stephanie Rolin-Poarch Band Creek Indians Initiate Consultation for Vogtle COLA. 6 Page(s)	Letter	NRC/NRO/DSER/RAP1	Poarch Band of Creek Nation	05200025 05200026
12/10/2009	ML092730059	Chadwick Smith- Vogtle Initiate Consultation to the Cherokee nation of Oklahoma. 7 Page(s)	Letter	NRC/NRO/DSER/RAP1	Cherokee Nation	05200025 05200026
12/10/2009	ML092730092	Environmental Impact Statement for Southern Nuclear Operating Company's Combined License Application for the Propose construction and Operation of Units 3 and 4 at the Vogtle Power Plant in Waynesboro, Georgia. 7 Page(s)	Letter	NRC/NRO/DSER/RAP1	Cherokee Nation	05200025 05200026
12/10/2009	ML092730147	Letter re: Environmental Impact Statement for Southern Nuclear Operating Company's Combined License Application for the Proposed Construction and Operating of Units 3 and 4 at the Vogtle Electric Generating Plant in Waynesboro, Georgia.	Letter	NRC/NRO/DSER/RAP1	Chickasaw Nation	05200025 05200026

Appendix B

Chronology of the Combined License Application for Vogtle Electric Generating Plant Units 3 and 4

This appendix contains a chronological listing of routine licensing correspondence between the staff of the U.S. Nuclear Regulatory Commission (NRC) regarding the review of the Vogtle Electric Generating Plant, Units 3 and 4 plant design under Docket Nos. 052-000025 and 052-000026

Document Date	Accession Number	Title & Estimated Page Count	Document Type	Author Affiliation	Addressee Affiliation	Docket Number
12/10/2009	ML092730177	Environmental Impact Statement for Southern Nuclear Operating Company's Combined License Application for the Proposed Construction and Operation of Units 3 and 4 at the Vogtle Power Plant in Waynesboro, Georgia. 7 Page(s)	Letter	NRC/NRO/DSER/RAP1	Chickasaw Nation	05200025 05200026
12/10/2009	ML092730208	Carleton-Mississippi Band of Choctaw Indians Initiate Consultation to the Tribe for Vogtle COLA 7 Page(s)	Letter	NRC/NRO/DSER/RAP1	Mississippi Band of Choctaw Indians	05200025 05200026
12/10/2009	ML092730252	Debbie Thomas - Alabama-Coushatta Tribe of Texas Initiate Consultation to the Tribes for Vogtle COLA. 7 Page(s)	Letter	NRC/NRO/DSER/RAP1	Alabama-Coushatta Tribe of Texas	05200025 05200026
12/10/2009	ML092730274	Yargee- Alabama-Quassarte Tribal Town Initiate Consultation to the Tribes for Vogtle COLA. 7 Page(s)	Letter	NRC/NRO/DSER/RAP1	Alabama-Quassarte Tribal Town	05200025 05200026
12/10/2009	ML092730283	Kaniatobe-Absentee-Shawnee Tribe of Oklahoma Initiate Consultation to the Tribes for Vogtle COLA. 5 Page(s)	Letter	NRC/NRO/DSER/RAP1	Absentee-Shawnee Tribe of Oklahoma	05200025 05200026
12/10/2009	ML092730292	Zachary- Coushatta Tribe of	Letter	NRC/NRO/DSER/RAP1	Coushatta Tribe of	05200025

Appendix B

Chronology of the Combined License Application for Vogtle Electric Generating Plant Units 3 and 4

This appendix contains a chronological listing of routine licensing correspondence between the staff of the U.S. Nuclear Regulatory Commission (NRC) regarding the review of the Vogtle Electric Generating Plant, Units 3 and 4 plant design under Docket Nos. 052-000025 and 052-000026

Document Date	Accession Number	Title & Estimated Page Count	Document Type	Author Affiliation	Addressee Affiliation	Docket Number
		Louisiana Initiate Consultation to the Tribes of Vogtle COLa. 7 Page(s)			Louisiana	05200026
12/10/2009	ML092730317	McCoy-Eastern Band of Cherokee Indians - Initiate Consultation to the Tribes for Vogtle COLA. 7 Page(s)	Letter	NRC/NRO/DSER/RAP1	Eastern Band of Cherokee Indians	05200025 05200026
12/10/2009	ML092730321	Blue - Catawba Indian Tribe Initiate Consultation to the Tribes for Vogtle COLA. 7 Page(s)	Letter	NRC/NRO/DSER/RAP1	Catawba Indian Nation	05200025 05200026
12/10/2009	ML092730350	Ellis-Muscogee (Creek) Nation Initiate Consultation to the Tribes for Vogtle COLA. 7 Page(s)	Letter	NRC/NRO/DSER/RAP1	Muscogee (Creek) Nation	05200025 05200026
12/10/2009	ML092740375	Terry - Miccosukee Tribe of Indians of Florida Initiate Consultation to the Tribes for Vogtle COLA. 7 Page(s)	Letter	NRC/NRO/DSER/RAP1	Miccosukee Indian Tribe	05200025 05200026
12/10/2009	ML092740388	Bucktrot-Kialegee Tribal Town Initiate Consultation to the Tribes for Vogtle COLA. 7 Page(s)	Letter	NRC/NRO/DSER/RAP1	Kialegee Tribal Town	05200025 05200026
12/10/2009	ML092740393	Proctor-United Keetoowah Band of Cherokee Indians Initiate Consultation to the Tribes for Vogtle COLA. 7 Page(s)	Letter	NRC/NRO/DSER/RAP1	United Keetoowah Band of Cherokee Indians	05200025 05200026
12/10/2009	ML092740546	Holland-United Keetoowah Band of	Letter	NRC/NRO/DSER/RAP1	United Keetoowah Band of	05200025

Appendix B

Chronology of the Combined License Application for Vogtle Electric Generating Plant Units 3 and 4

This appendix contains a chronological listing of routine licensing correspondence between the staff of the U.S. Nuclear Regulatory Commission (NRC) regarding the review of the Vogtle Electric Generating Plant, Units 3 and 4 plant design under Docket Nos. 052-000025 and 052-000026

Document Date	Accession Number	Title & Estimated Page Count	Document Type	Author Affiliation	Addressee Affiliation	Docket Number
		Cherokee Indians Initiate Consultation to the Tribes for Vogtle COLA. 7 Page(s)			Cherokee Indians	05200026
12/10/2009	ML092740554	McGertt-Thlopthlocco Tribal Town Initiate Consultation to the Tribes for Vogtle COLA. 7 Page(s)	Letter	NRC/NRO/DSER/RAP1	Thlopthlocco Tribal Town	05200025 05200026
12/10/2009	ML092920488	Letter to Mr. W. Steele - Seminole Tribe of Florida - Vogtle Electric Generating Plant EIS. 8 Page(s)	Letter	NRC/NRO/DSER/RAP1	Seminole Tribe of Florida	05200025 05200026
12/10/2009	ML092920490	Letter to Joyce A. Bear - Muscogee (Creek) Nation of Oklahoma - Vogtle - EIS. 7 Page(s)	Letter	NRC/NRO/DSER/RAP1	Muscogee (Creek) Nation of Oklahoma	05200025 05200026
12/10/2009	ML092930629	Early Site Permit Review for the Vogtle Site. 7 Page(s)	Letter	NRC/NRO/DSER/RAP1	Seminole Nation of Oklahoma	05200025 05200026

Appendix B

Chronology of the Combined License Application for Vogtle Electric Generating Plant Units 3 and 4

This appendix contains a chronological listing of routine licensing correspondence between the staff of the U.S. Nuclear Regulatory Commission (NRC) regarding the review of the Vogtle Electric Generating Plant, Units 3 and 4 plant design under Docket Nos. 052-000025 and 052-000026

Document Date	Accession Number	Title & Estimated Page Count	Document Type	Author Affiliation	Addressee Affiliation	Docket Number
12/10/2009	ML092940250	U.S. Nuclear Regulatory Commission's Supplemental Environmental Impact Statement For Southern Nuclear Operating Company's Combined License Application For The Proposed Construction and Operation of Units 3 and 4 at The Vogtle Electric Generating Plant... 6 Page(s)	Letter	NRC/NRO/DSER/RAP3	Eastern Band of Cherokee Indians	05200025 05200026
12/10/2009	ML093140059	Memo-Request for Additional Information Regarding the Environmental Review of the Combined License Application for Vogtle Electric Generating Plant, 3 and 4. 13 Page(s)	Letter Request for Additional Information (RAI)	NRC/NRO/DSER/RAP1	Southern Nuclear Operating Co, Inc	05200025 05200026
12/10/2009	ML093441525	2009/12/10 Vogtle COL Review - Courtesy Email Copy of NRC RAIs for Vogtle COL to SNC 15 Page(s)	E-Mail	NRC/NRO	NRC/NRO/DNRL/NWE1	05200025 05200026
12/10/2009	ML093490905	2009/12/10 Vogtle COL Review - FW: Fwd: Plant Vogtle Comment Letter for Dan 4 Page(s)	E-Mail	Battelle Memorial Institute, Pacific Northwest National Lab	Battelle Memorial Institute, Pacific Northwest National Lab NRC/NRO	05200025 05200026
12/10/2009	ML093520276	Kuntzleman Vogtle Call Record Bob Perry SCDNR. 2 Page(s)	Note to File incl Telcon Record, Verbal Comm	Battelle Memorial Institute, Pacific Northwest National Lab	NRC/NRO	05200025 05200026
12/11/2009	ML093500016	2009/12/11 Vogtle COL Review -	E-Mail	NRC/NRO	NRC/NRO/DNRL/NWE1	05200025

Appendix B

Chronology of the Combined License Application for Vogtle Electric Generating Plant Units 3 and 4

This appendix contains a chronological listing of routine licensing correspondence between the staff of the U.S. Nuclear Regulatory Commission (NRC) regarding the review of the Vogtle Electric Generating Plant, Units 3 and 4 plant design under Docket Nos. 052-000025 and 052-000026

Document Date	Accession Number	Title & Estimated Page Count	Document Type	Author Affiliation	Addressee Affiliation	Docket Number
		RAI Letter No. 167 Related SRP Section 9.1.2 for Bellefonte Units 3 and 4 10 Page(s)				05200026
12/11/2009	ML093570043	Vogtle, Units 3 & 4, Combined License Application Submittal No. 5. 8 Page(s)	Letter	Southern Nuclear Operating Co, Inc	NRC/Document Control Desk NRC/NRO	05200025 05200026
12/11/2009	ML100080892	2009/12/11 Vogtle COL Review - SNC Letter ND-09-2001 - SNC VEGP Units 3&4 Combined License Application Submittal No. 5 12 Page(s)	E-Mail	- No Known Affiliation	NRC/NRO/DNRL/NWE1	05200025 05200026
12/14/2009	ML093491035	Vogtle Electric Generating Plant Units 3 and 4 Combined License Application Response to Bellefonte Units 3 and 4 Safety Evaluation Report Open Items for Chapter 3. 16 Page(s)	Letter	Southern Nuclear Operating Co, Inc	NRC/Document Control Desk NRC/NRO	05200025 05200026
12/14/2009	ML100080832	2009/12/14 Vogtle COL Review - RAI Letter No. 168 Related to SRP section 6.4 for Bellefonte Units 3 and 4 9 Page(s)	E-Mail	NRC/NRO	NRC/NRO/DNRL/NWE1	05200025 05200026
12/14/2009	ML100080899	2009/12/14 Vogtle COL Review - FW: Plant Vogtle Expansion 3 Page(s)	E-Mail	NRC/OPA	NRC/NRO/DNRL/NWE1	05200025 05200026
12/14/2009	ML100080914	2009/12/14 Vogtle COL Review - SNC Letter ND-09-2015 - SNC	E-Mail	- No Known Affiliation	NRC/NRO/DNRL/NWE1	05200025 05200026

Chronology of the Combined License Application for Vogtle Electric Generating Plant Units 3 and 4

This appendix contains a chronological listing of routine licensing correspondence between the staff of the U.S. Nuclear Regulatory Commission (NRC) regarding the review of the Vogtle Electric Generating Plant, Units 3 and 4 plant design under Docket Nos. 052-000025 and 052-000026

Document Date	Accession Number	Title & Estimated Page Count	Document Type	Author Affiliation	Addressee Affiliation	Docket Number
		VEGP Units 3&4 COL Response to Bellefonte Units 3&4 Safety Evaluation Report Open Items for Chapter 3 20 Page(s)				
12/14/2009	ML100081532	2009/12/14 Vogtle COL Review - RE: Vogtle RAI data for Tuesday's call, 12/15/09 27 Page(s)	E-Mail	NRC/NRO	NRC/NRO/DNRL/NWE1	05200025 05200026
12/15/2009	ML093491132	2009/12/15 Vogtle COL Review - FW: SC State Threatened and Endangered Species in the Vicinity of Vogtle Electric Generating Plant 3 Page(s)	E-Mail	NRC/NRO	NRC/NRO/DNRL/NWE1	05200025 05200026
12/15/2009	ML093491138	2009/12/15 Vogtle COL Review - Call records for GaDNR and FWS 5 Page(s)	E-Mail	- No Known Affiliation	NRC/NRO/DNRL/NWE1	05200025 05200026
12/15/2009	ML100080844	2009/12/15 Vogtle COL Review - January 14 meeting 2 Page(s)	E-Mail	NRC/NRO	NRC/NRO/DNRL/NWE1	05200025 05200026
12/15/2009	ML100080859	2009/12/15 Vogtle COL Review - preliminary plan for 2010 ACRS interactions.doc 4 Page(s)	E-Mail	NRC/NRO		05200025 05200026
12/16/2009	ML093500211	2009/12/16 Vogtle COL Review - RE: FW: GDNR email Vogtle COL 3 Page(s)	E-Mail	NRC/NRO	NRC/NRO/DNRL/NWE1	05200025 05200026
12/16/2009	ML093500215	2009/12/16 Vogtle COL Review - FW: FW: GDNR email Vogtle COL	E-Mail	NRC/NRO	NRC/NRO/DNRL/NWE1	05200025 05200026

Chronology of the Combined License Application for Vogtle Electric Generating Plant Units 3 and 4

This appendix contains a chronological listing of routine licensing correspondence between the staff of the U.S. Nuclear Regulatory Commission (NRC) regarding the review of the Vogtle Electric Generating Plant, Units 3 and 4 plant design under Docket Nos. 052-000025 and 052-000026

Document Date	Accession Number	Title & Estimated Page Count	Document Type	Author Affiliation	Addressee Affiliation	Docket Number
12/17/2009	ML093520011	2009/12/17 Vogtle COL Review - Submittal Letter ND-09-2031 Involving Periodic Update of ITAAC Schedule Information for Vogtle 3 & 4 7 Page(s) 9 Page(s)	E-Mail	- No Known Affiliation	NRC/NRO/DNRL/NWE1	05200025 05200026
12/21/2009	ML093580195	Vogtle, Units 3 and 4 Combined License Application, Submittal No. 5 Roadmap. 30 Page(s)	Graphics incl Charts and Tables Letter Report, Miscellaneous	Southern Nuclear Operating Co, Inc	NRC/Document Control Desk NRC/NRO	05200025 05200026
12/21/2009	ML100080875	2009/12/21 Vogtle COL Review - Transmittal of Southern Nuclear Letter ND-09-2064, VEGP Units 3 & 4 COL Application Submittal No. 5 Roadmap 33 Page(s)	E-Mail	- No Known Affiliation	NRC/NRO/DNRL/NWE1	05200025 05200026
12/21/2009	ML100081506	2009/12/21 Vogtle COL Review - RE: Vogtle RAI data for Tuesday's call, 12/15/09 27 Page(s)	E-Mail	NRC/NRO	NRC/NRO/DNRL/NWE1	05200025 05200026
12/21/2009	ML100490042	2009/12/21 Vogtle COL Review - Re: FW: GDNR email Vogtle COL 9 Page(s)	E-Mail	- No Known Affiliation	NRC/NRO/DNRL/NWE1	05200025 05200026
12/22/2009	ML093370165	Memo, Vogtle Electric Generating Plant (VEGP), Final Safety Evaluation Report Input Regarding the Vogtle Combined License Final	Memoranda Safety	NRC/NRO/DCIP/CQVP	NRC/NRO/DNRL	05200025 05200026

Appendix B

Chronology of the Combined License Application for Vogtle Electric Generating Plant Units 3 and 4

This appendix contains a chronological listing of routine licensing correspondence between the staff of the U.S. Nuclear Regulatory Commission (NRC) regarding the review of the Vogtle Electric Generating Plant, Units 3 and 4 plant design under Docket Nos. 052-000025 and 052-000026

Document Date	Accession Number	Title & Estimated Page Count	Document Type	Author Affiliation	Addressee Affiliation	Docket Number
		Safety Analysis Report, Section 14.2. 46 Page(s)				
12/22/2009	ML100080935	2009/12/22 Vogtle COL Review - SNC Letter ND-09-2071 - SNC VEGP Units 3&4 COL Supplemental Response to RAI Letter No. 042 Loss of Large Areas of the Plant Due to Explosions or Fire 8 Page(s)	E-Mail	- No Known Affiliation	NRC/NRO/DNRL/NWE1	05200025 05200026
12/23/2009	ML092600785	Vogtle – Historic Preservation – Request for Information on Historic Properties Within the Area Under Evaluation for the Plant Vogtle Units 3 and 4. 6 Page(s)	Letter	NRC/NRO/DSER/RAP1	US Advisory Council On Historic Preservation	05200025 05200026
12/23/2009	ML093410022	Summary of 10/21/2009 Teleconference Held with Southern Nuclear Operating Company Regarding Vogtle Electric Generating Plant Site for a Combined License Application. 2 Page(s)	Meeting Summary Memoranda	NRC/NRO/DSER/RAP1	NRC/NRO/DSER/RAP1	05200025 05200026
12/23/2009	ML093630679	Vogtle, Units 3 & 4, Combined License Application, Supplemental Response to Request for Additional Information Letter No. 042, Loss of Large Areas of the Plant Due to	Letter	Southern Nuclear Operating Co, Inc	NRC/Document Control Desk NRC/NRO	05200025 05200026

Appendix B

Chronology of the Combined License Application for Vogtle Electric Electric Generating Plant Units 3 and 4

This appendix contains a chronological listing of routine licensing correspondence between the staff of the U.S. Nuclear Regulatory Commission (NRC) regarding the review of the Vogtle Electric Generating Plant, Units 3 and 4 plant design under Docket Nos. 052-000025 and 052-000026

Document Date	Accession Number	Title & Estimated Page Count	Document Type	Author Affiliation	Addressee Affiliation	Docket Number
12/23/2009	ML100080950	Explosions or Fire. 4 Page(s) 2009/12/23 Vogtle COL Review - SNC Letter ND-09-2071 - SNC VEGP Units 3&4 COL Supplemental Response to RAI Letter No. 042 Loss of Large Areas of the Plant Due to Explosions or Fire (corrected copy) 8 Page(s)	E-Mail	- No Known Affiliation	NRC/NRO/DNRL/NWE1	05200025 05200026
12/28/2009	ML100080966	2009/12/28 Vogtle COL Review - Phone Call Summary Vogtle RCOL Chap 12 December 17, 2009 3 Page(s)	E-Mail	NRC/NRO	NRC/NRO/DNRL/NWE1	05200025 05200026
12/29/2009	ML093631261	12/16/2009 Summary of Category 1 Public Meeting with Southern Nuclear Operating Company, Inc. Regarding Vogtle Electric Generating Plant, Units 3 & 4, 9 Page(s)	Letter Meeting Summary	NRC/RGN-II/DCP	Southern Nuclear Operating Co, Inc	05200011 05200025 05200026
12/30/2009	ML100040142	Southern Nuclear Operating Company Vogtle Electric Generating Plant Units 3 and 4 Combined License Application Response to Bellefonte Units 3 and 4 Safety Evaluation Report Open Items for Chapter 9. 11 Page(s)	Letter	Southern Nuclear Operating Co, Inc	NRC/Document Control Desk NRC/NRO	05200025 05200026
12/30/2009	ML100081144	2009/12/30 Vogtle COL Review -	E-Mail	- No Known Affiliation	NRC/NRO/DNRL/NWE1	05200025

B-122

Appendix B

Chronology of the Combined License Application for Vogtle Electric Generating Plant Units 3 and 4

This appendix contains a chronological listing of routine licensing correspondence between the staff of the U.S. Nuclear Regulatory Commission (NRC) regarding the review of the Vogtle Electric Generating Plant, Units 3 and 4 plant design under Docket Nos. 052-000025 and 052-000026

Document Date	Accession Number	Title & Estimated Page Count	Document Type	Author Affiliation	Addressee Affiliation	Docket Number
		SNC Letter ND-09-2078 transmitting Response to Bellefonte Units 3 and 4 Safety Evaluation Report Open Items for Chapter 9 14 Page(s)				05200026
12/31/2009	ML100050270	Vogtle Electric Generating Plant Units 3 & 4 Combined License Application Supplemental Response to Bellefonte Units 3 & 4, Safety Evaluation Report Open Items for Chapter 17. 79 Page(s)	Letter	Southern Nuclear Operating Co, Inc	NRC/NRO	05200014 05200015 05200025 05200026
12/31/2009	ML100081148	2009/12/31 Vogtle COL Review - SNC Letter ND-09-2082 transmitting VEGP Units 3 & 4 COLA Supplemental Response to Bellefonte Units 3 & 4 Safety Evaluation Report Open Items for Chapter 17 82 Page(s)	E-Mail	Southern Nuclear Operating Co, Inc	NRC/NRO/DNRL/NWE1	05200025 05200026
01/04/2010	ML100081141	2010/01/04 Vogtle COL Review - FW: SNC Letter ND-09-2071 - SNC VEGP Units 3&4 COL Supplemental Response to RAI Letter No. 042 Loss of Large Areas of the Plant Due to Explosions or Fire (corrected copy) 2 Page(s)	E-Mail	NRC/NRO	NRC/NRO/DNRL/NWE1	05200025 05200026
01/04/2010	ML100081473	2010/01/04 Vogtle COL Review -	E-Mail	NRC/NRO	NRC/NRO/DNRL/NWE1	05200025

Appendix B

Chronology of the Combined License Application for Vogtle Electric Generating Plant Units 3 and 4

This appendix contains a chronological listing of routine licensing correspondence between the staff of the U.S. Nuclear Regulatory Commission (NRC) regarding the review of the Vogtle Electric Generating Plant, Units 3 and 4 plant design under Docket Nos. 052-000025 and 052-000026

Document Date	Accession Number	Title & Estimated Page Count	Document Type	Author Affiliation	Addressee Affiliation	Docket Number
		RE: Vogtle RAI data for Tuesday's call, 1/5/2010 27 Page(s)				05200026
01/07/2010	ML092600684	Vogtle Electric Generating Plant, Units 3 & 4 COL, Request for List of Protected Species Within the Area Under Evaluation (FWS). 7 Page(s)	Letter	NRC/NRO/DSER/RAP1	US Dept of Interior, Fish & Wildlife Service	05200025 05200026
01/07/2010	ML100060777	Letter from G. P. Hatchett to D. Rodgers re: U.S. Nuclear Regulatory Commission's Supplemental Environmental Impact Statement For Southern's Combined License Application for the Proposed Construction and Operation of Units 3 and 4 at Vogtle Plant. 7 Page(s)	Letter	NRC/NRO/DSER/RAP1	Catawba Indian Nation	05200025 05200026
01/07/2010	ML100070349	SRM-M100107 - Affirmation Session: I. - SECY-09-0117 (Summer); II. 09-0135 (Levy County); III. 09-0139 (Fermi); IV. 09-0141 (GE-Hitachi); V. 09-0142 (Vogtle); VI. 09-0145 (Bellefonte); VII. 09-0158 (Bell Bend); VIII. 09-0171 (Shieldalloy). 3 Page(s)	Commission Staff Requirements Memo (SRM)	NRC/SECY	NRC/OCAA NRC/OGC	04007102 05200011 05200014 05200015 05200025 05200026 05200027 05200028 05200029 05200030 05200039

Appendix B

Chronology of the Combined License Application for Vogtle Electric Generating Plant Units 3 and 4

This appendix contains a chronological listing of routine licensing correspondence between the staff of the U.S. Nuclear Regulatory Commission (NRC) regarding the review of the Vogtle Electric Generating Plant, Units 3 and 4 plant design under Docket Nos. 052-000025 and 052-000026

Document Date	Accession Number	Title & Estimated Page Count	Document Type	Author Affiliation	Addressee Affiliation	Docket Number
01/07/2010	ML100070449	M100107 - Affirmation Session: I. - SECY-09-0117 (Summer); II. 09-0135 (Levy County); III. 09-0139 (Fermi); IV. 09-0141 (GE-Hitachi); V. 09-0142 (Vogtle); VI. 09-0145 (Bellefonte); VII. 09-0158 (Bell Bend); VIII. 09-0171 (Shieldalloy). 8 Page(s)	Commission Meeting Transcript/Exhibit	NRC/OCM		07007016 07200072 05000438 05000439 05200025 05200026 05200027 05200028 05200029 05200030 05200033 05200039 07001113
01/07/2010	ML100081142	2010/01/07 Vogtle COL Review - RAI Letter No. 169 Related to SRP section 6.4 for Bellefonte Units 3 and 4 9 Page(s)	E-Mail	NRC/NRO	NRC/NRO/DNRL/NWE1	05200025 05200026
01/07/2010	ML100081151	2010/01/07 Vogtle COL Review - SNC Letter ND-10-0008 - SNC VEGP Units 3&4 Combined License Application Response to Request for Additional Information Letter No. 044 12 Page(s)	E-Mail	- No Known Affiliation	NRC/NRO/DNRL/NWE1	05200025 05200026
01/07/2010	ML100110073	Vogtle Electric Generating Plant, Units 3 and 4 - Combined License Application Response to Request for Additional Information Letter No.	Letter	Southern Nuclear Operating Co, Inc	NRC/Document Control Desk NRC/NRO	05200025 05200026

Appendix B

Chronology of the Combined License Application for Vogtle Electric Generating Plant Units 3 and 4

This appendix contains a chronological listing of routine licensing correspondence between the staff of the U.S. Nuclear Regulatory Commission (NRC) regarding the review of the Vogtle Electric Generating Plant, Units 3 and 4 plant design under Docket Nos. 052-000025 and 052-000026

Document Date	Accession Number	Title & Estimated Page Count	Document Type	Author Affiliation	Addressee Affiliation	Docket Number
01/08/2010	ML100081152	044. 8 Page(s)				05200025 05200026
01/08/2010	ML100081152	2010/01/08 Vogtle COL Review - Conference Call Summary - January 7, 2010 - Vogtle Units 3 and 4 COLA - RAI 4184 related to SRP Section 19.0– Loss of large area Fires	E-Mail	NRC/NRO	NRC/NRO/DNRL/NWE1	05200025 05200026
01/08/2010	ML100081481	2010/01/08 Vogtle COL Review - FW: Vogtle RAI data for Tuesday's call, 1/5/2010 2 Page(s)	E-Mail	NRC/NRO	NRC/NRO/DNRL/NWE1	05200025 05200026
01/08/2010	ML100120291	Vogtle Electric Generating Plant, Units 3 & 4 - Combined License Application, Supplemental Information Addressing Bellefonte, Units 3 & 4, Safety Evaluation Report of Chapter 04, Confirmatory Item 4.4-1. 6 Page(s)	Letter	Southern Nuclear Operating Co, Inc	NRC/Document Control Desk NRC/NRO	05200014 05200015 05200025 05200026
01/08/2010	ML100120479	Vogtle Electric Generating Plant Units 3 & 4 Combined License Application Response to Request for Additional Information Letter on Environmental Issues. 106 Page(s)	Letter	Southern Nuclear Operating Co, Inc	NRC/Document Control Desk NRC/NRO	05200025 05200026
01/08/2010	ML100120840	2010/01/08 Vogtle COL Review - SNC Letter ND-10-0023	E-Mail	- No Known Affiliation	NRC/NRO/DNRL/NWE1	05200025 05200026

B-126

Chronology of the Combined License Application for Vogtle Electric Generating Plant Units 3 and 4

This appendix contains a chronological listing of routine licensing correspondence between the staff of the U.S. Nuclear Regulatory Commission (NRC) regarding the review of the Vogtle Electric Generating Plant, Units 3 and 4 plant design under Docket Nos. 052-000025 and 052-000026

Document Date	Accession Number	Title & Estimated Page Count	Document Type	Author Affiliation	Addressee Affiliation	Docket Number
		Transmitting Responses to NRC Environmental RAIs (Transmittal # 2) 46 Page(s)				
01/08/2010	ML100130017	2010/01/08 Vogtle COL Review - SNC Letter ND-10-0006 - SNC VEGP Units 3&4 Combined License Application Supplemental Information Addressing Bellefonte Units 3 and 4 Safety Evaluation Report Chapter 04, Confirmatory Item 4.4-1 10 Page(s)	E-Mail	- No Known Affiliation	NRC/NRO/DNRL/NWE1	05200025 05200026
01/08/2010	ML100130022	2010/01/08 Vogtle COL Review - SNC Letter ND-10-0023 - SNC VEGP Units 3&4 Combined License Application Response to Request for Additional Information Letter on Environmental Issues 6 Page(s)	E-Mail	- No Known Affiliation	NRC/NRO/DNRL/NWE1	05200025 05200026
01/08/2010	ML100130078	2010/01/08 Vogtle COL Review - SNC Letter ND-10-0023 Transmitting Responses to NRC Environmental RAIs (Transmittal # 1) 49 Page(s)	E-Mail	- No Known Affiliation	NRC/NRO/DNRL/NWE1	05200025 05200026
01/08/2010	ML100130089	2010/01/08 Vogtle COL Review - SNC Letter ND-10-0023 Transmitting Responses to NRC	E-Mail	- No Known Affiliation	NRC/NRO/DNRL/NWE1	05200025 05200026

Appendix B

Chronology of the Combined License Application for Vogtle Electric Generating Plant Units 3 and 4

This appendix contains a chronological listing of routine licensing correspondence between the staff of the U.S. Nuclear Regulatory Commission (NRC) regarding the review of the Vogtle Electric Generating Plant, Units 3 and 4 plant design under Docket Nos. 052-000025 and 052-000026

Document Date	Accession Number	Title & Estimated Page Count	Document Type	Author Affiliation	Addressee Affiliation	Docket Number
01/11/2010		Environmental RAIs (Transmittal # 5) 3 Page(s)				
01/11/2010	ML100050136	01/14/2010 Notice of Meeting with the Southern Nuclear Operating Company on the Vogtle Cyber Security Plan Submittal. 8 Page(s)	Meeting Agenda Meeting Notice Memoranda	NRC/NRO/DNRL/NWE1	NRC/NRO/DNRL/NWE1	05200025 05200026
01/11/2010	ML100110012	2010/01/11 Vogtle COL Review - FW: Courtesy Copy of RAI Response 2 Page(s)	E-Mail	NRC/NRO	NRC/NRO/DNRL/NWE1	05200025 05200026
01/11/2010	ML100130082	2010/01/11 Vogtle COL Review - Proposed Agenda for Closed Cyber Security Plan Meeting 2 Page(s)	E-Mail	NRC/NRO	NRC/NRO/DNRL/NWE1	05200025 05200026
01/11/2010	ML100130085	2010/01/11 Vogtle COL Review - January 14 meeting attendees 2 Page(s)	E-Mail	NRC/NRO	NRC/NRO/DNRL/NWE1	05200025 05200026
01/12/2010	ML100141434	2010/01/12-Memorandum from M. Zobler - Licensing Board Memorandum and Orders in Southern Nuclear Operating Co. (Vogtle Electric Generating Plant, Units 3 and 4) (LBP-10-1 & unpublished order). 2 Page(s)	Memoranda	NRC/OGC	NRC/NRO	05200025 05200026
01/12/2010	ML100141734	Vogtle, Units 3 and 4 Combined License Application, Response to	Letter	Southern Nuclear Operating Co, Inc	NRC/Document Control Desk	05200014 05200015

Appendix B

Chronology of the Combined License Application for Vogtle Electric Generating Plant Units 3 and 4

This appendix contains a chronological listing of routine licensing correspondence between the staff of the U.S. Nuclear Regulatory Commission (NRC) regarding the review of the Vogtle Electric Generating Plant, Units 3 and 4 plant design under Docket Nos. 052-000025 and 052-000026

Document Date	Accession Number	Title & Estimated Page Count	Document Type	Author Affiliation	Addressee Affiliation	Docket Number
		Bellefonte, Units 3 and 4, Safety Evaluation Report Open Items for Chapter 3. 9 Page(s)			NRC/NRO	05200025 05200026
01/13/2010	ML100130283	2010/01/13 Vogtle COL Review - SNC Letter ND-10-0064 - SNC VEGP Units 3&4 COL Response to Bellefonte Units 3&4 Safety Evaluation Report Open Items for Chapter 3 13 Page(s)	E-Mail	- No Known Affiliation	NRC/NRO/DNRL/NWE1	05200025 05200026
01/14/2010	ML100450016	2010/01/14 Vogtle COL Review - ACRS Meeting February 2 - 3 2 Page(s)	E-Mail	NRC/NRO	NRC/NRO/DNRL/NWE1	05200025 05200026
01/14/2010	ML100450017	2010/01/14 Vogtle COL Review - RE: Vogtle RAI data for Tuesday's call, 1/12/2010 4 Page(s)	E-Mail	- No Known Affiliation	NRC/NRO/DNRL/NWE1	05200025 05200026
01/15/2010	ML100450018	2010/01/15 Vogtle COL Review - RE: Vogtle RAI data for Tuesday's call, 1/20/2010 27 Page(s)	E-Mail	NRC/NRO	NRC/NRO/DNRL/NWE1	05200025 05200026
01/15/2010	ML100450019	2010/01/15 Vogtle COL Review - RE: Vogtle RAI data for Tuesday's call, 1/12/2010 4 Page(s)	E-Mail	- No Known Affiliation	NRC/NRO/DNRL/NWE1	05200025 05200026
01/15/2010	ML100450020	2010/01/15 Vogtle COL Review - Info: Westinghouse's plans to address Calorimetric Uncertainty	E-Mail	NRC/NRO	NRC/NRO/DNRL/NWE1	05200025 05200026

Appendix B

Chronology of the Combined License Application for Vogtle Electric Generating Plant Units 3 and 4

This appendix contains a chronological listing of routine licensing correspondence between the staff of the U.S. Nuclear Regulatory Commission (NRC) regarding the review of the Vogtle Electric Generating Plant, Units 3 and 4 plant design under Docket Nos. 052-000025 and 052-000026

Document Date	Accession Number	Title & Estimated Page Count	Document Type	Author Affiliation	Addressee Affiliation	Docket Number
		through DCA ITAAC 4 Page(s)				
01/19/2010	ML100190604	2010/01/19 Vogtle COL Review - Federal Register Notice for Atlantic Sturgeon, 90-day Petition Finding, Request for Info 2 Page(s)	E-Mail	NRC/NRO	NRC/NRO/DNRL/NWE1	05200025 05200026
01/19/2010	ML100191980	2010/01/19 Vogtle COL Review - FW: Atlantic Sturgeon - NMFS 90-day petition finding; request for info 2 Page(s)	E-Mail	NRC/NRO	NRC/NRO/DNRL/NWE1	05200025 05200026
01/19/2010	ML100192005	2010/01/19 Vogtle COL Review - FW: Atlantic Sturgeon - NMFS 90-day petition finding; request for info 2 Page(s)	E-Mail	NRC/NRO	NRC/NRO/DNRL/NWE1	05200025 05200026
01/19/2010	ML100450021	2010/01/19 Vogtle COL Review - RE: NI15 RAI Responses 2 Page(s)	E-Mail	NRC/NRO	NRC/NRO/DNRL/NWE1	05200025 05200026
01/21/2010	ML100450022	2010/01/21 Vogtle COL Review - Call at 3:30 Eastern (2:30 Central) 2 Page(s)	E-Mail	- No Known Affiliation	NRC/NRO/DNRL/NWE1	05200025 05200026
01/21/2010	ML100200305	Memo: Final Safety Analysis Report Section 2.3 Safety Evaluation Report Input for Vogtle Units 3 & 4 Combined License Application. 2 Page(s)	Memoranda	NRC/NRO/DSER/RSAC	NRC/NRO/DNRL/NWE1	05200025 05200026
01/21/2010	ML100200361	Enclosure: Vogtle SCOL Safety Evaluation Report, Chapter 2. 14 Page(s)	Safety Evaluation Report	NRC/NRO/DSER/RSAC	NRC/NRO/DNRL/NWE1	05200025 05200026

Appendix B

Chronology of the Combined License Application for Vogtle Electric Generating Plant Units 3 and 4

This appendix contains a chronological listing of routine licensing correspondence between the staff of the U.S. Nuclear Regulatory Commission (NRC) regarding the review of the Vogtle Electric Generating Plant, Units 3 and 4 plant design under Docket Nos. 052-000025 and 052-000026

Document Date	Accession Number	Title & Estimated Page Count	Document Type	Author Affiliation	Addressee Affiliation	Docket Number
01/21/2010	ML100210213	2010/01/21 Vogtle COL Review - NS Information for energy alternatives 3 Page(s)	E-Mail	NRC/NRO	NRC/NRO/DNRL/NWE1	05200025 05200026
01/21/2010	ML100210569	2010/01/21 Vogtle COL Review - FW: Disturbed Area Clarification 2 Page(s)	E-Mail	NRC/NRO	NRC/NRO/DNRL/NWE1	05200025 05200026
01/22/2010	ML100110122	Memo, Vogtle Units 3 & 4, Draft Safety Evaluation Report. 26 Page(s)	Draft Safety Evaluation Report (DSER) Memoranda	NRC/NRO/DSER/RGS1 NRC/NRO/DSER/RGS2	NRC/NRO/DNRL/NWE1	05200025 05200026
01/22/2010	ML100260127	Vogtle Electric Generating Plant, Units 3 and 4 - Combined License Application, Response to Bellefonte Units 3 and 4 Safety Evaluation Report Open Items for Chapter 15. 7 Page(s)	Letter	Southern Nuclear Operating Co, Inc	NRC/Document Control Desk NRC/NRO	05200025 05200026
01/22/2010	ML100450013	2010/01/22 Vogtle COL Review - AP1000 Cyber security plan 2 Page(s)	E-Mail	NRC/NRO	NRC/NRO/DNRL/NWE1	05200025 05200026
01/22/2010	ML100450049	2010/01/22 Vogtle COL Review - SNC Letter ND-10-0004 - SNC VEGP Units 3&4 COL Response to Bellefonte Units 3&4 Safety Evaluation Report Open Items for Chapter 15 11 Page(s)	E-Mail	- No Known Affiliation	NRC/NRO/DNRL/NWE1	05200025 05200026

Appendix B

Chronology of the Combined License Application for Vogtle Electric Generating Plant Units 3 and 4

This appendix contains a chronological listing of routine licensing correspondence between the staff of the U.S. Nuclear Regulatory Commission (NRC) regarding the review of the Vogtle Electric Generating Plant, Units 3 and 4 plant design under Docket Nos. 052-000025 and 052-000026

Document Date	Accession Number	Title & Estimated Page Count	Document Type	Author Affiliation	Addressee Affiliation	Docket Number
01/24/2010	ML100450023	2010/01/24 Vogtle COL Review - FW: SNC Letter ND-10-0120 - SNC VEGP Units 3&4 COL Summary of NRC Clarifications from January 14th Cyber Security Plan Meeting 10 Page(s)	E-Mail	NRC/NRO	NRC/NRO/DNRL/NWE1	05200025 05200026
01/25/2010	ML100450024	2010/01/25 Vogtle COL Review - RE: Vogtle RAI data for Tuesday's call, 1/26/2010 27 Page(s)	E-Mail	NRC/NRO	NRC/NRO/DNRL/NWE1	05200025 05200026
01/25/2010	ML100450037	2010/01/25 Vogtle COL Review - Vogtle ACRS Draft Slides for Chapter 15 COL Standard Content 8 Page(s)	E-Mail	NRC/NRO	NRC/NRO/DNRL/NWE1	05200025 05200026
01/25/2010	ML100450038	2010/01/25 Vogtle COL Review - Vogtle Draft RAIs for 13.6 Physical Security and a 14.03.12 Physical Security Hardware ITAAC 8 Page(s)	E-Mail	NRC/NRO	NRC/NRO/DNRL/NWE1	05200025 05200026
01/25/2010	ML100450039	2010/01/25 Vogtle COL Review - RE: Vogtle Draft RAIs for 13.6 Physical Security and a 14.03.12 Physical Security Hardware ITAAC 2 Page(s)	E-Mail	- No Known Affiliation	NRC/NRO/DNRL/NWE1	05200025 05200026
01/25/2010	ML100450040	2010/01/25 Vogtle COL Review - RE: Vogtle RAI data for Tuesday's call, 1/26/2010 4 Page(s)	E-Mail	- No Known Affiliation	NRC/NRO/DNRL/NWE1	05200025 05200026
01/25/2010	ML100450078	2010/01/25 Vogtle COL Review -	E-Mail	- No Known Affiliation	NRC/NRO/DNRL/NWE1	05200025

Appendix B

Chronology of the Combined License Application for Vogtle Electric Generating Plant Units 3 and 4

This appendix contains a chronological listing of routine licensing correspondence between the staff of the U.S. Nuclear Regulatory Commission (NRC) regarding the review of the Vogtle Electric Generating Plant, Units 3 and 4 plant design under Docket Nos. 052-000025 and 052-000026

Document Date	Accession Number	Title & Estimated Page Count	Document Type	Author Affiliation	Addressee Affiliation	Docket Number
		AP1000 COLA Draft Presentation - February 2-3 48 Page(s)				05200026
01/26/2010	ML100450041	2010/01/26 Vogtle COL Review - FW: Summary - AP1000 IST open items phone call held 1/21/10 at 2 p.m. 3 Page(s)	E-Mail	NRC/NRO	NRC/NRO/DNRL/NWE1	05200025 05200026
01/26/2010	ML100450042	2010/01/26 Vogtle COL Review - RE: Vogtle Draft RAIs for 13.6 Physical Security and a 14.03.12 Physical Security Hardware ITAAC 3 Page(s)	E-Mail	NRC/NRO	NRC/NRO/DNRL/NWE1	05200025 05200026
01/26/2010	ML100450043	2010/01/26 Vogtle COL Review - Vogtle 13.7 FFD Draft RAIs 3 Page(s)	E-Mail	NRC/NRO	NRC/NRO/DNRL/NWE1	05200025 05200026
01/26/2010	ML100450077	2010/01/26 Vogtle COL Review - Southern Nuclear Letter ND-10-0118, Periodic Update of Construction Schedule Information Data Files 6 Page(s)	E-Mail	- No Known Affiliation	NRC/NRO/DNRL/NWE1	05200025 05200026
01/26/2010	ML100450079	2010/01/26 Vogtle COL Review - REVISED SCANNED COPY of SNC Letter ND-10-0120 - SNC VEGP Units 3&4 COL Summary of NRC Clarifications from January 14th Cyber Security Plan Meeting 14 Page(s)	E-Mail	- No Known Affiliation	NRC/NRO/DNRL/NWE1	05200025 05200026

Chronology of the Combined License Application for Vogtle Electric Generating Plant Units 3 and 4

This appendix contains a chronological listing of routine licensing correspondence between the staff of the U.S. Nuclear Regulatory Commission (NRC) regarding the review of the Vogtle Electric Generating Plant, Units 3 and 4 plant design under Docket Nos. 052-000025 and 052-000026

Document Date	Accession Number	Title & Estimated Page Count	Document Type	Author Affiliation	Addressee Affiliation	Docket Number
01/27/2010	ML100290015	Vogtle, Units 3 and 4, Nuclear Development Executive Management Additions. 4 Page(s)	Letter	Southern Nuclear Operating Co, Inc	NRC/Document Control Desk NRC/NRO	05200025 05200026
01/27/2010	ML100450044	2010/01/27 Vogtle COL Review - ACRS Feb 2nd - Chapter 3 NRC Presentation 27 Page(s)	E-Mail	NRC/NRO	NRC/NRO/DNRL/NWE1	05200025 05200026
01/27/2010	ML100450045	2010/01/27 Vogtle COL Review - RE: Draft RAI 4281 - 6.4 Control Room Habitability System 5 Page(s)	E-Mail	NRC/NRO	NRC/NRO/DNRL/NWE1	05200025 05200026
01/27/2010	ML100450046	2010/01/27 Vogtle COL Review - RE: Draft RAI 4281 - 6.4 Control Room Habitability System 6 Page(s)	E-Mail	NRC/NRO	NRC/NRO/DNRL/NWE1	05200025 05200026
01/27/2010	ML100450050	2010/01/27 Vogtle COL Review - SNC Letter ND-10-0115 - SNC VEGP Units 3&4 Nuclear Development Executive Management Additions 8 Page(s)	E-Mail	- No Known Affiliation	NRC/NRO/DNRL/NWE1	05200025 05200026
01/27/2010	ML100450076	2010/01/27 Vogtle COL Review - Rev 2: ACRS Feb 2nd - Chapter 3 NRC presentation 27 Page(s)	E-Mail	NRC/NRO	NRC/NRO/DNRL/NWE1	05200025 05200026
01/28/2010	ML092630002	Issuance of the Environmental Review Schedule for the Combined License Application Review for	Letter	NRC/NRO/DNRL	Southern Nuclear Operating Co, Inc	05200025 05200026

Appendix B

Chronology of the Combined License Application for Vogtle Electric Generating Plant Units 3 and 4

This appendix contains a chronological listing of routine licensing correspondence between the staff of the U.S. Nuclear Regulatory Commission (NRC) regarding the review of the Vogtle Electric Generating Plant, Units 3 and 4 plant design under Docket Nos. 052-000025 and 052-000026

Document Date	Accession Number	Title & Estimated Page Count	Document Type	Author Affiliation	Addressee Affiliation	Docket Number
		Vogtle Electric Generating Plant, Units 3 & 4. 11 Page(s)				
01/28/2010	ML100280586	2010/01/28 Vogtle COL Review - FW: SNC Letter ND-10-0115 - SNC VEGP Units 3&4 Nuclear Development Executive Management Additions 6 Page(s)	E-Mail	NRC/NRO	NRC/NRO/DNRL/NWE1	05200025 05200026
01/28/2010	ML100450074	2010/01/28 Vogtle COL Review - fyi - attached please find latest agenda for ACRS Feb meetings 4 Page(s)	E-Mail	NRC/NRO		05200025 05200026
01/28/2010	ML100450075	2010/01/28 Vogtle COL Review - AP1000 NRO Chapter 9 ACRS COLA Draft Presentation - February 2-3 14 Page(s)	E-Mail	NRC/NRO	NRC/NRO/DNRL/NWE1	05200025 05200026
01/29/2010	ML100290538	2010/01/29-Notification of NRC Staff Intentions to Issue the Draft COL Supplement to ESP Final Environmental Impact Statement on 05/07/2010 in the Matter of Vogtle, Units 3 and 4. 4 Page(s)	Legal-Correspondence/Miscellaneous	NRC/OGC	NRC/ASLBP	05200025 05200026
01/29/2010	ML100300004	2010/01/29 Vogtle COL Review - SNC Letter ND-10-0134 - SNC VEGP Units 3&4 COL Voluntary Revision to FSAR Chapter 17	E-Mail	- No Known Affiliation	NRC/NRO/DNRL/NWE1	05200025 05200026

Appendix B

Chronology of the Combined License Application for Vogtle Electric Generating Plant Units 3 and 4

This appendix contains a chronological listing of routine licensing correspondence between the staff of the U.S. Nuclear Regulatory Commission (NRC) regarding the review of the Vogtle Electric Generating Plant, Units 3 and 4 plant design under Docket Nos. 052-000025 and 052-000026

Document Date	Accession Number	Title & Estimated Page Count	Document Type	Author Affiliation	Addressee Affiliation	Docket Number
01/29/2010	ML100300005	2010/01/29 Vogtle COL Review - Response to Request for Additional Information Letter No. 021 Related to Radiation Protection - SNC Letter ND-10-0145 13 Page(s) 10 Page(s)	E-Mail	- No Known Affiliation	NRC/NRO/DNRL/NWE1	05200025 05200026
01/29/2010	ML100300006	2010/01/29 Vogtle COL Review - SNC Letter - ND-10-0177 SNC VEGP Units 3&4 COL Replacement DVD for Letter ND-09-1673 10 Page(s)	E-Mail	- No Known Affiliation	NRC/NRO/DNRL/NWE1	05200025 05200026
01/29/2010	ML100330387	Vogtle, Units 3 & 4 Combined License Application, Supplemental response 2 to Request for Additional Information Letter No. 021 Related to Radiation Protection. 9 Page(s)	Letter	Southern Nuclear Operating Co, Inc	NRC/Document Control Desk NRC/NRO	05200025 05200026
01/29/2010	ML100330720	Vogtle, Units 3 & 4, Combined License Application Voluntary Revision to Final Safety Analysis Report Chapter 17. 6 Page(s)	Letter	Southern Nuclear Operating Co, Inc	NRC/Document Control Desk NRC/NRO	05200025 05200026
01/29/2010	ML100630202	Vogtle, Units 3 & 4, Combined License Application, Replacement DVD for Letter ND-09-1673. 6 Page(s)	Letter	Southern Nuclear Operating Co, Inc	NRC/Document Control Desk NRC/NRO	05200025 05200026
01/30/2010	ML100450073	2010/01/30 Vogtle COL Review -	E-Mail	NRC/NRO	NRC/NRO/DNRL/NWE1	05200025

Appendix B

Chronology of the Combined License Application for Vogtle Electric Generating Plant Units 3 and 4

This appendix contains a chronological listing of routine licensing correspondence between the staff of the U.S. Nuclear Regulatory Commission (NRC) regarding the review of the Vogtle Electric Generating Plant, Units 3 and 4 plant design under Docket Nos. 052-000025 and 052-000026

Document Date	Accession Number	Title & Estimated Page Count	Document Type	Author Affiliation	Addressee Affiliation	Docket Number
02/01/2010		RE: Comments on NRC Staff's draft slides for Feb. 3 ACRS Presentation on LOLA 3 Page(s)				05200026
02/01/2010	ML100320769	2010/02/01 Vogtle RAI for SER – Request for Additional Information Letter No. 045 Related to SRP Section 2.5.4 for the Vogtle Electric Generating Plant Units 3 and 4 Combined License Application 6 Page(s)	E-Mail Request for Additional Information (RAI)	NRC/NRO	NRC/NRO/DNRL/NWE1	05200025 05200026
02/012010	ML100320770	2010/02/01 Vogtle RAI for SER – Request for Additional Information Letter No. 46 Related to SRP Section 3.12 for the Vogtle Electric Generating Plant Units 3 and 4 Combined License Application 6 Page(s)	E-Mail Request for Additional Information (RAI)	NRC/NRO	NRC/NRO/DNRL/NWE1	05200025 05200026
02/01/2010	ML100450070	2010/02/01 Vogtle COL Review - RAI Letter number 046 related SRP Section 3.12 for Vogtle Units 3 and 4 8 Page(s)	E-Mail	NRC/NRO	NRC/NRO/DNRL/NWE1	05200025 05200026
02/01/2010	ML100450071	2010/02/01 Vogtle COL Review - RAI letter number 45 related SRP section 2.5.4 for Vogtle Units 3 and 4 8 Page(s)	E-Mail	NRC/NRO	NRC/NRO/DNRL/NWE1	05200025 05200026
02/01/2010	ML100450072	2010/02/01 Vogtle COL Review -	E-Mail	NRC/NRO	NRC/NRO/DNRL/NWE1	05200025

Appendix B

Chronology of the Combined License Application for Vogtle Electric Generating Plant Units 3 and 4

This appendix contains a chronological listing of routine licensing correspondence between the staff of the U.S. Nuclear Regulatory Commission (NRC) regarding the review of the Vogtle Electric Generating Plant, Units 3 and 4 plant design under Docket Nos. 052-000025 and 052-000026

Document Date	Accession Number	Title & Estimated Page Count	Document Type	Author Affiliation	Addressee Affiliation	Docket Number
		RE: Vogtle RAI data for Tuesday's call, 2/1/2010 27 Page(s)				05200026
02/02/2010	ML100450069	2010/02/02 Vogtle COL Review - RE: Vogtle RAI data for Tuesday's call, 2/1/2010 4 Page(s)	E-Mail	- No Known Affiliation	NRC/NRO/DNRL/NWE1	05200025 05200026
02/03/2010	ML100450064	2010/02/03 Vogtle COL Review - FW: ACRS presenters 2 Page(s)	E-Mail	NRC/NRO	NRC/NRO/DNRL/NWE1	05200025 05200026
02/03/2010	ML100450065	2010/02/03 Vogtle COL Review - FW: today's 3:00 Touchpoint call has been cancelled 2 Page(s)	E-Mail	NRC/NRO	NRC/NRO/DNRL/NWE1	05200025 05200026
02/03/2010	ML100450066	2010/02/03 Vogtle COL Review - FW: ACRS agenda 3 Page(s)	E-Mail	NRC/NRO	NRC/NRO/DNRL/NWE1	05200025 05200026
02/03/2010	ML100450067	2010/02/03 Vogtle COL Review - RE: Vogtle Fitness for Duty RAI Call 2 Page(s)	E-Mail	- No Known Affiliation	NRC/NRO/DNRL/NWE1	05200025 05200026
02/03/2010	ML100450068	2010/02/03 Vogtle COL Review - Vogtle Fitness for Duty RAI Call 2 Page(s)	E-Mail	NRC/NRO	NRC/NRO/DNRL/NWE1	05200025 05200026
02/042010	ML100280034	Request for Additional Information Regarding the Environmental Review of the Limited Work Authorization for the Vogtle Electric Generating Plant, Units 3 and 4.	Letter Request for Additional Information (RAI)	NRC/NRO/DSER/RAP1	Southern Nuclear Operating Co, Inc	05200025 05200026

Appendix B

Chronology of the Combined License Application for Vogtle Electric Generating Plant Units 3 and 4

This appendix contains a chronological listing of routine licensing correspondence between the staff of the U.S. Nuclear Regulatory Commission (NRC) regarding the review of the Vogtle Electric Generating Plant, Units 3 and 4 plant design under Docket Nos. 052-000025 and 052-000026

Document Date	Accession Number	Title & Estimated Page Count	Document Type	Author Affiliation	Addressee Affiliation	Docket Number
		9 Page(s)				
02/04/2010	ML100350179	2010/02/04 Vogtle RAI for SER – Request for Additional Information Letter No. 047 Related to SRP Section 13.6 for the Vogtle Electric Generating Plant Units 3 and 4 Combined License Application 9 Page(s)	Request for Additional Information (RAI)	NRC/NRO	NRC/NRO/DNRL/NWE1	05200025 05200026
02/04/2010	ML100450061	2010/02/04 Vogtle COL Review - FW: FFD RAI Phone call participants 2 Page(s)	E-Mail	NRC/NRO	NRC/NRO/DNRL/NWE1	05200025 05200026
02/04/2010	ML100450062	2010/02/04 Vogtle COL Review - RE: Request for Additional Information Letter No. 047 Related to SRP Section 13.6 for the Vogtle Units 3 and 4 Combined License Application 2 Page(s)	E-Mail	- No Known Affiliation	NRC/NRO/DNRL/NWE1	05200025 05200026
02/04/2010	ML100450063	2010/02/04 Vogtle COL Review – Request for Additional Information Letter No. 047 Related to SRP Section 13.6 for the Vogtle Units 3 and 4 Combined License Application 11 Page(s)	E-Mail	NRC/NRO	NRC/NRO/DNRL/NWE1	05200025 05200026
02/05/2010	ML100360026	2010/02/05 Vogtle RAI for SER – Request for Additional Information Letter No. 048 Re;ated tp SRP	E-Mail	NRC/NRO	NRC/NRO/DNRL/NWE1	05200025 05200026

Appendix B

Chronology of the Combined License Application for Vogtle Electric Generating Plant Units 3 and 4

This appendix contains a chronological listing of routine licensing correspondence between the staff of the U.S. Nuclear Regulatory Commission (NRC) regarding the review of the Vogtle Electric Generating Plant, Units 3 and 4 plant design under Docket Nos. 052-000025 and 052-000026

Document Date	Accession Number	Title & Estimated Page Count	Document Type	Author Affiliation	Addressee Affiliation	Docket Number
		Section 6.4 for the Vogtle Electric Generting Plant, Units 3 AND 4 Combined License Application 6 Page(s)				
02/05/2010	ML100360823	2010/02/05 Vogtle RAI for SER – Request for Additional Information Letter No. 049 Related to SRP Section 13.6, Fitness for Duty, for the Vogtle Electric Generating Plant Units 3 and 4 Combined License Application 6 Page(s)	Request for OMB Review	NRC/NRO	NRC/NRO/DNRL/NWE1	05200025 05200026
02/05/2010	ML100450051	2010/02/05 Vogtle COL Review - Supplemental Response to Request for Additional Information Letter No. 042 Loss of Large Areas of the Plant due to Explosions of Fire - SNC Letter ND-10-0243 8 Page(s)	E-Mail	- No Known Affiliation	NRC/NRO/DNRL/NWE1	05200025 05200026
02/05/2010	ML100450052	2010/02/05 Vogtle COL Review - SNC Letter ND-10-0187 - SNC VEGP Units 3&4 COL Response to Bellefonte Units 3&4 Safety Evaluation Report Open Items for Chapter 3 63 Page(s)	E-Mail	- No Known Affiliation	NRC/NRO/DNRL/NWE1	05200025 05200026
02/05/2010	ML100450055	2010/02/05 Vogtle COL Review - RE: SER OI closure procedure 7 Page(s)	E-Mail	NRC/NRO	NRC/NRO/DNRL/NWE1	05200025 05200026

Appendix B

Chronology of the Combined License Application for Vogtle Electric Generating Plant Units 3 and 4

This appendix contains a chronological listing of routine licensing correspondence between the staff of the U.S. Nuclear Regulatory Commission (NRC) regarding the review of the Vogtle Electric Generating Plant, Units 3 and 4 plant design under Docket Nos. 052-000025 and 052-000026

Document Date	Accession Number	Title & Estimated Page Count	Document Type	Author Affiliation	Addressee Affiliation	Docket Number
02/05/2010	ML100450056	2010/02/05 Vogtle COL Review - RE: Request for Additional Information Letter No. 049 Related to SRP Section 13.6, Fitness for Duty, for the Vogtle Units 3 and 4 Combined License Application 2 Page(s)	E-Mail	- No Known Affiliation	NRC/NRO/DNRL/NWE1	05200025 05200026
02/05/2010	ML100450057	2010/02/05 Vogtle COL Review – Request for Addition Information Letter No. 049 Related to SRP Section 13.6, Fitness for Duty for the Vogtle Units 3 AND 4 Combined License Application 8 Page(s)	E-Mail	NRC/NRO	NRC/NRO/DNRL/NWE1	05200025 05200026
02/05/2010	ML100450059	2010/02/05 Vogtle COL Review - Phone Call Summary Vogtle RCOL 020410 - SRP 6.4 Control Room Habitability 3 Page(s)	E-Mail	NRC/NRO	NRC/NRO/DNRL/NWE1	05200025 05200026
02/05/2010	ML100450060	2010/02/05 Vogtle COL Review - RAI Letter 048 Related to SRP 6.4 Control Room Habitability for Vogtle Units 3 and 4 COL Application 8 Page(s)	E-Mail	NRC/NRO	NRC/NRO/DNRL/NWE1	05200025 05200026

Appendix B

Chronology of the Combined License Application for Vogtle Electric Generating Plant Units 3 and 4

This appendix contains a chronological listing of routine licensing correspondence between the staff of the U.S. Nuclear Regulatory Commission (NRC) regarding the review of the Vogtle Electric Generating Plant, Units 3 and 4 plant design under Docket Nos. 052-000025 and 052-000026

Document Date	Accession Number	Title & Estimated Page Count	Document Type	Author Affiliation	Addressee Affiliation	Docket Number
02/05/2010	ML100470600	Vogtle, Units 3 and 4, Combined License Application Environmental Report to Support Revision 1 to Part 6, LWA Request. 28 Page(s)	Letter	Southern Nuclear Operating Co, Inc	NRC/Document Control Desk NRC/NRO	05200025 05200026
02/05/2010	ML100480513	Vogtle, Units 3 & 4, Combined License Application, Response to Bellefonte Units 3 and 4 Safety Evaluation Report Open Items for Chapter 3. 59 Page(s)	Letter	Southern Nuclear Operating Co, Inc	NRC/Document Control Desk NRC/NRO	05200014 05200015 05200025 05200026
02/05/2010	ML100481140	Vogtle, Units 3 and 4 - Combined License Application Supplemental Response to Request for Additional Information Letter No. 042 Loss of Large Areas of the Plant due to Explosions or Fire. 4 Page(s)	Letter	Southern Nuclear Operating Co, Inc	NRC/Document Control Desk NRC/NRO	05200025 05200026
02/06/2010	ML100450054	2010/02/06 Vogtle COL Review - RE: SER OI closure procedure 2 Page(s)	E-Mail	- No Known Affiliation	NRC/NRO/DNRL/NWE1	05200025 05200026
02/07/2010	ML100380015	2010/02/07 Vogtle COL Review - Vogtle Chapter 9 Open Item 2 Page(s)	E-Mail	NRC/NRO	NRC/NRO/DNRL/NWE1	05200025 05200026
02/08/2010	ML100450014	2010/02/08 Vogtle COL Review - RE: Vogtle RAI data for Tuesday's call, 2/9/2010 24 Page(s)	E-Mail	NRC/NRO	NRC/NRO/DNRL/NWE1	05200025 05200026
02/08/2010	ML100450015	2010/02/08 Vogtle COL Review -	E-Mail	- No Known Affiliation	NRC/NRO/DNRL/NWE1	05200025

Appendix B

Chronology of the Combined License Application for Vogtle Electric Generating Plant Units 3 and 4

This appendix contains a chronological listing of routine licensing correspondence between the staff of the U.S. Nuclear Regulatory Commission (NRC) regarding the review of the Vogtle Electric Generating Plant, Units 3 and 4 plant design under Docket Nos. 052-000025 and 052-000026

Document Date	Accession Number	Title & Estimated Page Count	Document Type	Author Affiliation	Addressee Affiliation	Docket Number
		Southern Nuclear Submittal Letter ND-10-0227, Environmental Report in Support of Revision 1 to VEGP Units 3 & 4 COL Application Part 6 31 Page(s)				05200026
02/08/2010	ML100450053	2010/02/08 Vogtle COL Review - RE: Vogtle RAI data for Tuesday's call, 2/9/2010 4 Page(s)	E-Mail	- No Known Affiliation	NRC/NRO/DNRL/NWE1	05200025 05200026
02/11/2010	ML100470517	G20100031/EDATS: OEDO-2010-0023, Email Closeout for Briefing Package for EDO's Visit to Vogtle and MOX on Feb. 23-26, 2010. 2 Page(s)	E-Mail	NRC/RGN-II	NRC/EDO/AO	05000424 05000425 05200025
02/12/2010	ML100350998	Vogtle Electric Generating Plants, Units 3 and 4 Safety Evaluation Report (Phase 2) Input for Section 8 Referencing AP 1000 Design (Project No. 0755). 33 Page(s)	Memoranda Safety Evaluation Report	NRC/NRO/DE/EEB	NRC/NRO/DNRL/NWE1	05200025 05200026 PROJ0755
02/12/2010	ML100450048	2010/02/12 Vogtle COL Review - RE: Vogtle RAI data for Tuesday's call, 2/16/2010 28 Page(s)	E-Mail	NRC/NRO	NRC/NRO/DNRL/NWE1	05200025 05200026
02/12/2010	ML100470632	2010/02/12 Vogtle COL Review - NAIP image of site 3 Page(s)	E-Mail	Battelle Memorial Institute, Pacific Northwest National Lab	Battelle Memorial Institute, Pacific Northwest National Lab NRC/NRO	05200025 05200026
02/16/2010	ML100470809	2010/02/16 Vogtle COL Review -	E-Mail	NRC/NRO	NRC/NRO/DNRL/NWE1	05200025

Appendix B

Chronology of the Combined License Application for Vogtle Electric Generating Plant Units 3 and 4

This appendix contains a chronological listing of routine licensing correspondence between the staff of the U.S. Nuclear Regulatory Commission (NRC) regarding the review of the Vogtle Electric Generating Plant, Units 3 and 4 plant design under Docket Nos. 052-000025 and 052-000026

Document Date	Accession Number	Title & Estimated Page Count	Document Type	Author Affiliation	Addressee Affiliation	Docket Number
		Emailing: Smith Vogtle COL SEIS new references 2 Page(s)				05200026
02/16/2010	ML100500302	2010/02/16 Vogtle COL Review - SHPO MOU 4 Page(s)	E-Mail	- No Known Affiliation	NRC/NRO/DNRL/NWE1	05200025 05200026
02/16/2010	ML100710662	2010/02/16 Vogtle COL Review - RE: fyi - press conference today at 10am - DOE Loan Guarantee for Southern Nuclear 2 Page(s)	E-Mail	NRC/NRO	NRC/NRO/DNRL/NWE1	05200025 05200026
02/16/2010	ML100710667	2010/02/16 Vogtle COL Review - RE: fyi - press conference today at 10am - DOE Loan Guarantee for Southern Nuclear 2 Page(s)	E-Mail	NRC/NRO	NRC/NRO/DNRL/NWE1	05200025 05200026
02/16/2010	ML100710672	2010/02/16 Vogtle COL Review - RE: fyi - press conference today at 10am - DOE Loan Guarantee for Southern Nuclear 2 Page(s)	E-Mail	NRC/OPA	NRC/NRO/DNRL/NWE1	05200025 05200026
02/16/2010	ML100710674	2010/02/16 Vogtle COL Review - fyi - press conference today at 10am - DOE Loan Guarantee for Southern Nuclear 2 Page(s)	E-Mail	NRC/NRO	NRC/NRO/DNRL/NWE1	05200025 05200026
02/17/2010	ML100430800	03/03/2010 Notice of Public Meeting With Southern Nuclear Operating Company (SNC) to	Meeting Agenda Meeting Notice	NRC/NRO/DNRL/NWE1	NRC/NRO/DNRL/NWE1	05200025 05200026

Appendix B

Chronology of the Combined License Application for Vogtle Electric Generating Plant Units 3 and 4

This appendix contains a chronological listing of routine licensing correspondence between the staff of the U.S. Nuclear Regulatory Commission (NRC) regarding the review of the Vogtle Electric Generating Plant, Units 3 and 4 plant design under Docket Nos. 052-000025 and 052-000026

Document Date	Accession Number	Title & Estimated Page Count	Document Type	Author Affiliation	Addressee Affiliation	Docket Number
		Discuss SNC's Plans to Submit An Exemption Request From Requirements of 10 CFR 50.10(D) for Its Combined License Application the Vogtle Electric Generating Plant. 4 Page(s)				
02/17/2010	ML100490040	2010/02/17 Vogtle COL Review - FW: SNC Letter ND-10-0023 Transmitting Responses to NRC Environmental RAIs (Transmittal # 5) 2 Page(s)	E-Mail	NRC/NRO	NRC/NRO/DNRL/NWE1	05200025 05200026
02/17/2010	ML100490044	2010/02/17 Vogtle COL Review - Impingement Study 53 Page(s)	E-Mail	- No Known Affiliation	NRC/NRO/DNRL/NWE1	05200025 05200026
02/17/2010	ML100770582	2010/02/17 Vogtle COL Review - FW: Southern Nuclear Submittal Letter ND-10-0227, Environmental Report in Support of Revision 1 to VEGP Units 3 & 4 COL Application Part 6 30 Page(s)	E-Mail	- No Known Affiliation	NRC/NRO/DNRL/NWE1	05200025 05200026
02/18/2010	ML100492136	2010/02/18-NRC Staff Submittal of Vogtle COL Hearing File Index, Update 6 - February 18, 2010. 14 Page(s)	Legal-Hearing File	NRC/OGC	NRC/ASLBP	05200025 05200026
02/19/2010	ML100539569	2010/02/19 Vogtle COL Review - Transmittal of Southern Nuclear	E-Mail	- No Known Affiliation	NRC/NRO/DNRL/NWE1	05200025 05200026

Appendix B

Chronology of the Combined License Application for Vogtle Electric Generating Plant Units 3 and 4

This appendix contains a chronological listing of routine licensing correspondence between the staff of the U.S. Nuclear Regulatory Commission (NRC) regarding the review of the Vogtle Electric Generating Plant, Units 3 and 4 plant design under Docket Nos. 052-000025 and 052-000026

Document Date	Accession Number	Title & Estimated Page Count	Document Type	Author Affiliation	Addressee Affiliation	Docket Number
		Submittal Letter ND-10-0300, Cyber Security Plan, Rev. 2 7 Page(s)				
02/19/2010	ML100540545	Vogtle, Units 3 & 4 Combined License Application Cyber Security Plan, Revision 2. 7 Page(s)	Letter	Southern Nuclear Operating Co, Inc	NRC/Document Control Desk NRC/NRO	05200025 05200026
02/19/2010	ML100540547	Vogtle, Units 3 & 4, Enclosure 2 to ND-10-0300, Cyber Security Plan, Revision 2. 43 Page(s)	Security Plan	Southern Nuclear Operating Co, Inc	NRC/NRO	05200025 05200026
02/19/2010	ML100550033	Vogtle, Units 3 & 4, Combined License Application re: Large Component Transportation Method Decision. 5 Page(s)	Letter	Southern Nuclear Operating Co, Inc	NRC/Document Control Desk NRC/NRO	05200025 05200026
02/19/2010	ML100700031	2010/02/19 Vogtle COL Review - Transmittal of Southern Nuclear Letter ND-10-0261, Large Component Transportation Method Decision 8 Page(s)	E-Mail	- No Known Affiliation	NRC/NRO/DNRL/NWE1	05200025 05200026
02/19/2010	ML100710661	2010/02/19 Vogtle COL Review - RE: Notice for the Public meeting on March 3, 2010 10 Page(s)	E-Mail	NRC/NRO	NRC/NRO/DNRL/NWE1	05200025 05200026
02/22/2010	ML100539570	2010/02/22 Vogtle COL Review - FW: Transmittal of Southern Nuclear Submittal Letter ND-10-	E-Mail	NRC/NRO	NRC/NRO/DNRL/NWE1	05200025 05200026

Appendix B

Chronology of the Combined License Application for Vogtle Electric Generating Plant Units 3 and 4

This appendix contains a chronological listing of routine licensing correspondence between the staff of the U.S. Nuclear Regulatory Commission (NRC) regarding the review of the Vogtle Electric Generating Plant, Units 3 and 4 plant design under Docket Nos. 052-000025 and 052-000026

Document Date	Accession Number	Title & Estimated Page Count	Document Type	Author Affiliation	Addressee Affiliation	Docket Number
02/22/2010	ML100541589	0300, Cyber Security Plan, Rev. 2 2 Page(s)	E-Mail	- No Known Affiliation	NRC/NRO/DNRL/NWE1	05200025 05200026
02/22/2010	ML100560114	2010/02/22 Vogtle COL Review - DSM Supporting Information 4 Page(s)	E-Mail	- No Known Affiliation	NRC/NRO/DNRL/NWE1	05200025 05200026
02/22/2010	ML100560114	Turkey Point, Units 6 & 7, Vogtle, Units 3 & 4, Submittal of AP1000 R-COLA (VEGP) Supplemental Evaluation of Safety Evaluation Report Open Items for Standard Applicability. 4 Page(s)	Letter	Florida Power & Light Co	NRC/Document Control Desk NRC/NRO	05200025 05200026 05200040 05200041
02/22/2010	ML100710656	2010/02/22 Vogtle COL Review - RE: Vogtle RAI data for Tuesday's call, 2/23/2010 28 Page(s)	E-Mail	NRC/NRO	NRC/NRO/DNRL/NWE1	05200025 05200026
02/23/2010	ML100541586	2010/02/23 Vogtle COL Review - Telcon Summary - Telcon with Vogtle 2/22/10 2 Page(s)	E-Mail	NRC/NRO	NRC/NRO/DNRL/NWE1	05200025 05200026
02/23/2010	ML100710647	2010/02/23 Vogtle COL Review - RE: Vogtle RAI data for Tuesday's call, 2/23/2010 4 Page(s)	E-Mail	- No Known Affiliation	NRC/NRO/DNRL/NWE1	05200025 05200026
02/24/2010	ML100560166	2010/02/24 Vogtle COL Review - FW: predecisional information - Vogtle thermal plume 7 Page(s)	E-Mail	- No Known Affiliation	NRC/NRO/DNRL/NWE1	05200025 05200026
02/24/2010	ML100560286	2010/02/24 Vogtle COL Review -	E-Mail	- No Known Affiliation	NRC/NRO/DNRL/NWE1	05200025

Appendix B

Chronology of the Combined License Application for Vogtle Electric Generating Plant Units 3 and 4

This appendix contains a chronological listing of routine licensing correspondence between the staff of the U.S. Nuclear Regulatory Commission (NRC) regarding the review of the Vogtle Electric Generating Plant, Units 3 and 4 plant design under Docket Nos. 052-000025 and 052-000026

Document Date	Accession Number	Title & Estimated Page Count	Document Type	Author Affiliation	Addressee Affiliation	Docket Number
		Southern Nuclear Letters Involving Vogtle 3 and 4 Project Schedule/Status Information 15 Page(s)				05200026
02/26/2010	ML100700030	2010/02/26 Vogtle COL Review - DRAFT Presentation for Next Wednesday 14 Page(s)	E-Mail	- No Known Affiliation	NRC/NRO/DNRL/NWE1	05200025 05200026
02/26/2010	ML100700039	2010/02/26 Vogtle COL Review - FW: DRAFT Presentation for Next Wednesday 14 Page(s)	E-Mail	NRC/NRO	NRC/NRO/DNRL/NWE1	05200025 05200026
03/01/2010	ML100570038	09/21/09 Discussion with US Army Corps of Engineers, Savannah District, Concerning Participation in Development of Supplemental Environmental Impact Statement for Combined Operating License for Vogtle Electric Generating Plant, Units 3 & 4. 3 Page(s)	Meeting Summary Memoranda	NRC/NRO/DSER/RAP1	NRC/NRO/DSER/RAP1	05200025 05200026
03/01/2010	ML100600818	Georgia Power's Application for The Certification of Units 3 and 4 at Plant Vogtle and Updated Integrated Resource Plan. 233 Page(s)	Report, Miscellaneous	Georgia Power Co	NRC/NRO	05200025 05200026

Appendix B

Chronology of the Combined License Application for Vogtle Electric Generating Plant Units 3 and 4

This appendix contains a chronological listing of routine licensing correspondence between the staff of the U.S. Nuclear Regulatory Commission (NRC) regarding the review of the Vogtle Electric Generating Plant, Units 3 and 4 plant design under Docket Nos. 052-000025 and 052-000026

Document Date	Accession Number	Title & Estimated Page Count	Document Type	Author Affiliation	Addressee Affiliation	Docket Number
03/01/2010	ML100610021	2010/03/01 Vogtle COL Review - SNC Letter ND-10-0393 - SNC VEGP Units 3&4 COL Response to Bellefonte Units 3&4 Safety Evaluation Report Open Items for Chapter 3 19 Page(s)	E-Mail	- No Known Affiliation	NRC/NRO/DNRL/NWE1	05200025 05200026
03/01/2010	ML100620825	Vogtle, Units 3 and 4 Combined License Application Revised Response to Bellefonte Units 3 and 4 Safety Evaluation Report Open Items for Chapter 16. 6 Page(s)	Letter	Southern Nuclear Operating Co, Inc	NRC/Document Control Desk NRC/NRO	05200014 05200015 05200025 05200026
03/01/2010	ML100620826	Vogtle, Units 3 & 4 Combined License Application Response to Bellefonte Units 3 & 4 Safety Evaluation Report Open Items for Chapter 3. 15 Page(s)	Letter	Southern Nuclear Operating Co, Inc	NRC/Document Control Desk NRC/NRO	05200014 05200015 05200025 05200026
03/01/2010	ML100700029	2010/03/01 Vogtle COL Review - SNC Letter ND-10-0383 - SNC VEGP Units 3&4 COL Revised Response to Bellefonte Units 3&4 Safety Evaluation Report Open Items for Chapter 16 10 Page(s)	E-Mail	- No Known Affiliation	NRC/NRO/DNRL/NWE1	05200025 05200026
03/01/2010	ML100700036	2010/03/01 Vogtle COL Review - RE: Vogtle RAI data for Tuesday's call, 3/2/2010	E-Mail	- No Known Affiliation	NRC/NRO/DNRL/NWE1	05200025 05200026

Appendix B

Chronology of the Combined License Application for Vogtle Electric Generating Plant Units 3 and 4

This appendix contains a chronological listing of routine licensing correspondence between the staff of the U.S. Nuclear Regulatory Commission (NRC) regarding the review of the Vogtle Electric Generating Plant, Units 3 and 4 plant design under Docket Nos. 052-000025 and 052-000026

Document Date	Accession Number	Title & Estimated Page Count	Document Type	Author Affiliation	Addressee Affiliation	Docket Number
03/01/2010	ML100700038	2010/03/01 Vogtle COL Review - RE: Vogtle RAI data for Tuesday's call, 3/2/2010 4 Page(s)	E-Mail	NRC/NRO	NRC/NRO/DNRL/NWE1	05200025 05200026
03/02/2010	ML100620052	2010/03/02 Vogtle COL Review - SAS-2007-01837 Southern Nuclear Operating Company, Inc. 28 Page(s)	E-Mail	- No Known Affiliation	NRC/NRO/DNRL/NWE1	05200025 05200026
03/02/2010	ML100630207	Vogtle Electric Generating Plant Units 3 & 4 Combined License Application Response to Request for Additional Information Letter No. 046. 6 Page(s)	Letter	Southern Nuclear Operating Co, Inc	NRC/Document Control Desk NRC/NRO	05200025 05200026
03/02/2010	ML100630208	Vogtle, Units 3 & 4 Combined License Application, Response to Request for Additional Information Letter No. 045. 6 Page(s)	Letter	Southern Nuclear Operating Co, Inc	NRC/Document Control Desk NRC/NRO	05200025 05200026
03/02/2010	ML100700027	2010/03/02 Vogtle COL Review - SNC Letter ND-10-0325 - SNC VEGP Units 3&4 COL Response to RAI Letter No. 46 10 Page(s)	E-Mail	- No Known Affiliation	NRC/NRO/DNRL/NWE1	05200025 05200026
03/02/2010	ML100700028	2010/03/02 Vogtle COL Review - SNC Letter ND-10-0324 - SNC VEGP Units 3&4 COL Response to RAI Letter No. 045	E-Mail	- No Known Affiliation	NRC/NRO/DNRL/NWE1	05200025 05200026

Appendix B

Chronology of the Combined License Application for Vogtle Electric Generating Plant Units 3 and 4

This appendix contains a chronological listing of routine licensing correspondence between the staff of the U.S. Nuclear Regulatory Commission (NRC) regarding the review of the Vogtle Electric Generating Plant, Units 3 and 4 plant design under Docket Nos. 052-000025 and 052-000026

Document Date	Accession Number	Title & Estimated Page Count	Document Type	Author Affiliation	Addressee Affiliation	Docket Number
		10 Page(s)				
03/02/2010	ML100700035	2010/03/02 Vogtle COL Review - Cyber call on draft RAIs 2 Page(s)	E-Mail	NRC/NRO	NRC/NRO/DNRL/NWE1	05200025 05200026
03/03/2010	ML100251515	Revision 2 of The Vogtle Combined License Application. 3 Page(s)	Letter	NRC/NSIR/DPR	US Dept of Homeland Security US Federal Emergency Mgmt Agency (FEMA)	05200025 05200026
03/03/2010	ML100700025	2010/03/03 Vogtle COL Review - Exemption Request Meeting Attendees 2 Page(s)	E-Mail	- No Known Affiliation	NRC/NRO/DNRL/NWE1	05200025 05200026
03/03/2010	ML100700026	2010/03/03 Vogtle COL Review - RE: Southern Meeting 3 Page(s)	E-Mail	- No Known Affiliation	NRC/NRO/DNRL/NWE1	05200025 05200026
03/03/2010	ML101180339	03/03/2010 Southern Nuclear Meeting Presentation on Vogtle & 4 Project. 7 Page(s)	Meeting Agenda Meeting Briefing Package/Handouts Slides and Viewgraphs	Southern Nuclear Operating Co, Inc	NRC/NRO	05200025 05200026
03/04/2010	ML100640702	Vogtle, Units 3 and 4 - Combined License Application, Mitigative Strategies Description - Updated Reviewer's Aid. 7 Page(s)	Letter	Southern Nuclear Operating Co, Inc	NRC/Document Control Desk NRC/NRO	05200025 05200026
03/05/2010	ML100640579	Vogtle COL Cyber Security RAIs. 6 Page(s)	- No Document Type Applies	NRC/NSIR		05200025 05200026
03/05/2010	ML100680206	Vogtle, Units 3 and 4 Combined License Application, Response to	Letter	Southern Nuclear Operating Co, Inc	NRC/Document Control Desk	05200025 05200026

Appendix B

Chronology of the Combined License Application for Vogtle Electric Generating Plant Units 3 and 4

This appendix contains a chronological listing of routine licensing correspondence between the staff of the U.S. Nuclear Regulatory Commission (NRC) regarding the review of the Vogtle Electric Generating Plant, Units 3 and 4 plant design under Docket Nos. 052-000025 and 052-000026

Document Date	Accession Number	Title & Estimated Page Count	Document Type	Author Affiliation	Addressee Affiliation	Docket Number
		Request for Additional Information Letter No. 048, Control Room Habitability Analyses. 14 Page(s)			NRC/NRO	
03/05/2010	ML100680207	Vogtle, Units 3 and 4 Combined License Application, Response to Request for Additional Information Letter No. 049, Fitness for Duty Program - Construction Phase. 14 Page(s)	Letter	Southern Nuclear Operating Co, Inc	NRC/Document Control Desk NRC/NRO	05200025 05200026
03/05/2010	ML100680210	Vogtle, Units 3 and 4 Combined License Application, Response to Request for Additional Information Letter No. 047, Physical Security Plan. 14 Page(s)	Letter	Southern Nuclear Operating Co, Inc	NRC/Document Control Desk NRC/NRO	05200025 05200026
03/05/2010	ML100680212	Vogtle, Units 3 and 4 Combined License Application, Response to Request for Additional Information Letter No. 047, Physical Security Inspections, Tests, Analyses, and Acceptance Criteria. 31 Page(s)	Letter	Southern Nuclear Operating Co, Inc	NRC/Document Control Desk NRC/NRO	05200025 05200026
03/06/2010	ML100700024	2010/03/06 Vogtle COL Review - RE: Letter 2 Page(s)	E-Mail	NRC/NRO	NRC/NRO/DNRL/NWE1	05200025 05200026
03/08/2010	ML100700023	2010/03/08 Vogtle COL Review - RE: Vogtle RAI data for Tuesday's call, 3/9/2010	E-Mail	NRC/NRO	NRC/NRO/DNRL/NWE1	05200025 05200026

Appendix B

Chronology of the Combined License Application for Vogtle Electric Generating Plant Units 3 and 4

This appendix contains a chronological listing of routine licensing correspondence between the staff of the U.S. Nuclear Regulatory Commission (NRC) regarding the review of the Vogtle Electric Generating Plant, Units 3 and 4 plant design under Docket Nos. 052-000025 and 052-000026

Document Date	Accession Number	Title & Estimated Page Count	Document Type	Author Affiliation	Addressee Affiliation	Docket Number
03/11/2010	ML100630453	Final Safety Evaluation Report Input Regarding the Vogtle Combined License Final Safety Analysis Report, Sections 17.0, 17.1, 17.2, 17.3, AND 17.5. 27 Page(s) 28 Page(s)	Final Safety Evaluation Report (FSER) Memoranda	NRC/NRO/DCIP/CQVP	NRC/NRO/DNRL/NWE1	05200025 05200026
03/11/2010	ML100740441	Vogtle, Units 3 and 4, Combined License Application Supplemental to Environmental Report in Support of Revision 1 to Part 6, LWA Request. 7 Page(s)	Environmental Report Letter	Southern Nuclear Operating Co, Inc	NRC/Document Control Desk NRC/NRO	05200025 05200026
03/11/2010	ML100750657	2010/03/11 Vogtle COL Review - SNC Letter ND-10-0520 - SNC VEGP Units 3&4 COL Supplement to Environmental Report in Support of Revision 1 to Part 6, LWA Request 12 Page(s)	E-Mail	- No Known Affiliation	NRC/NRO/DNRL/NWE1	05200025 05200026
03/11/2010	ML100750662	2010/03/11 Vogtle COL Review - LWA ER Section 11 4 Page(s)	E-Mail	- No Known Affiliation	NRC/NRO/DNRL/NWE1	05200025 05200026
03/11/2010	ML100760545	2010/03/11 Vogtle COL Review - Documents to ADAMize 196 Page(s)	E-Mail	- No Known Affiliation	NRC/NRO/DNRL/NWE1	05200025 05200026
03/12/2010	ML100710642	2010/03/12 Vogtle COL Review - FW: Southern Nuclear Letter ND-10-0459, Response to NRC RAI	E-Mail	NRC/NRO	NRC/NRO/DNRL/NWE1	05200025 05200026

Appendix B

Chronology of the Combined License Application for Vogtle Electric Generating Plant Units 3 and 4

This appendix contains a chronological listing of routine licensing correspondence between the staff of the U.S. Nuclear Regulatory Commission (NRC) regarding the review of the Vogtle Electric Generating Plant, Units 3 and 4 plant design under Docket Nos. 052-000025 and 052-000026

Document Date	Accession Number	Title & Estimated Page Count	Document Type	Author Affiliation	Addressee Affiliation	Docket Number
		Letter No. 48 on the VEGP Units 3 and 4 COL Application 16 Page(s)				
03/12/2010	ML100750038	Vogtle, Units 3 and 4 Combined License Application, Supporting Information for Environmental Report Review. 143 Page(s)	Letter	Southern Nuclear Operating Co, Inc	NRC/Document Control Desk NRC/NRO	05200025 05200026
03/12/2010	ML100750319	2010/03/12 Vogtle COL Review - SNC Letter ND-10-0510 - SNC VEGP Units 3&4 COL New and Significant Evaluation for Rail Transport 51 Page(s)	E-Mail	- No Known Affiliation	NRC/NRO/DNRL/NWE1	05200025 05200026
03/12/2010	ML100750322	2010/03/12 Vogtle COL Review - SNC Letter ND-10-0526 - SNC VEGP Units 3&4 COL Supporting Information for Environmental Report Review 8 Page(s)	E-Mail	- No Known Affiliation	NRC/NRO/DNRL/NWE1	05200025 05200026
03/12/2010	ML100750689	Vogtle, Units 3 and 4 Combined License Application, New and Significant Evaluation for Rail Transport. 37 Page(s)	Letter	Southern Nuclear Operating Co, Inc	NRC/Document Control Desk NRC/NRO	05200025 05200026
03/12/2010	ML100750690	Vogtle, Units 3 and 4, SVP_SVO_000254, "Rail Delivery Mode for AP1000 Major Components."	- No Document Type Applies	Southern Nuclear Operating Co, Inc	NRC/NRO	05200025 05200026

Appendix B

Chronology of the Combined License Application for Vogtle Electric Generating Plant Units 3 and 4

This appendix contains a chronological listing of routine licensing correspondence between the staff of the U.S. Nuclear Regulatory Commission (NRC) regarding the review of the Vogtle Electric Generating Plant, Units 3 and 4 plant design under Docket Nos. 052-000025 and 052-000026

Document Date	Accession Number	Title & Estimated Page Count	Document Type	Author Affiliation	Addressee Affiliation	Docket Number
03/14/2010	ML100750705	2010/03/14 Vogtle COL Review - RE: Vogtle RAI data for Tuesday's call, 3/16/2010 17 Page(s) 38 Page(s)	E-Mail	NRC/NRO	NRC/NRO/DNRL/NWE1	05200025 05200026
03/15/2010	ML100740671	2010/03/15 Vogtle COL Review - Draft RAI 4497 Related SRP Section 9.3.3 for Vogtle Units 3 and 4 3 Page(s)	E-Mail	NRC/NRO	NRC/NRO/DNRL/NWE1	05200025 05200026
03/15/2010	ML100740672	2010/03/15 Vogtle COL Review - Telcon Summary - Telcon with Vogtle 3/11/10 2 Page(s)	E-Mail	NRC/NRO	NRC/NRO/DNRL/NWE1	05200025 05200026
03/15/2010	ML100750099	2010/03/15 Vogtle COL Review - SNC Letter ND-10-0559 - SNC VEGP Units 3&4 COL Supplemental Response 2 to RAI Letter No. 005 12 Page(s)	E-Mail	- No Known Affiliation	NRC/NRO/DNRL/NWE1	05200025 05200026
03/15/2010	ML100750706	2010/03/15 Vogtle COL Review - RE: Conference call next week 2 Page(s)	E-Mail	NRC/NRO	NRC/NRO/DNRL/NWE1	05200025 05200026
03/15/2010	ML100750707	2010/03/15 Vogtle COL Review - RE: Vogtle RAI data for Tuesday's call, 3/16/2010 3 Page(s)	E-Mail	- No Known Affiliation	NRC/NRO/DNRL/NWE1	05200025 05200026
03/15/2010	ML100760095	Vogtle, Units 3 and 4 Combined License Application, Supplemental	Letter	Southern Nuclear Operating Co, Inc	NRC/Document Control Desk	05200025 05200026

Appendix B

Chronology of the Combined License Application for Vogtle Electric Generating Plant Units 3 and 4

This appendix contains a chronological listing of routine licensing correspondence between the staff of the U.S. Nuclear Regulatory Commission (NRC) regarding the review of the Vogtle Electric Generating Plant, Units 3 and 4 plant design under Docket Nos. 052-000025 and 052-000026

Document Date	Accession Number	Title & Estimated Page Count	Document Type	Author Affiliation	Addressee Affiliation	Docket Number
03/16/2011	ML100630372	Response 2 to Request for Additional Information Letter No. 005. 8 Page(s)			NRC/NRO	
03/16/2010	ML100750300	Vogtle LAF Review Phase 2 (Additional). 4 Page(s)	Request for Additional Information (RAI)	NRC/NRO		05200025 05200026
03/16/2010	ML100750480	2010/03/16 Vogtle RAI for SER - Request for Additional Information Letter No. 050 Related to SRP Section 09.03.03 for the Vogtle Units 3 and 4 Combined License Application 6 Page(s)	E-Mail	NRC/NRO	NRC/NRO/DNRL/NWE1	05200025 05200026
03/16/2010	ML100950020	2010/03/16 Vogtle COL Review - RAI Letter No. 050 Related to SRP Section 9.3.3 for Vogtle Units 3 and 4 8 Page(s)	E-Mail	NRC/NRO	NRC/NRO/DNRL/NWE1	05200025 05200026
03/16/2010	ML100950021	2010/03/16 Vogtle COL Review - Cyber RAI call in # 2 Page(s)	E-Mail	NRC/NRO	NRC/NRO/DNRL/NWE1	05200025 05200026
03/16/2010	ML100950022	2010/03/16 Vogtle COL Review - FW: LOLA MSD Reviewer's Aid 2 Page(s)	E-Mail	NRC/NRO	NRC/NRO/DNRL/NWE1	05200025 05200026
03/16/2010	ML100950022	2010/03/16 Vogtle COL Review - FW: Participants in the LOLA Follow-up Call - January 7, 2010 3 Page(s)	E-Mail	NRC/NRO	NRC/NRO/DNRL/NWE1	05200025 05200026

Appendix B

Chronology of the Combined License Application for Vogtle Electric Generating Plant Units 3 and 4

This appendix contains a chronological listing of routine licensing correspondence between the staff of the U.S. Nuclear Regulatory Commission (NRC) regarding the review of the Vogtle Electric Generating Plant, Units 3 and 4 plant design under Docket Nos. 052-000025 and 052-000026

Document Date	Accession Number	Title & Estimated Page Count	Document Type	Author Affiliation	Addressee Affiliation	Docket Number
03/16/2010	ML101090238	2010/03/16 Vogtle COL Review - FW: LOLA Draft RAI information 8 Page(s)	E-Mail	NRC/NRO	NRC/NRO/DNRL/NWE1	05200025 05200026
03/17/2010	ML100760357	2010/03/17 Vogtle RAI for SER – Request for Additional Information Letter No. 051 Related to SRP Section 13.6, Cyber Security, for the Vogtle Electric Generating Plant Units 3 and 4 Combined License Application 6 Page(s)	Request for Additional Information (RAI)	NRC/NRO	NRC/NRO/DNRL/NWE1	05200025 05200026
03/17/2010	ML101090241	2010/03/17 Vogtle COL Review - FW: Courtesy copy - RAI Letter 169 - Habitability Systems (BLN Chapter 6) 9 Page(s)	E-Mail	NRC/NRO	NRC/NRO/DNRL/NWE1	05200025 05200026
03/18/2010	ML100780393	Vogtle, Units 3 and 4 Combined License Application, Response to Bellefonte Units 3 and 4 Safety Evaluation Report Open Items for Chapter 3. 11 Page(s)	Letter	Southern Nuclear Operating Co, Inc	NRC/Document Control Desk NRC/NRO	05200014 05200015 05200025 05200026
03/18/2010	ML101090242	2010/03/18 Vogtle COL Review - FW: Request for Additional Information Letter No. 051 Related to SRP Section 3.6, Cyber Security, for the Vogtle Electric Generating Plant Units 3 and 4 Combined License Application	E-Mail	NRC/NRO	NRC/NRO/DNRL/NWE1	05200025 05200026

Appendix B

Chronology of the Combined License Application for Vogtle Electric Generating Plant Units 3 and 4

This appendix contains a chronological listing of routine licensing correspondence between the staff of the U.S. Nuclear Regulatory Commission (NRC) regarding the review of the Vogtle Electric Generating Plant, Units 3 and 4 plant design under Docket Nos. 052-000025 and 052-000026

Document Date	Accession Number	Title & Estimated Page Count	Document Type	Author Affiliation	Addressee Affiliation	Docket Number
03/18/2010	ML101090246	2010/03/18 Vogtle COL Review - SNC Letter ND-10-0585 - SNC VEGP Units 3&4 COL Response to Bellefonte Units 3 & 4 Safety Evaluation Report Open Items for Chapter 3 8 Page(s)	E-Mail	- No Known Affiliation	NRC/NRO/DNRL/NWE1	05200025 05200026
03/19/2010	ML100780131	2010/03/19 Vogtle COL Review - Draft RAI 4525 Related SRP Section 8.2 for Vogtle Units 3 and 4 3 Page(s)	E-Mail	NRC/NRO	NRC/NRO/DNRL/NWE1	05200025 05200026
03/19/2010	ML100850120	Request for Additional Information Letter No. 052 Related to SRP Section 19.0 for the Vogtle Electric Generating Plant Units 3 and 4 Combined License Application. 7 Page(s)	Letter Request for Additional Information (RAI)	NRC/NRO/DNRL/NWE1	Southern Nuclear Operating Co, Inc	05200025 05200026
03/22/2010	ML100810311	2010/03/22-NRC Staff Letter Regarding Notice of Issuance of Chapter 11 of Vogtle Advanced SER. 3 Page(s)	Legal-Correspondence/Miscellaneous	NRC/OGC	NRC/ASLBP	05200025 05200026
03/22/2010	ML101090243	2010/03/22 Vogtle COL Review - RE: Vogtle RAI data for Tuesday's call, 3/23/2010 30 Page(s)	E-Mail	NRC/NRO	NRC/NRO/DNRL/NWE1	05200025 05200026
03/22/2010	ML101090244	2010/03/22 Vogtle COL Review - RE: Vogtle RAI data for Tuesday's	E-Mail	- No Known Affiliation	NRC/NRO/DNRL/NWE1	05200025 05200026

Appendix B

Chronology of the Combined License Application for Vogtle Electric Generating Plant Units 3 and 4

This appendix contains a chronological listing of routine licensing correspondence between the staff of the U.S. Nuclear Regulatory Commission (NRC) regarding the review of the Vogtle Electric Generating Plant, Units 3 and 4 plant design under Docket Nos. 052-000025 and 052-000026

Document Date	Accession Number	Title & Estimated Page Count	Document Type	Author Affiliation	Addressee Affiliation	Docket Number
		call, 3/23/2010 4 Page(s)				
03/23/2010	ML100820580	2010/03/23 Vogtle COL Review - SNC Letter ND-10-0526, Env. Info 2 Page(s)	E-Mail	- No Known Affiliation	NRC/NRO/DNRL/NWE1	05200025 05200026
03/25/2010	ML100840706	2010/03/25 Vogtle RAI for SER - Request for Additional Information Letter No. 053 Related to SRP Section 08.02 for the Vogtle Units 3 and 4 Combined License Application 6 Page(s)	E-Mail	NRC/NRO	NRC/NRO/DNRL/NWE1	05200025 05200026
03/25/2010	ML100840708	2010/03/25 Vogtle COL Review - Telcon Summary - Telcon with Vogtle 3/25/10 2 Page(s)	E-Mail	NRC/NRO	NRC/NRO/DNRL/NWE1	05200025 05200026
03/25/2010	ML100840778	2010/03/25 Vogtle COL Review - RAI Letter No. 053 Related to SRP Section 8.2 for Vogtle Units 3 and 4 8 Page(s)	E-Mail	NRC/NRO	NRC/NRO/DNRL/NWE1	05200025 05200026
03/25/2010	ML100840779	2010/03/25 Vogtle COL Review - Telcon Summary - Telcon with Vogtle 3/18/10 2 Page(s)	E-Mail	NRC/NRO	NRC/NRO/DNRL/NWE1	05200025 05200026
03/29/2010	ML101090248	2010/03/29 Vogtle COL Review - RE: Vogtle RAI data for Tuesday's call, 3/30/2010 30 Page(s)	E-Mail	NRC/NRO	NRC/NRO/DNRL/NWE1	05200025 05200026
03/29/2010	ML101090272	2010/03/29 Vogtle COL Review -	E-Mail	- No Known Affiliation	NRC/NRO/DNRL/NWE1	05200025

Appendix B

Chronology of the Combined License Application for Vogtle Electric Generating Plant Units 3 and 4

This appendix contains a chronological listing of routine licensing correspondence between the staff of the U.S. Nuclear Regulatory Commission (NRC) regarding the review of the Vogtle Electric Generating Plant, Units 3 and 4 plant design under Docket Nos. 052-000025 and 052-000026

Document Date	Accession Number	Title & Estimated Page Count	Document Type	Author Affiliation	Addressee Affiliation	Docket Number
		RE: Vogtle RAI data for Tuesday's call, 3/30/2010 4 Page(s)				05200026
03/29/2010	ML101090273	2010/03/29 Vogtle COL Review - Draft Physical Security RAI 2 Page(s)	E-Mail	NRC/NRO	NRC/NRO/DNRL/NWE1	05200025 05200026
03/29/2010	ML101090277	2010/03/29 Vogtle COL Review - Southern Nuclear Letter Transmitting March's Construction Schedule Update 6 Page(s)	E-Mail	- No Known Affiliation	NRC/NRO/DNRL/NWE1	05200025 05200026
03/30/2010	ML101090274	2010/03/30 Vogtle COL Review - FW: Would it be possible to hold next Tuesday's Security scheduling call at 1:00 pm, instead of 10:00 am? 2 Page(s)	E-Mail	NRC/NRO	NRC/NRO/DNRL/NWE1	05200025 05200026
03/30/2010	ML101090275	2010/03/30 Vogtle COL Review - FW: Would it be possible to hold next Tuesday's Security scheduling call at 1:00 pm, instead of 10:00 am? 2 Page(s)	E-Mail	NRC/NRO	NRC/NRO/DNRL/NWE1	05200025 05200026
03/30/2010	ML101090276	2010/03/30 Vogtle COL Review - Public meeting at Vogtle 4 Page(s)	E-Mail	NRC/RGN-II	NRC/NRO/DNRL/NWE1	05200025 05200026
03/30/2010	ML101090280	2010/03/30 Vogtle COL Review - Southern Nuclear Letter Transmitting Module Mock-Up	E-Mail	- No Known Affiliation	NRC/NRO/DNRL/NWE1	05200025 05200026

Appendix B

Chronology of the Combined License Application for Vogtle Electric Generating Plant Units 3 and 4

This appendix contains a chronological listing of routine licensing correspondence between the staff of the U.S. Nuclear Regulatory Commission (NRC) regarding the review of the Vogtle Electric Generating Plant, Units 3 and 4 plant design under Docket Nos. 052-000025 and 052-000026

Document Date	Accession Number	Title & Estimated Page Count	Document Type	Author Affiliation	Addressee Affiliation	Docket Number
		Schedule Information 7 Page(s)				
03/31/2010	ML100810331	04/06/2010 Notice of Forthcoming Closed Conference Call with the Southern Nuclear Operating Company on the Vogtle Cyber Security Plan Submittal. 9 Page(s)	Meeting Agenda Meeting Notice	NRC/NRO/DNRL/NWE1	NRC/NRO/DNRL/NWE1	05200025 05200026
04/02/2010	ML100920455	Press Release-II-10-014: NRC Names First Resident Inspectors for Vogtle Nuclear Plant Construction Activities. 2 Page(s)	Press Release	NRC/OPA/RGN-II/FO		05200025 05200026
04/02/2010	ML100960108	Vogtle, Units 3 & 4 - Combined License Application, Voluntary Revision to Final Safety Analysis Report Chapter 17. 6 Page(s)	Final Safety Analysis Report (FSAR) Letter	Southern Nuclear Operating Co, Inc	NRC/Document Control Desk NRC/NRO	05200025 05200026
04/02/2010	ML100960109	Vogtle Electric Generating Plant, Units 3 & 4 Combined License Application, Response to Bellefonte Units 3 and 4 Safety Evaluation Report Open Items for Chapter 3. 7 Page(s)	Letter	Southern Nuclear Operating Co, Inc	NRC/Document Control Desk NRC/NRO	05200025 05200026
04/02/2010	ML101090281	2010/04/02 Vogtle COL Review - SNC Letter ND-10-0702 - SNC VEGP Units 3&4 COL Voluntary Revision to FSAR Chapter 17	E-Mail	- No Known Affiliation	NRC/NRO/DNRL/NWE1	05200025 05200026

Appendix B

Chronology of the Combined License Application for Vogtle Electric Generating Plant Units 3 and 4

This appendix contains a chronological listing of routine licensing correspondence between the staff of the U.S. Nuclear Regulatory Commission (NRC) regarding the review of the Vogtle Electric Generating Plant, Units 3 and 4 plant design under Docket Nos. 052-000025 and 052-000026

Document Date	Accession Number	Title & Estimated Page Count	Document Type	Author Affiliation	Addressee Affiliation	Docket Number
04/02/2010	ML101090283	2010/04/02 Vogtle COL Review - SNC Letter ND-10-0703 - SNC VEGP Units 3&4 COL Response to Bellefonte Units 3 & 4 Safety Evaluation Report Open Items for Chapter 3 11 Page(s)	E-Mail	- No Known Affiliation	NRC/NRO/DNRL/NWE1	05200025 05200026
04/05/2010	ML101090284	2010/04/05 Vogtle COL Review - RE: Vogtle RAI data for Tuesday's call, 4/6/2010 29 Page(s)	E-Mail	NRC/NRO	NRC/NRO/DNRL/NWE1	05200025 05200026
04/05/2010	ML101090285	2010/04/05 Vogtle COL Review - RE: Vogtle RAI data for Tuesday's call, 4/6/2010 4 Page(s)	E-Mail	- No Known Affiliation	NRC/NRO/DNRL/NWE1	05200025 05200026
04/05/2010	ML101090334	2010/04/05 Vogtle COL Review - Draft RAI question related to SRP Section 13.6 for the Vogtle Units 3 and 4 combined license application 3 Page(s)	E-Mail	NRC/NRO	NRC/NRO/DNRL/NWE1	05200025 05200026
04/06/2010	ML09631125	Summary of the Environmental New and Significant Site Audit Related to the Review of the Combined License Application for Vogtle Electric Generating Plant Site. 6 Page(s)	Memoranda	NRC/NRO/DSER/RAP1	NRC/NRO/DSER/RAP1	05200025 05200026

Appendix B

Chronology of the Combined License Application for Vogtle Electric Generating Plant Units 3 and 4

This appendix contains a chronological listing of routine licensing correspondence between the staff of the U.S. Nuclear Regulatory Commission (NRC) regarding the review of the Vogtle Electric Generating Plant, Units 3 and 4 plant design under Docket Nos. 052-000025 and 052-000026

Document Date	Accession Number	Title & Estimated Page Count	Document Type	Author Affiliation	Addressee Affiliation	Docket Number
04/06/2010	ML093631186	Enclosure - Vogtle Electric Generating Plant Combined License New and Significant Information Site Audit Trip Report September 28-October 1, 2009. 30 Page(s)	Audit Report Trip Report	NRC/NRO/DSER/RAP1	NRC/NRO/DSER/RAP1	05200025 05200026
04/6/2010	ML100970602	04/06/2010 Meeting Handouts for Meeting With Southern Nuclear to Discuss Backfill Operations at Vogtle Units 3 & 4. 44 Page(s)	Meeting Briefing Package/Handouts	Southern Nuclear Operating Co, Inc	NRC/NRO	05200011 05200025 05200026
04/06/2010	ML101090335	2010/04/06 Vogtle COL Review - RE: Draft RAI question related to SRP Section 13.6 for the Vogtle Units 3 and 4 combined license application 3 Page(s)	E-Mail	NRC/NRO	NRC/NRO/DNRL/NWE1	05200025 05200026
04/07/2010	ML101090337	2010/04/07 Vogtle COL Review - FW: April 13 Physical Security Conference Call 2 Page(s)	E-Mail	NRC/NRO	NRC/NRO/DNRL/NWE1	05200025 05200026
04/09/2010	ML100980390	Request for Additional Information (RAI) 4543, Vogtle Units 3 & 4, Southern Nuclear Operating Co. SRP Section 19.0-Loss of Large Areas of the Plant Due to Explosions or Fire - Mitigative Strategies Description and Plans. 5 Page(s)	Request for Additional Information (RAI)	NRC/NRO	Southern Nuclear Operating Co, Inc	05200025 05200026

Appendix B

Chronology of the Combined License Application for Vogtle Electric Generating Plant Units 3 and 4

This appendix contains a chronological listing of routine licensing correspondence between the staff of the U.S. Nuclear Regulatory Commission (NRC) regarding the review of the Vogtle Electric Generating Plant, Units 3 and 4 plant design under Docket Nos. 052-000025 and 052-000026

Document Date	Accession Number	Title & Estimated Page Count	Document Type	Author Affiliation	Addressee Affiliation	Docket Number
04/09/2010	ML100980600	Request for Additional Information Letter No. 054 Related to SRP Section 19.0 Related to Loss Of Large Areas of the Plant Due to Explosions or Fire-Mitigative Strategies Description & Plans For the Vogtle Electric Generating Plant Units 3 & 4 Combined... 4 Page(s)	Letter	NRC/NRO/DNRL/NWE2	Southern Nuclear Operating Co, Inc	05200025 05200026
04/12/2010	ML101090342	2010/04/12 Vogtle COL Review - RE: Vogtle RAI data for Tuesday's call, 4/13/2010 30 Page(s)	E-Mail	NRC/NRO	NRC/NRO/DNRL/NWE1	05200025 05200026
04/12/2010	ML101090348	2010/04/12 Vogtle COL Review - RE: Vogtle RAI data for Tuesday's call, 4/13/2010 3 Page(s)	E-Mail	- No Known Affiliation	NRC/NRO/DNRL/NWE1	05200025 05200026
04/14/2010	ML101090351	2010/04/14 Vogtle COL Review - Phone Call summary for Vogtle LOLA April 7, 2010 3 Page(s)	E-Mail	NRC/NRO	NRC/NRO/DNRL/NWE1	05200025 05200026
04/15/2010	ML101090146	Vogtle, Units 3 and 4, Combined License Application Response to Request for Additional Information Letter No. 050 Waste Water System. 11 Page(s)	Letter	Southern Nuclear Operating Co, Inc	NRC/Document Control Desk NRC/NRO	05200025 05200026
04/15/2010	ML101090352	2010/04/15 Vogtle COL Review - Southern Nuclear Submittal Letter	E-Mail	- No Known Affiliation	NRC/NRO/DNRL/NWE1	05200025 05200026

Appendix B

Chronology of the Combined License Application for Vogtle Electric Generating Plant Units 3 and 4

This appendix contains a chronological listing of routine licensing correspondence between the staff of the U.S. Nuclear Regulatory Commission (NRC) regarding the review of the Vogtle Electric Generating Plant, Units 3 and 4 plant design under Docket Nos. 052-000025 and 052-000026

Document Date	Accession Number	Title & Estimated Page Count	Document Type	Author Affiliation	Addressee Affiliation	Docket Number
		Responding to NRC RAI Letter No. 050 14 Page(s)				
04/16/2010	ML101090295	Vogtle, Units 3 and 4 Combined License Application, Submittal of 10 CFR 50.46 Annual Report. 3 Page(s)	Letter	Southern Nuclear Operating Co, Inc	NRC/Document Control Desk NRC/NRO	05200025 05200026
04/16/2010	ML101090296	Vogtle Electric Generating Plant Units 3 and 4 Combined License Application Response to Request for Additional Information Letter No. 051 Cyber Security Plan. 4 Page(s)	Letter	Southern Nuclear Operating Co, Inc	NRC/Document Control Desk NRC/RGN-II	05200025 05200026
04/16/2010	ML101090297	Response to NRC Request for Additional Information Letter No. 051 on the Vogtle Electric Generating Plants Units 3 & 4 COL Application Involving the Cyber Security Plan. 41 Page(s)	- No Document Type Applies	Southern Nuclear Operating Co, Inc	NRC/NRO	05200025 05200026
04/16/2010	ML101090353	2010/04/16 Vogtle COL Review - FW: Southern Nuclear Submittal Letter Responding to NRC RAI Letter No. 050 2 Page(s)	E-Mail	NRC/NRO	NRC/NRO/DNRL/NWE1	05200025 05200026

Appendix B

Chronology of the Combined License Application for Vogtle Electric Generating Plant Units 3 and 4

This appendix contains a chronological listing of routine licensing correspondence between the staff of the U.S. Nuclear Regulatory Commission (NRC) regarding the review of the Vogtle Electric Generating Plant, Units 3 and 4 plant design under Docket Nos. 052-000025 and 052-000026

Document Date	Accession Number	Title & Estimated Page Count	Document Type	Author Affiliation	Addressee Affiliation	Docket Number
04/16/2010	ML101340032	2010/04/16 Vogtle COL Review - SNC Letter ND-10-0773 - SNC VEGP Units 3&4 COL Response to RAI Letter No. 051 Cyber Security Plan 8 Page(s)	E-Mail	- No Known Affiliation	NRC/NRO/DNRL/NWE1	05200025 05200026
04/16/2010	ML101340035	2010/04/16 Vogtle COL Review - Southern Nuclear Submittal Letter ND-10-0690 Annual 50.46 Report for VEGP Units 3 and 4 COL Application 7 Page(s)	E-Mail	- No Known Affiliation	NRC/NRO/DNRL/NWE1	05200025 05200026
04/19/2010	ML101090162	2010/04/19-NRC Staff Letter to the Board Informing that Staff does not Anticipate that Chapter 11 of the Vogtle Advanced SER will be Issued Before June 2010. 3 Page(s)	Legal-Correspondence/Miscellaneous	NRC/OGC	NRC/ASLBP	05200025 05200026
04/19/2010	ML101100050	2010/04/19 Vogtle COL Review - RE: Vogtle RAI data for Tuesday's call, 4/20/2010 31 Page(s)	E-Mail	NRC/NRO	NRC/NRO/DNRL/NWE1	05200025 05200026
04/19/2010	ML101100051	2010/04/19 Vogtle COL Review - RE: Vogtle RAI data for Tuesday's call, 4/20/2010 4 Page(s)	E-Mail	- No Known Affiliation	NRC/NRO/DNRL/NWE1	05200025 05200026
04/20/2010	ML101120089	Southern Nuclear Operating Co. Early Site Permit Site Safety Analysis Report Change Request	Letter Report, Miscellaneous	Southern Nuclear Operating Co, Inc	NRC/Document Control Desk NRC/NRO	05200011 05200025 05200026

B-166

Appendix B

Chronology of the Combined License Application for Vogtle Electric Generating Plant Units 3 and 4

This appendix contains a chronological listing of routine licensing correspondence between the staff of the U.S. Nuclear Regulatory Commission (NRC) regarding the review of the Vogtle Electric Generating Plant, Units 3 and 4 plant design under Docket Nos. 052-000025 and 052-000026

Document Date	Accession Number	Title & Estimated Page Count	Document Type	Author Affiliation	Addressee Affiliation	Docket Number
		Vogtle Electric Generating Plant Units 3 and 4 Use of Category 1 and 2 Backfill Material from Additional Onsite Areas On An Exigent Basis for Units 3 and 4. 17 Page(s)				
04/23/2010	ML101130316	2010/04/23 Vogtle COL Review - Telcon Summary - Telcon with Vogtle 8/28/09 2 Page(s)	E-Mail	NRC/NRO	NRC/NRO/DNRL/NWE1	05200025 05200026
04/23/2010	ML101160532	Vogtle, Units 3 and 4 Combined License Application - Response to Bellefonte Units 3 and 4 Safety Evaluation Report Open Items for Chapter 9. 7 Page(s)	Letter	Southern Nuclear Operating Co, Inc	NRC/Document Control Desk NRC/NRO	05200014 05200015 05200025 05200026
04/23/2010	ML101160533	Vogtle, Units 3 and 4 Combined License Application - Response to Bellefonte Units 3 and 4 Safety Evaluation Report Open Items for Chapter 3. 11 Page(s)	Letter	Southern Nuclear Operating Co, Inc	NRC/Document Control Desk NRC/NRO	05200014 05200015 05200025 05200026
04/23/2010	ML101340030	2010/04/23 Vogtle COL Review - RE: SNC Letter ND-10-0781 - SNC VEGP Units 3&4 COL Response to Bellefonte Units 3 & 4 Safety Evaluation Report Open Items for Chapter 9 11 Page(s)	E-Mail	- No Known Affiliation	NRC/NRO/DNRL/NWE1	05200025 05200026

Chronology of the Combined License Application for Vogtle Electric Generating Plant Units 3 and 4

This appendix contains a chronological listing of routine licensing correspondence between the staff of the U.S. Nuclear Regulatory Commission (NRC) regarding the review of the Vogtle Electric Generating Plant, Units 3 and 4 plant design under Docket Nos. 052-000025 and 052-000026

Document Date	Accession Number	Title & Estimated Page Count	Document Type	Author Affiliation	Addressee Affiliation	Docket Number
04/23/2010	ML101340031	2010/04/23 Vogtle COL Review - SNC Letter ND-10-0801 - SNC VEGP Units 3&4 COL Response to Bellefonte Units 3 & 4 Safety Evaluation Report Open Items for Chapter 3 15 Page(s)	E-Mail	- No Known Affiliation	NRC/NRO/DNRL/NWE1	05200025 05200026
04/23/2010	ML101340069	2010/04/23 Vogtle COL Review - DRAFT RAI - SRP section 3.12 - Vogtle Units 3 and 4 Combined License Application 3 Page(s)	E-Mail	NRC/NRO	NRC/NRO/DNRL/NWE1	05200025 05200026
04/23/2010	ML101340070	2010/04/23 Vogtle COL Review - DRAFT RAIs - SRP section 3.9.6 - Vogtle Units 3 and 4 Combined License Application 3 Page(s)	E-Mail	NRC/NRO	NRC/NRO/DNRL/NWE1	05200025 05200026
04/26/2010	ML101340037	2010/04/26 Vogtle COL Review - RE: Please reschedule Security scheduling meeting 2 Page(s)	E-Mail	NRC/NRO	NRC/NRO/DNRL/NWE1	05200025 05200026
04/26/2010	ML101340038	2010/04/26 Vogtle COL Review - FW: Please reschedule Security scheduling meeting 2 Page(s)	E-Mail	NRC/NRO	NRC/NRO/DNRL/NWE1	05200025 05200026
04/26/2010	ML101340067	2010/04/26 Vogtle COL Review - RE: Vogtle RAI data for Tuesday's call, 4/27/2010 4 Page(s)	E-Mail	- No Known Affiliation	NRC/NRO/DNRL/NWE1	05200025 05200026

Appendix B

Chronology of the Combined License Application for Vogtle Electric Electric Generating Plant Units 3 and 4

This appendix contains a chronological listing of routine licensing correspondence between the staff of the U.S. Nuclear Regulatory Commission (NRC) regarding the review of the Vogtle Electric Generating Plant, Units 3 and 4 plant design under Docket Nos. 052-000025 and 052-000026

Document Date	Accession Number	Title & Estimated Page Count	Document Type	Author Affiliation	Addressee Affiliation	Docket Number
04/26/2010	ML101340068	2010/04/26 Vogtle COL Review - RE: Vogtle RAI data for Tuesday's call, 4/27/2010 31 Page(s)	E-Mail	NRC/NRO	NRC/NRO/DNRL/NWE1	05200025 05200026
04/27/2010	ML093560768	United States Nuclear Regulatory Commission, Southern Nuclear Operating Company, ET. Al: Supplementary Notice Of Hearing And Opportunity To Petition For Leave To Intervene On A Combined License Application For The Vogtle Electric Generating Plant Units 9 Page(s)	Federal Register Notice	NRC/SECY		05200025 05200026
04/27/2010	ML101180058	2010/04/27 Vogtle COL Review - Safe Harbor Agreement 13 Page(s)	E-Mail	- No Known Affiliation	NRC/NRO/DNRL/NWE1	05200025 05200026
04/27/2010	ML101340036	2010/04/27 Vogtle COL Review - Vogtle EALs supplemental RAI response 2 Page(s)	E-Mail	NRC/NRO	NRC/NRO/DNRL/NWE1	05200025 05200026
04/28/2010	ML093560656	Southern Nuclear Operating Company - Supplementary Notice Of Hearing And Opportunity To Petition For Leave To Intervene On A Supplement To The Combined License Application For The Vogtle Electric Generating Plant Units 3 And 4 To Include A Limited	Letter	NRC/NRO/DNRL/NWE1	Southern Nuclear Operating Co, Inc	05200025 05200026

Appendix B

Chronology of the Combined License Application for Vogtle Electric Generating Plant Units 3 and 4

This appendix contains a chronological listing of routine licensing correspondence between the staff of the U.S. Nuclear Regulatory Commission (NRC) regarding the review of the Vogtle Electric Generating Plant, Units 3 and 4 plant design under Docket Nos. 052-000025 and 052-000026

Document Date	Accession Number	Title & Estimated Page Count	Document Type	Author Affiliation	Addressee Affiliation	Docket Number
04/28/2010	ML101160362	04/06/2010 Summary of Meeting with Southern Nuclear Operating Company to Discuss Plans for Potential License Amendments Regarding Safety Related Backfill for its Early Site Permit for the Vogtle Site. 6 Page(s)	Meeting Agenda Meeting Summary	NRC/NRO/DNRL/NWE1	NRC/NRO	05200011 05200025 05200026
04/28/2010	ML101180234	2010/04/28 Vogtle COL Review - Telcon Summary - Telcon with Vogtle 4/28/10 9 Page(s)	E-Mail	NRC/NRO	NRC/NRO/DNRL/NWE1	05200025 05200026
04/28/2010	ML101180327	2010/04/28 Vogtle RAI for SER – Request for Additional Information Letter No. 055 Related to SRP Section 13.6 for the Vogtle Electric Plant Units 3 and 4 Combined License Application 3 Page(s)	E-Mail Request for Additional Information (RAI)	NRC/NRO	NRC/NRO/DNRL/NWE1	05200025 05200026
04/28/2010	ML101200570	Vogtle Electric Generating Plant Units 3 & 4 Combined License Application, Supplemental Response 2 to Request for Additional Information Letter No. 029. 6 Page(s)	Letter	Southern Nuclear Operating Co, Inc	NRC/Document Control Desk NRC/NRO	05200025 05200026
04/28/2010	ML101200571	Southern Nuclear Operating Company Vogtle Electric 9 Page(s)	Letter	Southern Nuclear Operating Co, Inc	NRC/Document Control Desk	05200025 05200026

Appendix B

Chronology of the Combined License Application for Vogtle Electric Generating Plant Units 3 and 4

This appendix contains a chronological listing of routine licensing correspondence between the staff of the U.S. Nuclear Regulatory Commission (NRC) regarding the review of the Vogtle Electric Generating Plant, Units 3 and 4 plant design under Docket Nos. 052-000025 and 052-000026

Document Date	Accession Number	Title & Estimated Page Count	Document Type	Author Affiliation	Addressee Affiliation	Docket Number
		Generating Plant Units 3 and 4 Combined License Application Revised Response to Request for Additional Information Numbers 08.02-2 and 08.02-7. 8 Page(s)			NRC/NRO	
04/28/2010	ML101340027	2010/04/28 Vogtle COL Review - SNC Letter ND-10-0613 - SNC VEGP Units 3&4 COL Supplemental Response 2 to RAI Letter No. 029 13 Page(s)	E-Mail	- No Known Affiliation	NRC/NRO/DNRL/NWE1	05200025 05200026
04/28/2010	ML101340028	2010/04/28 Vogtle COL Review - SNC Letter ND-10-0809 - SNC VEGP Units 3&4 COL Revised Response to RAI Numbers 08.02-2 and 08.02-7 12 Page(s)	E-Mail	- No Known Affiliation	NRC/NRO/DNRL/NWE1	05200025 05200026
04/29/2010	ML101190566	2010/04/29 Vogtle RAI for SER – Request for Additional Information Letter No. 056 Related to SRP Section 3.9.6 for the Vogtle Electric Generating Plant Untis 3 and 4 Combined License Application 6 Page(s)	Request for Additional Information (RAI)	NRC/NRO	NRC/NRO/DNRL/NWE1	05200025 05200026
04/29/2010	ML101340024	2010/04/29 Vogtle COL Review – Additional Information Letter No. 056 Related to SRP Section 3.9.6 for the Vogtle Electric Generating	E-Mail	NRC/NRO	NRC/NRO/DNRL/NWE1	05200025 05200026

Appendix B

Chronology of the Combined License Application for Vogtle Electric Generating Plant Units 3 and 4

This appendix contains a chronological listing of routine licensing correspondence between the staff of the U.S. Nuclear Regulatory Commission (NRC) regarding the review of the Vogtle Electric Generating Plant, Units 3 and 4 plant design under Docket Nos. 052-000025 and 052-000026

Document Date	Accession Number	Title & Estimated Page Count	Document Type	Author Affiliation	Addressee Affiliation	Docket Number
		Plant Units 3 and 4 Combined License Application 8 Page(s)				
04/30/2010	ML101200256	2010/04/30 Vogtle COL Review - VogtleBackfillAudit Exaction Plan 532010 mas final.doc 7 Page(s)	E-Mail	NRC/NRO	NRC/NRO/DNRL/NWE1	05200025 05200026
04/30/2010	ML101250145	2010/04/30-Southern Nuclear Operating Co. Disclosures Pursuant to 10 CFR Section 2.336 for Contention SAFETY-1 (COL for Vogtle Units 3 and 4), Docket Nos. 52-025-COL and 52-026-COL. 6 Page(s)	Legal-Correspondence	Balch & Bingham, LLP Southern Nuclear Operating Co, Inc	Emory Univ School of Law NRC/OGC Turner Environmental Law Clinic	05200025 05200026
05/03/2010	ML101230158	Press Release-10-079: NRC Announces Supplemental Opportunity to Participate in Hearing on New Reactor Application for Vogtle Site. 2 Page(s)	Press Release	NRC/OPA		05200025 05200026
05/03/2010	ML101340018	2010/05/03 Vogtle COL Review - Revised DRAFT RAI - SRP section 3.12 - Vogtle Units 3 and 4 Combined License Application 3 Page(s)	E-Mail	NRC/NRO	NRC/NRO/DNRL/NWE1	05200025 05200026
05/03/2010	ML101340019	2010/05/03 Vogtle COL Review - RE: Vogtle RAI data for Tuesday's call, 5/4/2010 4 Page(s)	E-Mail	- No Known Affiliation	NRC/NRO/DNRL/NWE1	05200025 05200026

Appendix B

Chronology of the Combined License Application for Vogtle Electric Generating Plant Units 3 and 4

This appendix contains a chronological listing of routine licensing correspondence between the staff of the U.S. Nuclear Regulatory Commission (NRC) regarding the review of the Vogtle Electric Generating Plant, Units 3 and 4 plant design under Docket Nos. 052-000025 and 052-000026

Document Date	Accession Number	Title & Estimated Page Count	Document Type	Author Affiliation	Addressee Affiliation	Docket Number
05/03/2010	ML101340021	2010/05/03 Vogtle COL Review - RE: Vogtle RAI data for Tuesday's call, 5/4/2010 30 Page(s)	E-Mail	NRC/NRO	NRC/NRO/DNRL/NWE1	05200025 05200026
05/05/2010	ML101340017	2010/05/05 Vogtle COL Review - SNC Letter ND-10-0854 - SNC VEGP Units 3&4 Response to RAI Letter No. 052 Loss of Large Areas of the Plant Due to Explosions or Fire - Mitigative Strategies Description 8 Page(s)	E-Mail	- No Known Affiliation	NRC/NRO/DNRL/NWE1	05200025 05200026
05/05/2010	ML101270075	Vogtle, Units 3 & 4 - Response to Request for Additional Information Letter No. 052 - Loss of Large Areas of the Plant Due to Explosions or Fire - Mitigative Strategies Description. 4 Page(s)	Letter	Southern Nuclear Operating Co, Inc	NRC/Document Control Desk NRC/NRO	05200025 05200026
05/05/2010	ML101270076	Vogtle, Units 3 & 4 - Response to Request for Additional Information Letter No. 052 - Loss of Large Areas of the Plant Due to Explosions or Fire - Mitigative Strategies Description. 47 Page(s)	- No Document Type Applies	Southern Nuclear Operating Co, Inc	NRC/NRO	05200025 05200026
05/06/2010	ML101300089	Vogtle, Units 3 and 4 Combined License Application, Response to Request for Additional Information	Letter	Southern Nuclear Operating Co, Inc	NRC/Document Control Desk NRC/NRO	05200025 05200026

B-173

Appendix B

Chronology of the Combined License Application for Vogtle Electric Generating Plant Units 3 and 4

This appendix contains a chronological listing of routine licensing correspondence between the staff of the U.S. Nuclear Regulatory Commission (NRC) regarding the review of the Vogtle Electric Generating Plant, Units 3 and 4 plant design under Docket Nos. 052-000025 and 052-000026

Document Date	Accession Number	Title & Estimated Page Count	Document Type	Author Affiliation	Addressee Affiliation	Docket Number
05/06/2010	ML101340016	Letter No. 053. 7 Page(s)				
		2010/05/06 Vogtle COL Review - SNC Letter ND-10-0813 - SNC VEGP Units 3&4 COLA Response to RAI Letter No. 053 11 Page(s)	E-Mail	- No Known Affiliation	NRC/NRO/DNRL/NWE1	05200025 05200026
05/07/2010	ML101270255	2010/05/07 Vogtle COL Review - Telcon Summary - Telcon with Vogtle 4/28/10 -UPDATE 6 Page(s)	E-Mail	NRC/NRO	NRC/NRO/DNRL/NWE1	05200025 05200026
05/07/2010	ML101270364	2010/05/07-NRC Staff Letter to ASLBP Board Regarding Environmental Review Schedule for Vogtle, Units 3 and 4. 3 Page(s)	Legal-Correspondence/Miscellaneous	NRC/OGC	NRC/ASLBP	05200025 05200026
05/10/2010	ML101250259	Summary of the March 3, 2010 Meeting with Southern Nuclear Operating Company to Discuss Plans to Request Exemption from Requirements of 10 CFR 50.10(d) for Its Combined License Application for the Vogtle Electric Generating Plant Proposed Units 3 and 4. 5 Page(s)	Meeting Summary	NRC/NRO/DNRL/NWE1	Southern Nuclear Operating Co, Inc	05200025 05200026
05/10/2010	ML101310517	2010/05/10 Vogtle COL Review - Southern Nuclear Letter ND-10-0923, Post New and Significant	E-Mail	Southern Nuclear Operating Co, Inc	NRC/NRO/DNRL/NWE1	05200025 05200026

Appendix B

Chronology of the Combined License Application for Vogtle Electric Generating Plant Units 3 and 4

This appendix contains a chronological listing of routine licensing correspondence between the staff of the U.S. Nuclear Regulatory Commission (NRC) regarding the review of the Vogtle Electric Generating Plant, Units 3 and 4 plant design under Docket Nos. 052-000025 and 052-000026

Document Date	Accession Number	Title & Estimated Page Count	Document Type	Author Affiliation	Addressee Affiliation	Docket Number
05/10/2010	ML101320256	Audit Supporting Information 7 Page(s) Vogtle, Units 3 & 4 - Combined License Application Post New and Significant Audit Supporting Information. 146 Page(s)	Letter	Southern Nuclear Operating Co, Inc	NRC/Document Control Desk NRC/NRO	05200025 05200026
05/10/2010	ML101330141	Vogtle, Units 3 and 4, Early Site Permit Site Safety Analysis Report Amendment Request, Revised Table Title. 6 Page(s)	Letter	Southern Nuclear Operating Co, Inc	NRC/Document Control Desk NRC/NRO	05200011 05200025 05200026
05/10/2010	ML101340012	2010/05/10 Vogtle COL Review - RE: Vogtle RAI data for Tuesday's call, 5/11/2010 4 Page(s)	E-Mail	- No Known Affiliation	NRC/NRO/DNRL/NWE1	05200025 05200026
05/10/2010	ML101340013	2010/05/10 Vogtle COL Review - RE: Vogtle RAI data for Tuesday's call, 5/11/2010 13 Page(s)	E-Mail	NRC/NRO	NRC/NRO/DNRL/NWE1	05200025 05200026
05/11/2010	ML101310333	2010/05/11 Vogtle COL Review - Southern Nuclear Letter ND-10-0923, Post New and Significant Audit Supporting Information 150 Page(s)	E-Mail	- No Known Affiliation	NRC/NRO/DNRL/NWE1	05200025 05200026
05/11/2010	ML101310627	2010/05/11 Vogtle COL Review - RWI Drawing 3 Page(s)	E-Mail	- No Known Affiliation	NRC/NRO/DNRL/NWE1	05200025 05200026
05/11/2010	ML101340009	2010/05/11 Vogtle COL Review -	E-Mail	NRC/NRO	NRC/NRO/DNRL/NWE1	05200025

Appendix B

Chronology of the Combined License Application for Vogtle Electric Generating Plant Units 3 and 4

This appendix contains a chronological listing of routine licensing correspondence between the staff of the U.S. Nuclear Regulatory Commission (NRC) regarding the review of the Vogtle Electric Generating Plant, Units 3 and 4 plant design under Docket Nos. 052-000025 and 052-000026

Document Date	Accession Number	Title & Estimated Page Count	Document Type	Author Affiliation	Addressee Affiliation	Docket Number
		Additional DRAFT RAI - SRP section 14.2 - Vogtle Units 3 and 4 Combined License Application 4 Page(s)				05200026
05/11/2010	ML101340010	2010/05/11 Vogtle COL Review - DRAFT RAI - SRP section 14.2 - Vogtle Units 3 and 4 Combined License Application 4 Page(s)	E-Mail	NRC/NRO	NRC/NRO/DNRL/NWE1	05200025 05200026
05/11/2010	ML101340011	2010/05/11 Vogtle COL Review - RE: Vogtle RAI data for Tuesday's call. 5/11/2010 21 Page(s)	E-Mail	NRC/NRO	NRC/NRO/DNRL/NWE1	05200025 05200026
05/12/2010	ML101320629	Vogtle Cyber Open Items. 5 Page(s)	Request for Additional Information (RAI)	NRC/NRO/DNRL/NWE1	NRC/NRO	05200025 05200026
05/13/2010	ML101340104	2010/05/13 Vogtle COL Review - ND-10-0960 - SNC VEGP Units 3&4 ESP Safety Analysis Report Amendment Request Revised Site Safety Analysis Report Markup for Onsite Sources of Backfill 32 Page(s)	E-Mail	- No Known Affiliation	NRC/NRO/DNRL/NWE1	05200025 05200026
05/13/2010	ML101340649	Vogtle Units, 3 & 4, Response to Requests for Additional Information on the License Amendment Request. 28 Page(s)	Letter	Southern Nuclear Operating Co, Inc	NRC/Document Control Desk NRC/NRO	05200011 05200025 05200026

Appendix B

Chronology of the Combined License Application for Vogtle Electric Generating Plant Units 3 and 4

This appendix contains a chronological listing of routine licensing correspondence between the staff of the U.S. Nuclear Regulatory Commission (NRC) regarding the review of the Vogtle Electric Generating Plant, Units 3 and 4 plant design under Docket Nos. 052-000025 and 052-000026

Document Date	Accession Number	Title & Estimated Page Count	Document Type	Author Affiliation	Addressee Affiliation	Docket Number
05/14/2010	ML101370303	2010/05/14 Vogtle COL Review - ND-10-0949 - SNC VEGP Units 3&4 COL Response to Bellefonte Units 3 and 4 Safety Evaluation Report Open Items for Chapter 3 19 Page(s)	E-Mail	- No Known Affiliation	NRC/NRO/DNRL/NWE1	05200025 05200026
05/14/2010	ML101370306	2010/05/14 Vogtle COL Review - ND-10-0962 - SNC VEGP Units 3&4 COL Response to RAI Letter No. 055 Safety / Security Interface 10 Page(s)	E-Mail	- No Known Affiliation	NRC/NRO/DNRL/NWE1	05200025 05200026
05/14/2010	ML101380120	Vogtle, Units 3 and 4 Combined License Application, Response to Bellefonte, Units 3 and 4, Safety Evaluation Report Open Items for Chapter 3. 15 Page(s)	Letter	Southern Nuclear Operating Co, Inc	NRC/Document Control Desk NRC/NRO	05200014 05200015 05200025 05200026
05/14/2010	ML101380352	Vogtle, Units 3 and 4, Combine License Application Response to Request for Additional Information Letter No. 055 Safety/Security Interface. 6 Page(s)	Letter	Southern Nuclear Operating Co, Inc	NRC/Document Control Desk NRC/NRO	05200025 05200026
05/17/2010	ML101320689	04/08/2010 Summary of a Closed Conference Call with Southern Nuclear Operating Company on the Vogtle Cyber Security Plan Submittal. 4 Page(s)	Meeting Agenda Meeting Summary Memoranda	NRC/NRO/DNRL/NWE1	NRC/NRO/DNRL/NWE1	05200025 05200026

Appendix B

Chronology of the Combined License Application for Vogtle Electric Generating Plant Units 3 and 4

This appendix contains a chronological listing of routine licensing correspondence between the staff of the U.S. Nuclear Regulatory Commission (NRC) regarding the review of the Vogtle Electric Generating Plant, Units 3 and 4 plant design under Docket Nos. 052-000025 and 052-000026

Document Date	Accession Number	Title & Estimated Page Count	Document Type	Author Affiliation	Addressee Affiliation	Docket Number
05/18/2010	ML101330353	Revision of the Environmental Review Schedule for the Combined License Application Review for Vogtle Electric Generating Plant, Units 3 & 4. 11 Page(s)	Letter	NRC/NRO/DNRL	Southern Nuclear Operating Co, Inc	05200016 05200025 05200026
05/18/2010	ML101380685	2010/05/18 Vogtle RAI for SER – Request for Additional Information Letter No. 057 Related to SRP Section SECTION 3.12 for the Vogtle Electric Generating Plant Units 3 and 4 Combined License Application 6 Page(s)	E-Mail Request for Additional Information (RAI)	NRC/NRO	NRC/NRO/DNRL/NWE1	05200025 05200026
05/20/2010	ML092720790	Vogtle Units 3 & 4, SER Chapter 10 – "Steam and Power Conversion System." 33 Page(s)	NRO Safety Evaluation Report (SER)-Delayed	NRC/NRO/DNRL/NWE1		05200025 05200026
05/20/2010	ML100550521	Ltr.- CH 10- Vogtle Electric Generating Plant Units 3 and 4 Advanced Final Safety Evaluation Report with No Open Items for Chapter 10 "Steam and Power Conversion System." 9 Page(s)	Letter	NRC/NRO/DNRL/NWE1	Southern Nuclear Operating Co, Inc	05200025 05200026
05/20/2010	ML101400600	2010/05/20 Vogtle RAI for SER - 41 Page(s)	E-Mail	NRC/NRO	NRC/NRO/DNRL/NWE1	05200025 05200026
05/20/2010	ML101440385	Vogtle, Units 3 & 4, Early Site Permit Site Safety Analysis Report	Letter	Southern Nuclear Operating Co, Inc	NRC/Document Control Desk	05200011 05200025

Appendix B

Chronology of the Combined License Application for Vogtle Electric Generating Plant Units 3 and 4

This appendix contains a chronological listing of routine licensing correspondence between the staff of the U.S. Nuclear Regulatory Commission (NRC) regarding the review of the Vogtle Electric Generating Plant, Units 3 and 4 plant design under Docket Nos. 052-000025 and 052-000026

Document Date	Accession Number	Title & Estimated Page Count	Document Type	Author Affiliation	Addressee Affiliation	Docket Number
		Amendment Request, Revised Site Safety Analysis Report Markup for Onsite Sources of Backfill Correction. 4 Page(s)			NRC/NRO	05200026
05/21/2010	ML092610415	VEGP SER - Vogtle Electric Generating Plant Units 3 And 4 Advanced Final Safety Evaluation Report With No Open Items For Chapter 4, "Reactor". 7 Page(s)	NRO Safety Evaluation Report (SER)-Delayed	NRC/NRO/DNRL/NWE1		05200025 05200026
05/21/2010	ML100280953	Letter- CH 4- Vogtle Electric Generating Plant Units 3 And 4 Advanced Final Safety Evaluation Report With No Open Items For Chapter 4, "Reactor". 6 Page(s)	Letter	NRC/NRO/DNRL/NWE1	Southern Nuclear Operating Co, Inc	05200025 05200026
05/21/2010	ML100540758	Memo- CH10- Vogtle Electric Generating Plant Units 3 and 4 Advanced Final Safety Evaluation Report with No Open Items for Chapter 10 "Steam and Power Conversion System." 3 Page(s)	Memoranda	NRC/NRO/DNRL	NRC/ACRS	05200025 05200026
05/21/2010	ML101460201	Vogtle Units 3 & 4 Combined License Application Response to Bellefonte Units 3 & 4 Safety Evaluation Report Open Items for Chapter 15.	Letter	Southern Nuclear Operating Co, Inc	NRC/Document Control Desk NRC/NRO	05200025 05200026

Appendix B

Chronology of the Combined License Application for Vogtle Electric Generating Plant Units 3 and 4

This appendix contains a chronological listing of routine licensing correspondence between the staff of the U.S. Nuclear Regulatory Commission (NRC) regarding the review of the Vogtle Electric Generating Plant, Units 3 and 4 plant design under Docket Nos. 052-000025 and 052-000026

Document Date	Accession Number	Title & Estimated Page Count	Document Type	Author Affiliation	Addressee Affiliation	Docket Number
05/21/2010	ML101460202	Vogtle, Units 3 & 4 Combined License Application, Revised Response to Bellefonte Units 3 & 4 Safety Evaluation Report Open Items for Chapter 16. 8 Page(s)	Letter	Southern Nuclear Operating Co, Inc	NRC/Document Control Desk NRC/NRO	05200014 05200015 05200025 05200026
05/21/2010	ML101460204	Vogtle, Units 3 & 4, Combined License Application Response to Request for Additional Information Letter No. 041, Supplement 1, Physical Security Plan. 6 Page(s)	Letter	Southern Nuclear Operating Co, Inc	NRC/Document Control Desk NRC/NRO	05200025 05200026
05/21/2010	ML101460298	Vogtle, Units 3 and 4, Combined License Application Response to Request for Additional Information Letter No. 041, Supplement 2 Physical Security Plan. 4 Page(s)	Letter	Southern Nuclear Operating Co, Inc	NRC/Document Control Desk NRC/NRO	05200025 05200026
05/21/2010	ML101460299	Enclosure to ND-10-1008 – Vogtle, Units 3 and 4, Supplemental Response to RAI 13.06-06 from NRC RAI Letter No. 041 Involving the Physical Security Plan. 2 Page(s)	Security Plan	Southern Nuclear Operating Co, Inc	NRC/NRO	05200025 05200026

Appendix B

Chronology of the Combined License Application for Vogtle Electric Generating Plant Units 3 and 4

This appendix contains a chronological listing of routine licensing correspondence between the staff of the U.S. Nuclear Regulatory Commission (NRC) regarding the review of the Vogtle Electric Generating Plant, Units 3 and 4 plant design under Docket Nos. 052-000025 and 052-000026

Document Date	Accession Number	Title & Estimated Page Count	Document Type	Author Affiliation	Addressee Affiliation	Docket Number
05/24/2010	ML092650039	VEGP SER - Vogtle Electric Generating Plant Units 3 and 4 Advanced Final Safety Evaluation Report With No Open Items for Chapter 12, "Radiation Protection." 40 Page(s)	NRO Safety Evaluation Report (SER)-Delayed	NRC/NRO/DNRL/NWE1		05200025 05200026
05/24/2010	ML100331243	Memo- CH 4- Vogtle Electric Generating Plant Units 3 and 4 Advanced Final Safety Evaluation Report with No Open Items for Chapter 4, "Reactor". 3 Page(s)	Memoranda	NRC/NRO/DNRL	NRC/ACRS	05200025 05200026
05/24/2010	ML100890231	Letter, Vogtle Electric Generating Plant Units 3 And 4 Advanced Final Safety Evaluation Report With No Open Items For Chapter 12, "Radiation Protection." 10 Page(s)	Letter	NRC/NRO/DNRL/NWE1	Southern Nuclear Operating Co, Inc	05200025 05200026
05/24/2010	ML100890389	Memo - Vogtle Electric Generating Plant Units 3 and 4 Advanced Final Safety Evaluation Report With No Open Items for Chapter 12, "Radiation Protection." 5 Page(s)	Memoranda	NRC/NRO/DNRL	NRC/ACRS	05200025 05200026
05/24/2010	ML101460211	Vogtle Electric Generating Plant, Units 3 and 4 - Combined License Application Voluntary Revision to Final Safety Analysis Report Chapter 6.	Final Safety Analysis Report (FSAR) Letter	Southern Nuclear Operating Co, Inc	NRC/Document Control Desk NRC/NRO	05200025 05200026

Appendix B

Chronology of the Combined License Application for Vogtle Electric Generating Plant Units 3 and 4

This appendix contains a chronological listing of routine licensing correspondence between the staff of the U.S. Nuclear Regulatory Commission (NRC) regarding the review of the Vogtle Electric Generating Plant, Units 3 and 4 plant design under Docket Nos. 052-000025 and 052-000026

Document Date	Accession Number	Title & Estimated Page Count	Document Type	Author Affiliation	Addressee Affiliation	Docket Number
05/24/2010	ML101460248	Memorandum re: Atomic Safety and Licensing Board Memorandum and Order in Southern Nuclear Operating Co. (Vogtle Electric Generating Plant Units 3 and 4)(LBP-10-08). 2 Page(s)	Memoranda	NRC/OGC	NRC/NRO	05200025 05200026
05/24/2010	ML101460301	Vogtle Electric Generating Plant Units 3 and 4 Combined License Application Response to Request for Additional Information Letter No. 054 Loss of Large Areas of the Plant Due to Explosions or Fire - Mitigative Strategies Description. 4 Page(s)	Letter	Southern Nuclear Operating Co, Inc	NRC/Document Control Desk NRC/NRO	05200025 05200026
05/26/2010	ML092610470	SER Chapter 11 - Vogtle Electric Generating Plant Units 3 And 4 Advanced Final Safety Evaluation Report With No Open Items For Chapter 11, "Radioactive Waste Management" 41 Page(s)	Safety Evaluation Report	NRC/NRO/DNRL/NWE1		05200025 05200026
05/26/2010	ML100350412	Letter re: Vogtle Electric Generating Plant Units 3 and 4 Advanced Final Safety Evaluation Report with No Open Items for Chapter 11, "Radioactive Waste Management". 6 Page(s)	Letter	NRC/NRO/DNRL/NWE1	Southern Nuclear Operating Co, Inc	05200025 05200026

Appendix B

Chronology of the Combined License Application for Vogtle Electric Generating Plant Units 3 and 4

This appendix contains a chronological listing of routine licensing correspondence between the staff of the U.S. Nuclear Regulatory Commission (NRC) regarding the review of the Vogtle Electric Generating Plant, Units 3 and 4 plant design under Docket Nos. 052-000025 and 052-000026

Document Date	Accession Number	Title & Estimated Page Count	Document Type	Author Affiliation	Addressee Affiliation	Docket Number
05/26/2010	ML101481017	Vogtle, Units 3 and 4 - Combined License Application, Nuclear Development Quality Assurance Manual - Revision 9. 65 Page(s)	Letter Manual Quality Assurance Program	Southern Nuclear Operating Co, Inc	NRC/Document Control Desk NRC/NRO	05200025 05200026
05/26/2010	ML101520041	2010/05/26 Vogtle RAI for SER – Request for Additonal Information Letter No. 059 Related to SRP Section 14.2 for the Vogtle Electric Generating Plant Units 3 and 4 Combined License Application 6 Page(s)	E-Mail	NRC/NRO	NRC/NRO/DNRL/NWE1	05200025 05200026
05/27/2010	ML100340674	Vogtle, Units 3 And 4, Advanced Final Safety Evaluation Report With No Open Items For Chapter 11, "Radioactive Waste Management." 3 Page(s)	Memoranda	NRC/NRO/DNRL/NWE1	NRC/ACRS	05200025 05200026
05/27/2010	ML101460568	05/27/2010 Notice of Forthcoming Closed Meeting With Southern Nuclear Operating Company Regarding the Vogtle Cyber Security Plan Submittal. 13 Page(s)	Meeting Agenda Meeting Notice Memoranda	NRC/NRO/DNRL/NWE1	NRC/NRO/DNRL/NWE1	05200025 05200026
05/27/2010	ML101520046	2010/05/27 Vogtle RAI for SER – Request for Additional Information Letter No. 058 Related to SRP Section 14.2 for the Vogtle Units 3 and 4 COL Application 6 Page(s)	E-Mail	NRC/NRO	NRC/NRO/DNRL/NWE1	05200025 05200026

Appendix B

Chronology of the Combined License Application for Vogtle Electric Generating Plant Units 3 and 4

This appendix contains a chronological listing of routine licensing correspondence between the staff of the U.S. Nuclear Regulatory Commission (NRC) regarding the review of the Vogtle Electric Generating Plant, Units 3 and 4 plant design under Docket Nos. 052-000025 and 052-000026

Document Date	Accession Number	Title & Estimated Page Count	Document Type	Author Affiliation	Addressee Affiliation	Docket Number
05/27/2010	ML101520053	2010/05/27 Vogtle RAI for SER – Request for Additional Information Letter No. 059 Related to SRP Section 14.2 for the Vogtle Units 3 and 4 COL Application 6 Page(s)	E-Mail	NRC/NRO	NRC/NRO/DNRL/NWE1	05200025 05200026
05/27/2010	ML101520182	Vogtle Electric Generating Plant, Units 3 and 4 - Combined License Application Voluntary Revision to Final Safety Analysis Report Chapter 13. 7 Page(s)	Letter	Southern Nuclear Operating Co, Inc	NRC/Document Control Desk NRC/NRO	05200025 05200026
05/27/2010	ML101520183	Vogtle, Units 3 and 4, Combined License Application, Response to Request for Additional Information Letter No. 056. 7 Page(s)	Letter	Southern Nuclear Operating Co, Inc	NRC/Document Control Desk NRC/NRO	05200025 05200026
05/28/2010	ML101530057	Vogtle, Units 3 & 4 - Response to Request for Additional Information Letter No. 047, Supplement 1 re Physical Security Plan. 4 Page(s)	Letter	Southern Nuclear Operating Co, Inc	NRC/Document Control Desk NRC/NRO	05200025 05200026
05/28/2010	ML101530058	Vogtle, Units 3 & 4 - Supplemental Response to Request for Additional Information 13.06-31 from NRC RAI Letter No. 047, re Physical Security Plan. 2 Page(s)	- No Document Type Applies	Southern Nuclear Operating Co, Inc	NRC/NRO	05200025 05200026

Appendix B

Chronology of the Combined License Application for Vogtle Electric Generating Plant Units 3 and 4

This appendix contains a chronological listing of routine licensing correspondence between the staff of the U.S. Nuclear Regulatory Commission (NRC) regarding the review of the Vogtle Electric Generating Plant, Units 3 and 4 plant design under Docket Nos. 052-000025 and 052-000026

Document Date	Accession Number	Title & Estimated Page Count	Document Type	Author Affiliation	Addressee Affiliation	Docket Number
05/28/2010	ML101530059	Vogtle, Units 3 and 4, Response to Request for Additional Information Letter No. 052, Supplement 1, Loss of Large Areas of the Plant Due to Explosions or Fire - Mitigative Strategies Description. 4 Page(s)	Letter	Southern Nuclear Operating Co, Inc	NRC/Document Control Desk NRC/NRO	05200025 05200026
05/28/2010	ML101530060	Enclosure to ND-10-1104, Vogtle, Units 3 and 4, Response to Request for Additional Information Letter No. 052, Supplement 1, Loss of Large Areas of the Plant Due to Explosions or Fire - Mitigative Strategies Description. 33 Page(s)	- No Document Type Applies	Southern Nuclear Operating Co, Inc	NRC/NRO	05200025 05200026
06/04/2010	ML101550367	2010/06/04 Vogtle COL Review - Audit Plan for Review of Calculation Associated with Control Room Habitability Chemical Releases SRP 6.4 4 Page(s)	E-Mail	NRC/NRO	NRC/NRO/DNRL/NWE1	05200025 05200026
06/04/2010	ML101590402	Vogtle, Units 3 and 4 - Combined License Application Voluntary Letter Addressing Additional Information for Chapter 07. 10 Page(s)	Letter	Southern Nuclear Operating Co, Inc	NRC/Document Control Desk NRC/NRO	05200025 05200026
06/04/2010	ML101590403	Vogtle, Units 3 and 4 - Combined License Application Response to Request for Additional Information	Letter	Southern Nuclear Operating Co, Inc	NRC/Document Control Desk NRC/NRO	05200025 05200026

Appendix B

Chronology of the Combined License Application for Vogtle Electric Generating Plant Units 3 and 4

This appendix contains a chronological listing of routine licensing correspondence between the staff of the U.S. Nuclear Regulatory Commission (NRC) regarding the review of the Vogtle Electric Generating Plant, Units 3 and 4 plant design under Docket Nos. 052-000025 and 052-000026

Document Date	Accession Number	Title & Estimated Page Count	Document Type	Author Affiliation	Addressee Affiliation	Docket Number
06/04/2010	ML101590404	Letter No. 054, Supplement 1 Loss of Large Areas of the Plant Due to Explosions or Fire - Mitigative Strategies Description. 4 Page(s)	Report, Miscellaneous			
06/04/2010	ML101590404	Enclosure to ND-10-1124, Supplemental Response to RAI Letter No. 054 on Vogtle, Units 3 and 4 COL Application. 9 Page(s)	Report, Miscellaneous	Southern Nuclear Operating Co, Inc	NRC/NRO	05200025 05200026
06/11/2010	ML101650541	Vogtle, Units 3 & 4, Combined License Application, Departure Report Update. 4 Page(s)	Letter	Southern Nuclear Operating Co, Inc	NRC/Document Control Desk NRC/NRO	05200025 05200026
06/11/2010	ML101660063	Vogtle Electric Generating Plant, Units 3 & 4 - Combined License Application Response to Request for Additional Information Letter No. 047, Supplement 2, Physical Security Inspections, Tests, Analyses, and Acceptance Criteria (PS-IT AAC). 31 Page(s)	Letter	Southern Nuclear Operating Co, Inc	NRC/Document Control Desk NRC/NRO	05200025 05200026
06/14/2010	ML101650301	Memo: FSAR Sections 2.1 And 2.2 Safety Evaluation Report Input For Vogtle Units 3 And 4 Combined License Application. 2 Page(s)	Memoranda	NRC/NRO/DSER/RSAC	NRC/NRO/DNRL	05200025 05200026
06/14/2010	ML101650311	Enclosure: FSAR Sections 2.1 And	- No Document	NRC/NRO/DSER/RSAC	NRC/NRO/DNRL	05200025

Appendix B

Chronology of the Combined License Application for Vogtle Electric Generating Plant Units 3 and 4

This appendix contains a chronological listing of routine licensing correspondence between the staff of the U.S. Nuclear Regulatory Commission (NRC) regarding the review of the Vogtle Electric Generating Plant, Units 3 and 4 plant design under Docket Nos. 052-000025 and 052-000026

Document Date	Accession Number	Title & Estimated Page Count	Document Type	Author Affiliation	Addressee Affiliation	Docket Number
		2.2 Safety Evaluation Report Input For Vogtle Units 3 And 4 Combined License Application. 5 Page(s)	Type Applies			05200026
06/14/2010	ML101670137	Vogtle, Units 3 and 4, Combined License Application, Cyber Security Plan, Revision 0. 4 Page(s)	Letter	Southern Nuclear Operating Co, Inc	NRC/Document Control Desk NRC/NRO	05200025 05200026
06/14/2010	ML101670138	Vogtle, Units 3 & 4, Cyber Security Plan, Revision 0. 48 Page(s)	Security Plan	Southern Nuclear Operating Co, Inc	NRC/NRO	05200025 05200026
06/16/2010	ML101550102	Site Audit Summary of the Environmental Site Audit Related to the Supplemental New and Significant Information for the Review of the Combined License Application for Vogtle Electric Generating Plant Site. 5 Page(s)	Memoranda	NRC/NRO/DSER/RAP1	NRC/NRO/DSER/RAP1	05200025 05200026
06/16/2010	ML101550128	Vogtle, Combined License Supplement New and Significant Information Site Audit Trip Report. May 3-6, 2010 9 Page(s)	Audit Report	NRC/NRO/DSER		05200025 05200026
06/17/2010	ML101670079	Summary of Teleconference Calls Held with Georgia Department of Natural Resources for the Vogtle Electric Generating Plant, Untis 3 and 4 Onsite Backfill Amendment.	Meeting Summary Memoranda	NRC/NRO/DSER/RAP1	NRC/NRO/DSER	05200025 05200026

Appendix B

Chronology of the Combined License Application for Vogtle Electric Generating Plant Units 3 and 4

This appendix contains a chronological listing of routine licensing correspondence between the staff of the U.S. Nuclear Regulatory Commission (NRC) regarding the review of the Vogtle Electric Generating Plant, Units 3 and 4 plant design under Docket Nos. 052-000025 and 052-000026

Document Date	Accession Number	Title & Estimated Page Count	Document Type	Author Affiliation	Addressee Affiliation	Docket Number
		4 Page(s)				
06/17/2010	ML101690456	Review of Bellefonte Response to Request for Additional Information Numbers 06.01.02-02, 06.04-07 & 06.04-08 for Applicability to Vogtle Electric Generating Plant Units 3 & 4 Combined License Application. 9 Page(s)	Letter	Southern Nuclear Operating Co, Inc	NRC/Document Control Desk NRC/NRO	05200025 05200026
06/18/2010	ML101720578	Vogtle, Units 3 & 4 - Response to Request for Additional Information Letter No. 058 Regarding Initial Testing Program. 7 Page(s)	Letter	Southern Nuclear Operating Co, Inc	NRC/Document Control Desk NRC/NRO	05200025 05200026
06/18/2010	ML101720640	Vogtle, Units 3 & 4 - Combined License Application Response to Request for Additional Information Letter No. 059. 7 Page(s)	Letter	Southern Nuclear Operating Co, Inc	NRC/Document Control Desk NRC/NRO	05200025 05200026
06/18/2010	ML101720641	Vogtle, Units 3 & 4 Combined License Application, Voluntary Revision to Final Safety Analysis Report Chapter 3. 6 Page(s)	Letter	Southern Nuclear Operating Co, Inc	NRC/Document Control Desk NRC/NRO	05200025 05200026
06/22/2010	ML101730045	06/23/2010 Notice of Closed Meeting with the Southern Nuclear Operating Company to Discuss Vogtle, Loss of Large Areas of the Plant Due to Explosions or Fire, Combined Licensing Agreement	Meeting Agenda Meeting Notice Memoranda	NRC/NRO/DNRL/NWE1	NRC/NRO/DNRL/NWE1	05200025 05200026

Appendix B

Chronology of the Combined License Application for Vogtle Electric Generating Plant Units 3 and 4

This appendix contains a chronological listing of routine licensing correspondence between the staff of the U.S. Nuclear Regulatory Commission (NRC) regarding the review of the Vogtle Electric Generating Plant, Units 3 and 4 plant design under Docket Nos. 052-000025 and 052-000026

Document Date	Accession Number	Title & Estimated Page Count	Document Type	Author Affiliation	Addressee Affiliation	Docket Number
06/22/2010	ML101740487	Vogtle, Units 3 and 4, Combined License Application, Response to Request for Additional Information Letter No. 049, Supplement 1, Fitness for Duty Program. 7 Page(s) Submittal. 11 Page(s)	Letter	Southern Nuclear Operating Co, Inc	NRC/Document Control Desk NRC/NRO	05200025 05200026
06/24/2010	ML101720375	Safety Evaluation Report Section 3.9.6 on Vogtle Units 3 and 4 Combined License Application. 17 Page(s)	Memoranda Safety Evaluation	NRC/NRO/DE/CIB1	NRC/NRO/DNRL/NWE1	05200025 05200026
06/24/2010	ML101730034	Mechanical Equipment Environmental Qualification Input for Safety Evaluation Report Section 3.11 on Revision 2 to Vogtle Units 3 and 4 Combined License Application. 9 Page(s)	Memoranda	NRC/NRO/DE/CIB1	NRC/NRO/DNRL/NGE1	05200025 05200026
06/24/2010	ML101930516	Transcripts on the Advisory Committee on Reactor Safeguards AP100 Subcommittee (Westinghouse AP1000 DCD and Vogtle Units 3 and 4 COL) June 24, 2010. Pages 1-241. 526 Page(s)	Meeting Transcript	NRC/ACRS		05200010 05200025 05200026
06/28/2010	ML092650055	Vogtle, Units 3 & 4, Safety Evaluation Report Chapter 16, Technical Specifications, Clean	Safety Evaluation	NRC/NRO/DNRL/NWE1		05200025 05200026

B-189

Appendix B

Chronology of the Combined License Application for Vogtle Electric Generating Plant Units 3 and 4

This appendix contains a chronological listing of routine licensing correspondence between the staff of the U.S. Nuclear Regulatory Commission (NRC) regarding the review of the Vogtle Electric Generating Plant, Units 3 and 4 plant design under Docket Nos. 052-000025 and 052-000026

Document Date	Accession Number	Title & Estimated Page Count	Document Type	Author Affiliation	Addressee Affiliation	Docket Number
06/25/2010	ML102250354	Master. 15 Page(s) 2010/06/25-Portion of Meeting Transcript of The Advisory Committee on Reactor Safeguards (ACRS) Subcommittee on Westinghouse AP1000 DCD and Vogtle, Units 3 and 4, Exhibit 5. 59 Page(s)	Legal-Exhibit Meeting Transcript	NRC/ACRS		05200025 05200026
06/28/2010	ML100900407	Memo -CH16 - Vogtle Electric Generating Plant Units 3 and 4 Advanced Final Safety Evaluation Report with No Open Items for Chapter 16, "Technical Specifications". 3 Page(s)	Memoranda	NRC/NRO/DNRL	NRC/ACRS	05200025 05200026
06/28/2010	ML100900435	Ltr. -CH16 - Vogtle Electric Generating Plant Units 3 and 4 Advanced Final Safety Evaluation Report with No Open Items for Chapter 16, "Technical Specifications." 10 Page(s)	Letter	NRC/NRO/DNRL/NWE1	Southern Nuclear Operating Co, Inc	05200025 05200026
06/29/2010	ML101740184	SEB1 Input to SER for Vogtle Nuclear Plant Units 3 and 4 Reference COLA (Phase 4). 32 Page(s)	Memoranda Safety Evaluation Report	NRC/NRO/DE/SEB1	NRC/NRO/DNRL/NGE1	05200025 05200026
07/01/2010	ML101830392	Vogtle Units 3 & 4 - Combined License Application Voluntary	Letter	Southern Nuclear Operating Co, Inc	NRC/Document Control Desk	05200025 05200026

Appendix B

Chronology of the Combined License Application for Vogtle Electric Generating Plant Units 3 and 4

This appendix contains a chronological listing of routine licensing correspondence between the staff of the U.S. Nuclear Regulatory Commission (NRC) regarding the review of the Vogtle Electric Generating Plant, Units 3 and 4 plant design under Docket Nos. 052-000025 and 052-000026

Document Date	Accession Number	Title & Estimated Page Count	Document Type	Author Affiliation	Addressee Affiliation	Docket Number
		Revision to Final Safety Analysis Report Chapters 1, 2, & 3. 6 Page(s)			NRC/NRO	
07/01/2010	ML101830393	Vogtle, Units 3 and 4, Combined License Application - Voluntary Revision to Final Safety Analysis Report Chapter 2. 9 Page(s)	Letter	Southern Nuclear Operating Co, Inc	NRC/Document Control Desk NRC/NRO	05200025 05200026
07/02/2010	ML101870659	Vogtle, Units 3 & 4, Combined License Application, Voluntary Revision to Final Safety Analysis Report Chapter 6. 7 Page(s)	Letter	Southern Nuclear Operating Co, Inc	NRC/Document Control Desk NRC/NRO	05200025 05200026
07/02/2010	ML101870660	Vogtle, Units 3 & 4 - Combined License Application Response to Request for Additional Information Letter No. 057. 8 Page(s)	Letter	Southern Nuclear Operating Co, Inc	NRC/Document Control Desk NRC/NRO	05200025 05200026
07/06/2010	ML100890413	Memo – Vogtle Electric Generating Plant Units 3 and 4 Advanced Final Safety Evaluation Report with No Open Items for Chapter 17, "Quality Assurance." 3 Page(s)	Memoranda	NRC/NRO/DNRL	NRC/ACRS	05200025 05200026
07/06/2010	ML100890449	Vogtle, Units 3 and 4, Advanced Final Safety Evaluation Report with No Open Items for Chapter 17, "Quality Assurance." 6 Page(s)	Letter	NRC/NRO/DNRL/NWE1	Southern Nuclear Operating Co, Inc	05200025 05200026

Appendix B

Chronology of the Combined License Application for Vogtle Electric Generating Plant Units 3 and 4

This appendix contains a chronological listing of routine licensing correspondence between the staff of the U.S. Nuclear Regulatory Commission (NRC) regarding the review of the Vogtle Electric Generating Plant, Units 3 and 4 plant design under Docket Nos. 052-000025 and 052-000026

Document Date	Accession Number	Title & Estimated Page Count	Document Type	Author Affiliation	Addressee Affiliation	Docket Number
07/06/2010	ML100950499	Memo - Vogtle, Units 3 and 4 - Advanced Final Safety Evaluation Report with No Open Items for Chapter 2, "Site Characteristics". 3 Page(s)	Memoranda	NRC/NRO/DNRL	NRC/ACRS	05200025 05200026
07/06/2010	ML100950519	Letter re: Vogtle Electric Generating Plant Units 3 and 4 Advanced Final Safety Evaluation Report with No Open Items for Chapter 2, "Sites Characteristics." 6 Page(s)	Letter	NRC/NRO/DNRL/NWE1	Southern Nuclear Operating Co, Inc	05200025 05200026
07/06/2010	ML101890107	Vogtle, Units 3 and 4 Combined License Application, Additional Information for Final Safety Analysis Report Chapter 07 Supplement 1. 11 Page(s)	Letter	Southern Nuclear Operating Co, Inc	NRC/Document Control Desk NRC/NRO	05200025 05200026
07/08/2010	ML101890191	2010/07/08 Vogtle RAI for SER - 6 Page(s)	E-Mail	NRC/NRO	NRC/NRO/DNRL/NWE1	05200025 05200026
07/08/2010	ML101890430	2010/07/08 Vogtle RAI for SER - 6 Page(s)	E-Mail	NRC/NRO	NRC/NRO/DNRL/NWE1	05200025 05200026
07/08/2010	ML101890431	2010/07/08 Vogtle RAI for SER - email capture 2 Page(s)	E-Mail	NRC/NRO	NRC/NRO/DNRL/NWE1	05200025 05200026
07/08/2010	ML101890433	2010/07/08 Vogtle RAI for SER - RAI Letter No. 059 Related to SRP Section 5.2.5 for the Vogtle Electric Generating Plant Units 3 and 4 Combined License Application	E-Mail	NRC/NRO	NRC/NRO/DNRL/NWE1	05200025 05200026

Appendix B

Chronology of the Combined License Application for Vogtle Electric Generating Plant Units 3 and 4

This appendix contains a chronological listing of routine licensing correspondence between the staff of the U.S. Nuclear Regulatory Commission (NRC) regarding the review of the Vogtle Electric Generating Plant, Units 3 and 4 plant design under Docket Nos. 052-000025 and 052-000026

Document Date	Accession Number	Title & Estimated Page Count	Document Type	Author Affiliation	Addressee Affiliation	Docket Number
		6 Page(s)				
07/08/2010	ML101890436	2010/07/08 Vogtle RAI for SER - RAI Letter No. 060 Related to SRP Section 5.2.5 for the Vogtle Electric Generating Plant Units 3 and 4 COLA 6 Page(s)	E-Mail	NRC/NRO	NRC/NRO/DNRL/NWE1	05200025 05200026
07/08/2010	ML101890438	2010/07/08 Vogtle RAI for SER - RAI Letter No. 060 Related to SRP Section 5.2.5 for the Vogtle Electric Generating Plant Units 3 and 4 COLA 6 Page(s)	E-Mail	NRC/NRO	NRC/NRO/DNRL/NWE1	05200025 05200026
07/09/2010	ML101940025	Vogtle, Units 3 & 4, Combined License Application Response to Bellefonte Safety Evaluation Report Open Items for Chapter 01. 12 Page(s)	Letter	Southern Nuclear Operating Co, Inc	NRC/Document Control Desk NRC/NRO	05200025 05200026
07/09/2010	ML101940029	Vogtle, Units 3 & 4, Combined License Application Voluntary Revision to Application Part 7 Involving Departure Reports. 7 Page(s)	Letter	Southern Nuclear Operating Co, Inc	NRC/Document Control Desk NRC/NRO	05200025 05200026
07/12/2010	ML101940268	E-mail from SHPO June 18 2010 Vogtle Units 3 and 4 Onsite Backfill Activities. 1 Page(s)	E-Mail	State of GA, Dept of Natural Resources	NRC/NRO/DSER/RAP1	05200025 05200026
07/13/2010	ML092650063	VEGP SER –Vogtle Electric Generating Plant Units 3 and 4	NRO Safety Evaluation	NRC/NRO		05200025 05200026

Appendix B

Chronology of the Combined License Application for Vogtle Electric Generating Plant Units 3 and 4

This appendix contains a chronological listing of routine licensing correspondence between the staff of the U.S. Nuclear Regulatory Commission (NRC) regarding the review of the Vogtle Electric Generating Plant, Units 3 and 4 plant design under Docket Nos. 052-000025 and 052-000026

Document Date	Accession Number	Title & Estimated Page Count	Document Type	Author Affiliation	Addressee Affiliation	Docket Number
07/13/2010		Advanced Final Safety Evaluation Report With No Open Items for Chapter 17, "Quality Assurance". 37 Page(s)				
07/13/2010	ML100430348	Vogtle Electric Generating Plant, Units 3 and 4, SER Chapter 2 Tables and Figures 1 (2.4-1 – 3). 5 Page(s)	Drawing Graphics incl Charts and Table	NRC/NRO		05200025 05200026
07/13/2010	ML100430350	Vogtle Electric Generating Plant, SER Chapter 2 Figures 2 (2.4-4 – 2.4-7) and Tables. 4 Page(s)	Drawing Graphics incl Charts and Table	NRC/NRO		05200025 05200026
07/14/2010	ML101890896	07/09/2010 - Notice of Forthcoming Closed Teleconference with Southern Nuclear Operating Company on Vogtle Loss of Large Areas of Plant Due to Explosions of Fire, COLA Submittal. 6 Page(s)	Meeting Agenda Meeting Notice Memoranda	NRC/NRO/DNRL/NWE1	NRC/NRO/DNRL/NWE1	05200025 05200026
07/14/2010	ML101950311	07/14/2010 - Notice of Forthcoming Closed Teleconference with the Southern Nuclear Operating Company on the Vogtle Loss of Large Areas Due to Explosions or Fire, COLA Submittal. 6 Page(s)	Meeting Agenda Meeting Notice	NRC/NRO/DNRL/NWE1	NRC/NRO/DNRL/NWE1	05200025 05200026
07/15/2010	ML101960367	Final Order 2010 IRP DNs 31081 Vogtle Units 3 and 4. 5 Page(s)	Legal-Correspondence Report,	State of GA, Public Service Commission	NRC/NRO	05200024 05200025

Appendix B

Chronology of the Combined License Application for Vogtle Electric Generating Plant Units 3 and 4

This appendix contains a chronological listing of routine licensing correspondence between the staff of the U.S. Nuclear Regulatory Commission (NRC) regarding the review of the Vogtle Electric Generating Plant, Units 3 and 4 plant design under Docket Nos. 052-000025 and 052-000026

Document Date	Accession Number	Title & Estimated Page Count	Document Type	Author Affiliation	Addressee Affiliation	Docket Number
07/16/2010	ML102010031	Vogtle, Units 3 and 4 Combined License Application, New and Significant Information Evaluation for the Transportation of Backfill from an Additional Offsite Source. 9 Page(s)	Miscellaneous Letter	Southern Nuclear Operating Co, Inc	NRC/Document Control Desk NRC/NRO	05200025 05200026
07/19/2010	ML102000604	2010/07/19 Vogtle COL Review - Draft RAI 4856 - 6.4 Control Room Habitability System 4 Page(s)	E-Mail	NRC/NRO	NRC/NRO/DNRL/NWE1	05200025 05200026
07/23/2010	ML102040703	2010/07/23 Vogtle RAI for SER – Request for Additional Information Letter No. 061 Related to SRP Section 6.4 for the Vogtle Electric Generating Plant, Units 3 and 4 Combined License Application 6 Page(s)	E-Mail	NRC/NRO	NRC/NRO/DNRL/NWE1	05200025 05200026
07/27/2010	ML102100204	Vogtle, Units 3 & 4 and Combined License Application Voluntary Revisions to Part 2 FSAR and Part 5 Emergency Plan. 7 Page(s)	Emergency Preparedness- Emergency Plan Final Safety Analysis Report (FSAR) Letter	Southern Nuclear Operating Co, Inc	NRC/Document Control Desk NRC/NRO	05200025 05200026
07/29/2010	ML102140339	Vogtle Electric Generating Plant Units 3 and 4 Combined License Application, CFR 50.46 Thirty Day Report.	Letter	Southern Nuclear Operating Co, Inc	NRC/Document Control Desk NRC/NRO	05200025 05200026

Appendix B

Chronology of the Combined License Application for Vogtle Electric Generating Plant Units 3 and 4

This appendix contains a chronological listing of routine licensing correspondence between the staff of the U.S. Nuclear Regulatory Commission (NRC) regarding the review of the Vogtle Electric Generating Plant, Units 3 and 4 plant design under Docket Nos. 052-000025 and 052-000026

Document Date	Accession Number	Title & Estimated Page Count	Document Type	Author Affiliation	Addressee Affiliation	Docket Number
07/30/2010	ML102150196	Vogtle Electric Generating Plant Units 3 and 4 Combined License Application Voluntary Revision to Final Safety Analysis Report Chapter 6. 6 Page(s)	Final Safety Analysis Report (FSAR) Letter 4 Page(s)	Southern Nuclear Operating Co, Inc	NRC/Document Control Desk NRC/NRO	05200025 05200026
07/30/2010	ML102160289	Vogtle, Units 3 and 4 Combined License Application, Submittal of Proposed Physical Security Plan, Revision 2. 4 Page(s)	Letter	Southern Nuclear Operating Co, Inc	NRC/Document Control Desk NRC/NRO	05200025 05200026
07/31/2009	ML101940362	Neel-Schaffer Traffic Study Vogtle Units 3 and 4. 76 Page(s)	Report, Miscellaneous	Neel-Schaffer, Inc	NRC/NRO Southern Co	05200025 05200026
08/03/2010	ML102100311	Revision of the Environmental Review Schedule for the Combined License Application Review for Vogtle Electric Generating Plant Units 3 and 4. 3 Page(s)	Letter	NRC/NRO/DSER/RAP1	Southern Nuclear Operating Co, Inc	05200025 05200026
08/03/2010	ML102230424	Edwin I. Hatch, Joseph M. Farley, Vogtle Units 1, 2, 3, & 4, Response to Regulatory Issue Summary 2010-08, Preparation and Scheduling of Operator License Examinations. 12 Page(s)	Letter License-Operator License Exam, Draft	Southern Nuclear Operating Co, Inc	NRC/Document Control Desk NRC/NRR	05000321 05000348 05000364 05000366 05000424 05000425 05200025 05200026

Appendix B

Chronology of the Combined License Application for Vogtle Electric Generating Plant Units 3 and 4

This appendix contains a chronological listing of routine licensing correspondence between the staff of the U.S. Nuclear Regulatory Commission (NRC) regarding the review of the Vogtle Electric Generating Plant, Units 3 and 4 plant design under Docket Nos. 052-000025 and 052-000026

Document Date	Accession Number	Title & Estimated Page Count	Document Type	Author Affiliation	Addressee Affiliation	Docket Number
08/04/2010	ML102110264	07/29/2010 - Notice of Forthcoming Closed Meeting with the Southern Nuclear Operating Company on the Vogtle Loss of Large Areas of the Plant Due to Explosions or Fire, COLA Submittal. 10 Page(s)	Meeting Agenda Meeting Notice	NRC/NRO/DNRL/NWE1	NRC/NRO/DNRL/NWE1	05200025 05200026
08/05/2010	ML102210129	Vogtle, Units 3 and 4 - Combined License Application, Response to Request for Additional Information Letter No. 060. 9 Page(s)	Letter	Southern Nuclear Operating Co, Inc	NRC/Document Control Desk NRC/NRO	05200025 05200026
08/06/2010	ML102210342	Vogtle, Units 3 and 4, Combined License Application, Supplemental Response to Request for Additional Information Letter No. 057 Involving Pressurizer Surge Line Monitoring. 9 Page(s)	Letter	Southern Nuclear Operating Co, Inc	NRC/Document Control Desk NRC/NRO	05200025 05200026
08/06/2010	ML102210343	Vogtle, Units 3 and 4 - Combined License Application, Response to Bellefonte Safety Evaluation Report Open Items for Chapter 15. 11 Page(s)	Letter	Southern Nuclear Operating Co, Inc	NRC/Document Control Desk NRC/NRO	05200025 05200026
08/06/2010	ML102220520	Vogtle, Units 3 and 4 Combined License Application - Submittal 6 of the Application. 8 Page(s)	Letter	Southern Nuclear Operating Co, Inc	NRC/Document Control Desk NRC/NRO	05200025 05200026
08/10/2010	ML102220217	08/11/2010 Notice of Forthcoming Closed Meeting With the Southern	Meeting Agenda Meeting Notice	NRC/NRO/DNRL/NWE1	NRC/NRO/DNRL/NWE1	05200025 05200026

Appendix B

Chronology of the Combined License Application for Vogtle Electric Generating Plant Units 3 and 4

This appendix contains a chronological listing of routine licensing correspondence between the staff of the U.S. Nuclear Regulatory Commission (NRC) regarding the review of the Vogtle Electric Generating Plant, Units 3 and 4 plant design under Docket Nos. 052-000025 and 052-000026

Document Date	Accession Number	Title & Estimated Page Count	Document Type	Author Affiliation	Addressee Affiliation	Docket Number
		Nuclear Operating Company on the Vogtle Loss of Large Areas of the Plant Due to Explosions or Fire, COLA Submittal. 7 Page(s)	Memoranda			
08/13/2010	ML102250357	2010/08/13–Declaration of Arnold Gundersen Supporting Blue Ridge Environmental Defense League's New Contention Regarding AP1000 Containment Integrity on the Vogtle Nuclear Power Plant, Units 3 and 4, Exhibit 1. 10 Page(s)	Legal-Exhibit	Blue Ridge Environmental Defense League	NRC/ASLBP	05200025 05200026
08/13/2010	ML102290036	Vogtle, Units 3 and 4 – Combined License Application Voluntary Revision to Final Safety Analysis Report Chapter 6. 6 Page(s)	Letter	Southern Nuclear Operating Co, Inc	NRC/Document Control Desk NRC/NRO	05200025 05200026
08/13/2010	ML102290038	Vogtle, Units 3 and 4, Combined License Application Response to Request for Additional Information Letter No. 042, Supplement 4 Loss of Large Areas of the Plant Due to Explosions or Fire – Mitigative Strategies Description. 4 Page(s)	Letter	Southern Nuclear Operating Co, Inc	NRC/Document Control Desk NRC/NRO	05200025 05200026
08/13/2010	ML102290200	Vogtle, Units 3 and 4, Combined License Application Submittal No. 6 Roadmap.	Letter	Southern Nuclear Operating Co, Inc	NRC/Document Control Desk NRC/NRO	05200025 05200026

Appendix B

Chronology of the Combined License Application for Vogtle Electric Generating Plant Units 3 and 4

This appendix contains a chronological listing of routine licensing correspondence between the staff of the U.S. Nuclear Regulatory Commission (NRC) regarding the review of the Vogtle Electric Generating Plant, Units 3 and 4 plant design under Docket Nos. 052-000025 and 052-000026

Document Date	Accession Number	Title & Estimated Page Count	Document Type	Author Affiliation	Addressee Affiliation	Docket Number
08/17/2010	ML102310236	Vogtle, Units 3 and 4 Combined License Application, Voluntary Revision to Final Safety Analysis Report. Chapter 3. 57 Page(s) 7 Page(s)	Letter	Southern Nuclear Operating Co, Inc	NRC/Document Control Desk NRC/NRO	05200025 05200026
08/18/2010	ML102300108	2010/08/18-Notice of Appearance for Stephanie N. Liaw on Behalf of the U.S. Nuclear Regulatory Commission Regarding Vogtle, Units 3 and 4. 3 Page(s)	Legal-Pleading	NRC/OGC	NRC/ASLBP	05200025 05200026
08/23/2010	ML102370782	Vogtle, Units 3 and 4, Combined License Application, Summary Identification of Reference Combined License Application Standard Content Submittals. 10 Page(s)	Letter	Southern Nuclear Operating Co, Inc	NRC/Document Control Desk NRC/NRO	05200025 05200026
08/25/2010	MI093000107	VEGP SER – Vogtle Electric Generating Plant Units 3 and 4 Advanced Final Safety Evaluation Report With No Open Items for Chapter 18 "Human Factors Engineering." 27 Page(s)	NRO Safety Evalution Report (SER)-Delayed	NRC/NRO/DNRL/NWE1		05200025 05200026

Chronology of the Combined License Application for Vogtle Electric Generating Plant Units 3 and 4

This appendix contains a chronological listing of routine licensing correspondence between the staff of the U.S. Nuclear Regulatory Commission (NRC) regarding the review of the Vogtle Electric Generating Plant, Units 3 and 4 plant design under Docket Nos. 052-000025 and 052-000026

Document Date	Accession Number	Title & Estimated Page Count	Document Type	Author Affiliation	Addressee Affiliation	Docket Number
08/25/2010	ML100910026	Ltr. –CH18 – Vogtle Electric Generating Plant units 3 and 4 Advanced Final Safety Evaulation Report with No Open Items for Chapter 18 "Human Factors Engineering." 6 Page(s)	Letter	NRC/NRO/DNRL/NWE1	Soutehr Nuclear Operating Co, Inc	05200025 05200026
08/25/2010	Ml100910031	CH18 – Vogtle Electric Generating Plant Units 3 and 4 Advanced Final Safety Evaulation Report with No Open Items for Chapte r18 "Human Factors Engineering." 3 Page(s)	Memoranda	NRC/NRO/DNRL	NRC/ACRS	05200025 05200026
08/26/2010	ML102070018	Draft Supplemental Environmental Impact Statement for the Combined License for Vogtle Electric Generating Plant Units 3 and 4. 13 Page(s)	Letter	NRC/NRO/DSER	US Environmental Protection Agency (EPA)	05200025 05200026
08/26/2010	ML102080062	Letter: Notice of Availability of the Draft Supplemental Environmental Impact Statement for the Combined License for Vogtle Electric Generating Plant Units 3 and 4. 6 Page(s)	Letter	NRC/NRO/DSER/RAP1	Southern Nuclear Operating Co, Inc	05200025 05200026
08/27/2010	ML093230696	SER Chapter 7 - Vogtle Electric Generating Plant Units 3 and 4 Advanced Final Safety Evaluation Report with No Open Items for Chapter 7, "Instrumentation and	NRO Safety Evaluation Report (SER)–Delayed Safety Evaluation Report	NRC/NRO/DNRL/NWE1		05200025 05200026

Appendix B

Chronology of the Combined License Application for Vogtle Electric Generating Plant Units 3 and 4

This appendix contains a chronological listing of routine licensing correspondence between the staff of the U.S. Nuclear Regulatory Commission (NRC) regarding the review of the Vogtle Electric Generating Plant, Units 3 and 4 plant design under Docket Nos. 052-000025 and 052-000026

Document Date	Accession Number	Title & Estimated Page Count	Document Type	Author Affiliation	Addressee Affiliation	Docket Number
08/27/2010	ML100350507	Vogtle Electric Generating Plant Units 3 and 4, Advanced Final Safety Evaluation Report With No Open Items for Chapter 7, "Instrumentation and Controls" (Letter). 8 Page(s)	Letter	NRC/NRO/DNRL/NWE1	Southern Nuclear Operating Co, Inc	05200025 05200026
08/27/2010	ML102070055	Notice of Availability of the Draft Supplemental Environmental Impact Statement for the Combined License for the Vogtle Electric Generating Plant Units 3 and 4. 2 Page(s)	Memoranda	NRC/NRO/DSER/RAP1	NRC/ADM/DAS/RDEB	05200025 05200026
08/27/2010	ML102080417	FRN: Southern Nuclear Operating Company Notice of Availability of the Draft Supplemental Environmental Impact Statement for Combined Licenses (COLs) for Vogtle Electric Generating Plant Units 3 and 4. 4 Page(s)	Federal Register Notice	NRC/NRO/DSER/RAP1		05200025 05200026
08/27/2010	ML102170028	Maintenance of Reference Materials at the Burke County Library for the Vogtle Electric Generating Plant, Units 3 and 4 COL Application. 7 Page(s)	Letter	NRC/NRO/DSER/RAP1	Burke County, GA	05200025 05200026

Appendix B

Chronology of the Combined License Application for Vogtle Electric Generating Plant Units 3 and 4

This appendix contains a chronological listing of routine licensing correspondence between the staff of the U.S. Nuclear Regulatory Commission (NRC) regarding the review of the Vogtle Electric Generating Plant, Units 3 and 4 plant design under Docket Nos. 052-000025 and 052-000026

Document Date	Accession Number	Title & Estimated Page Count	Document Type	Author Affiliation	Addressee Affiliation	Docket Number
08/27/2010	ML102430184	Vogtle, Units 3 and 4 - Combined License Application, Voluntary Revision to Final Safety Analysis Report Chapter 5. 6 Page(s)	Letter	Southern Nuclear Operating Co, Inc	NRC/Document Control Desk NRC/NRO	05200025 05200026
08/30/2010	ML100360236	Vogtle Electric Generating Plant Units 3 And 4 Advanced Final Safety Evaluation Report With No Open Items For Chapter 7, "Instrumentation And Controls." 3 Page(s)	Memoranda	NRC/NRO/DNRL	NRC/ACRS	05200025 05200026
08/30/2010	ML102420352	09/02/2010 Cancelled Notice of Public Meeting with Southern Nuclear Operating Company to Discuss Exemption Request from 10 CFR 52.39(e) for Non-Safety Early Site Permit Site Safety Analysis Report for Vogtle, Units 3 and 4. 11 Page(s)	Meeting Agenda Meeting Notice Memoranda	NRC/NRO/DNRL/NWE1	NRC/NRO/DNRL/NWE1	05200011 05200025 05200026
08/31/2010	ML102430229	2010/08/31 Vogtle COL Review - Telcon Summary - Telcon with Vogtle 8/31/10 2 Page(s)	E-Mail	NRC/NRO	NRC/NRO/DNRL/NWE1	05200025 05200026
09/02/2010	ML102000149	Letter to Mr. Tullis: Section 106 Consultation And Notification Of The Issuance And Request For Comments On The Draft Supplemental Environmental	Letter	NRC/NRO/DSER/RAP1	Poarch Band of Creek Nation	05200025 05200026

Appendix B

Chronology of the Combined License Application for Vogtle Electric Generating Plant Units 3 and 4

This appendix contains a chronological listing of routine licensing correspondence between the staff of the U.S. Nuclear Regulatory Commission (NRC) regarding the review of the Vogtle Electric Generating Plant, Units 3 and 4 plant design under Docket Nos. 052-000025 and 052-000026

Document Date	Accession Number	Title & Estimated Page Count	Document Type	Author Affiliation	Addressee Affiliation	Docket Number
		Impact Statement For The Vogtle Electric Generating Plant, Units 3 And 4 Combined License Application. 8 Page(s)				
09/02/2010	ML102000191	Letter to Ms. Holland: Section 106 Consultation and Notification of the Issuance and Request for Comments on the Draft Supplemental Environmental Impact Statement for the Vogtle Electric Generating Plant Units 3 and 4 Combined License Application. 7 Page(s)	Letter	NRC/NRO/DSER/RAP1	United Keetoowah Band of Cherokee Indians	05200025 05200026
09/02/2010	ML102000210	Letter to Ms. McCoy: Section 106 Consultation and Notification of the Issuance and Request for Comments on the Draft Supplemental Environmental Impact Statement for the Vogtle Electric Generation Plant Units 3 and 4 Combined License Application. 7 Page(s)	Letter	NRC/NRO/DSER/RAP1	Eastern Band of Cherokee Indians	05200025 05200026

Appendix B

Chronology of the Combined License Application for Vogtle Electric Generating Plant Units 3 and 4

This appendix contains a chronological listing of routine licensing correspondence between the staff of the U.S. Nuclear Regulatory Commission (NRC) regarding the review of the Vogtle Electric Generating Plant, Units 3 and 4 plant design under Docket Nos. 052-000025 and 052-000026

Document Date	Accession Number	Title & Estimated Page Count	Document Type	Author Affiliation	Addressee Affiliation	Docket Number
09/02/2010	ML102000219	Letter to Mr. Zachary: Section 106 Consultation and Notification of the Issuance and Request for Comments on the Draft Supplemental Environmental Impact Statement for the Vogtle Electric Generating Plant Units 3 and 4 Combined License Application. 8 Page(s)	Letter	NRC/NRO/DSER/RAP1	Coushatta Tribe of Louisiana	05200025 05200026
09/02/2010	ML102000224	Letter to Ms. Bucktrot: Section 106 Consultation and Notification of the Issuance and Request for Comments on the Draft Supplemental Environmental Impact Statement for the Vogtle Electric Generating Plant Units 3 and 4 Combined License Application. 7 Page(s)	Letter	NRC/NRO/DSER/RAP1	Kialegee Tribal Town	05200025 05200026
09/02/2010	ML102000228	Letter to Mr. Terry: Section 106 Consultation and Notification of the Issuance and Request of Comments on the Draft Supplemental Environmental Impact Statement for the Vogtle Electric Generating Plant Units 3 and 4 Combined License Application.	Letter	NRC/NRO/DSER/RAP1	Miccosukee Tribe of Indians of Florida	05200025 05200026

Appendix B

Chronology of the Combined License Application for Vogtle Electric Electric Generating Plant Units 3 and 4

This appendix contains a chronological listing of routine licensing correspondence between the staff of the U.S. Nuclear Regulatory Commission (NRC) regarding the review of the Vogtle Electric Generating Plant, Units 3 and 4 plant design under Docket Nos. 052-000025 and 052-000026

Document Date	Accession Number	Title & Estimated Page Count	Document Type	Author Affiliation	Addressee Affiliation	Docket Number
09/02/2010	ML102000233	Letter to Ms. Thrower: Section 106 Consultation and Notification of the Issuance and Request for Comments of the Draft Supplemental Environmental Impact Statement for the Vogtle Electric Generating Plant Units 3 and 4 Combined License Application. 3 Page(s) 7 Page(s)	Letter	NRC/NRO/DSER/RAP1	Poarch Band of Creek Nation	05200025 05200026
09/02/2010	ML102000240	Letter to Mr. McGertt: Section 106 Consultation and Notification of the Issuance and Request for Comments on the Draft Supplemental Environmental Impact Statement for the Vogtle Electric Generating Plant Units 3 and 4 Combined License Application. 8 Page(s)	Letter	NRC/NRO/DSER/RAP1	Thlopthlocco Tribal Town	05200025 05200026
09/02/2010	ML102000264	Letter to Chief Ellis: Section 106 Consultation and Notification of the Issuance and Request for Comments on the Draft Supplemental Environmental Impact Statement for the Vogtle Electric Generating Plant Units 3 and 4 Combined License	Letter	NRC/NRO/DSER/RAP1	Muscogee (Creek) Nation	05200025 05200026

Appendix B

Chronology of the Combined License Application for Vogtle Electric Generating Plant Units 3 and 4

This appendix contains a chronological listing of routine licensing correspondence between the staff of the U.S. Nuclear Regulatory Commission (NRC) regarding the review of the Vogtle Electric Generating Plant, Units 3 and 4 plant design under Docket Nos. 052-000025 and 052-000026

Document Date	Accession Number	Title & Estimated Page Count	Document Type	Author Affiliation	Addressee Affiliation	Docket Number
		Application. 8 Page(s)				
09/02/2010	ML102000287	Letter to Mr. Allen: Section 106 Consultation and Notification of the Issuance and Request for Comments on the Draft Supplemental Environmental Impact Statement for the Vogtle Electric Generating Plant Units 3 and 4 Combined License Application. 7 Page(s)	Letter	NRC/NRO/DSER/RAP1	Cherokee Nation of Oklahoma	05200025 05200026
09/02/2010	ML102000331	Letter to Ms. Haii: Section 106 Consultation and Notification of the Issuance and Request for Comments on the Draft Supplemental Environmental Impact Statement for the Vogtle Electric Generating Plant Units 3 and 4 Combined License Application. 7 Page(s)	Letter	NRC/NRO/DSER/RAP1	Chickasaw Nation	05200025 05200026

Appendix B

Chronology of the Combined License Application for Vogtle Electric Generating Plant Units 3 and 4

This appendix contains a chronological listing of routine licensing correspondence between the staff of the U.S. Nuclear Regulatory Commission (NRC) regarding the review of the Vogtle Electric Generating Plant, Units 3 and 4 plant design under Docket Nos. 052-000025 and 052-000026

Document Date	Accession Number	Title & Estimated Page Count	Document Type	Author Affiliation	Addressee Affiliation	Docket Number
09/02/2010	ML102000335	Letter to Governor Anoatubby: Section 106 Consultation and Notification of the Issuance and Request for Comments on the Draft Supplemental Environmental Impact Statement for the Vogtle Electric Generating Plant Units 3 and 4 Combined License Application. 7 Page(s)	Letter	NRC/NRO/DSER/RAP1	Chickasaw Nation of Oklahoma	05200025 05200026
09/02/2010	ML102000345	Letter to Mr. Thurmond: Section 106 Consultation and Notification of the Issuance and Request for Comments on the Draft Supplemental Environmental Impact Statement for the Vogtle Electric Generating Plant Units 3 and 4 Combined License Application. 8 Page(s)	Letter	NRC/NRO/DSER/RAP1	Georgia Tribe of Eastern Cherokee	05200025 05200026
09/02/2010	ML102000349	Letter to Mr. Yargee: Section 106 Consultation and Notification of the Issuance and Request for Comments on the Draft Supplemental Environmental Impact Statement for the Vogtle Electric Generating Plant Units 3 and 4 Combined License Application.	Letter	NRC/NRO/DSER/RAP1	Alabama-Quassarte Tribal Town	05200025 05200026

Appendix B

Chronology of the Combined License Application for Vogtle Electric Generating Plant Units 3 and 4

This appendix contains a chronological listing of routine licensing correspondence between the staff of the U.S. Nuclear Regulatory Commission (NRC) regarding the review of the Vogtle Electric Generating Plant, Units 3 and 4 plant design under Docket Nos. 052-000025 and 052-000026

Document Date	Accession Number	Title & Estimated Page Count	Document Type	Author Affiliation	Addressee Affiliation	Docket Number
09/02/2010	ML102000355	Letter to Mr. Bowlegs: Section 106 Consultation and Notification of the Issuance and Request for Comments on the Draft Supplemental Environmental Impact Statement for the Vogtle Electric Generating Plant Units 3 and 4 Combined License Application. 7 Page(s)	Letter	NRC/NRO/DSER/RAP1	Seminole Nation of Oklahoma	05200025 05200026
09/02/2010	ML102000358	Letter to Chief Hicks: Section 106 Consultation and Notification of the Issuance and Request for Comments on the Draft Supplemental Environmental Impact Statement for the Vogtle Electric Generating Plant Units 3 and 4 Combined License Application. 8 Page(s)	Letter	NRC/NRO/DSER/RAP1	Eastern Band of Cherokee Indians	05200025 05200026
09/02/2010	ML102000360	Letter to Chief Proctor: Section 106 Consultation and Notification of the Issuance and Request for Comments on the Draft Supplemental Environmental Impact Statement for the Vogtle Electric Generating Plant Units 3 and 4 Combined License	Letter	NRC/NRO/DSER/RAP1	United Keetoowah Band of Cherokee Indians	05200025 05200026

Appendix B

Chronology of the Combined License Application for Vogtle Electric Generating Plant Units 3 and 4

This appendix contains a chronological listing of routine licensing correspondence between the staff of the U.S. Nuclear Regulatory Commission (NRC) regarding the review of the Vogtle Electric Generating Plant, Units 3 and 4 plant design under Docket Nos. 052-000025 and 052-000026

Document Date	Accession Number	Title & Estimated Page Count	Document Type	Author Affiliation	Addressee Affiliation	Docket Number
		Application. 8 Page(s)				
09/02/2010	ML102000365	Letter to Ms. Kaniatobe: Section 106 Consultation and Notification of the Issuance and Request for Comments on the Draft Supplemental Environmental Impact Statement for the Vogtle Electric Generating Plant, Units 2 and 4 Combined License Application. 8 Page(s)	Letter	NRC/NRO/DSER/RAP1	Absentee-Shawnee Tribe of Oklahoma	05200025 05200026
09/02/2010	ML102000367	Letter to Ms. Thomas: Section 106 Consultation and Notification of the Issuance and Request for Comments on the Draft Supplemental Environmental Impact Statement for the Vogtle Electric Generating Plant Units 3 and 4 Combined License Application. 8 Page(s)	Letter	NRC/NRO/DSER/RAP1	Alabama-Coushatta Tribe of Texas	05200025 05200026

Appendix B

Chronology of the Combined License Application for Vogtle Electric Generating Plant Units 3 and 4

This appendix contains a chronological listing of routine licensing correspondence between the staff of the U.S. Nuclear Regulatory Commission (NRC) regarding the review of the Vogtle Electric Generating Plant, Units 3 and 4 plant design under Docket Nos. 052-000025 and 052-000026

Document Date	Accession Number	Title & Estimated Page Count	Document Type	Author Affiliation	Addressee Affiliation	Docket Number
09/02/2010	ML102000368	Letter to Mrs. Bear: Section 106 Consultation and Notification of the Issuance and Request for Comments on the Draft Supplemental Environmental Impact Statement for the Vogtle Electric Generating Plant Units 3 and 4 Combined License Application. 8 Page(s)	Letter	NRC/NRO/DSER/RAP1	Muscogee (Creek) Nation of Oklahoma	05200025 05200026
09/02/2010	ML102000375	Letter to Chief Smith: Section 106 Consultation and Notification of the Issuance and Request for Comments on the Draft Supplemental Environmental Impact Statement for the Vogtle Electric Generating Plant Units 3 and 4 Combined License Application. 8 Page(s)	Letter	NRC/NRO/DSER/RAP1	Cherokee Nation of Oklahoma	05200025 05200026
09/02/2010	ML102000382	Letter to Mr. Steele: Section 106 Consultation and Notification of the Issuance and Request for Comments on the Draft Supplemental Environmental Impact Statement for the Vogtle Electric Generating Plant units 3 and 4 Combined License Application.	Letter	NRC/NRO/DSER/RAP1	Seminole Tribe of Florida	05200025 05200026

Appendix B

Chronology of the Combined License Application for Vogtle Electric Generating Plant Units 3 and 4

This appendix contains a chronological listing of routine licensing correspondence between the staff of the U.S. Nuclear Regulatory Commission (NRC) regarding the review of the Vogtle Electric Generating Plant, Units 3 and 4 plant design under Docket Nos. 052-000025 and 052-000026

Document Date	Accession Number	Title & Estimated Page Count	Document Type	Author Affiliation	Addressee Affiliation	Docket Number
09/02/2010	ML102000384	Letter to Mr. Carleton: Section 106 Consultation and Notification of the Issuance and Request for Comments on the Draft Supplemental Environmental Impact Statement for the Vogtle Electric Generating Plant Units 3 and 4 Combined License Application. 8 Page(s)	Letter	NRC/NRO/DSER/RAP1	Mississippi Band of Choctaw Indians	05200025 05200026
09/02/2010	ML102000390	Letter to Ms. Rolin: Section 106 Consultation and Notification of the Issuance and Request for Comments on the Draft Supplemental Environmental Impact Statement for the Vogtle Electric Generating Plant Units 3 and 4 Combined License Application. 7 Page(s)	Letter	NRC/NRO/DSER/RAP1	Poarch Band of Creek Nation	05200025 05200026
09/02/2010	ML102320187	Letter to Ms. Bernstein USACE: Notification of the Issuance and Request for Comments on the Draft Supplemental Environmental Impact Statement for the Vogtle Electric Generating Plant, Units 3 and 4. 2 Page(s)	Letter	NRC/NRO/DSER/RAP1	US Dept of the Army, Corps of Engineers, Savannah District	05200025 05200026

Appendix B

Chronology of the Combined License Application for Vogtle Electric Generating Plant Units 3 and 4

This appendix contains a chronological listing of routine licensing correspondence between the staff of the U.S. Nuclear Regulatory Commission (NRC) regarding the review of the Vogtle Electric Generating Plant, Units 3 and 4 plant design under Docket Nos. 052-000025 and 052-000026

Document Date	Accession Number	Title & Estimated Page Count	Document Type	Author Affiliation	Addressee Affiliation	Docket Number
09/03/2010	ML102320162	Letter to Bernhart NMFS: Issuance of the Draft Environmental Impact Statement for the Vogtle Electric Generating Plant, Units 3 and 4 Combined License Application. 7 Page(s)	Letter	NRC/NRO/DSER/RAP1	State of FL, National Marine Fisheries Services	05200025 05200026
09/03/2010	ML102320174	Letter to Mr. Perry SCDNR: Notification of the Issuance and Request for Comments on the Draft Supplemental Environmental Impact Statement for the Vogtle Electric Generating Plant, Units 3 and 4 Combined License Application Review. 8 Page(s)	Letter	NRC/NRO/DSER/RAP1	State of SC, Dept of Natural Resources	05200025 05200026
09/03/2010	ML102320222	Letter to Ms. Tucker FWS: Notification of the Issuance and Request for Comments on the Draft Supplemental Environmental Impact Statement for the Vogtle Generating Plant, Units 3 and 4. 7 Page(s)	Letter	NRC/NRO/DSER	US Dept of Interior, Fish & Wildlife Service	05200025 05200026

Appendix B

Chronology of the Combined License Application for Vogtle Electric Generating Plant Units 3 and 4

This appendix contains a chronological listing of routine licensing correspondence between the staff of the U.S. Nuclear Regulatory Commission (NRC) regarding the review of the Vogtle Electric Generating Plant, Units 3 and 4 plant design under Docket Nos. 052-000025 and 052-000026

Document Date	Accession Number	Title & Estimated Page Count	Document Type	Author Affiliation	Addressee Affiliation	Docket Number
09/03/2010	ML102320234	Letter to Dr. Crass: Issuance of the Draft Supplemental Environmental Impact Statement for the Vogtle Electric Generating Plant, Units 3 and 4 Combined License Application. 7 Page(s)	Letter	NRC/NRO/DSER	State of GA, Dept of Natural Resources	05200025 05200026
09/03/2010	ML102320244	Letter to Director Klima SHPO: Issuance of the Draft Supplemental Environmental Impact Statement for the Vogtle Electric Generating Plant, Units 3 and 4 Combined License Application. 7 Page(s)	Letter	NRC/NRO/DSER	US Advisory Council On Historic Preservation	05200025 05200026
09/03/2010	ML102460405	Letter to Hardeman: Notification of the Issuance and Request for Comments on the Draft Supplemental Environmental Impact Statement for the Vogtle Electric Generating Plant, Units 3 and 4 Combined License Application. 7 Page(s)	Letter	NRC/NRO/DSER	State of GA, Dept of Natural Resources State of GA, Environmental Protection Div	05200025 05200026
09/03/2010	ML102500477	Vogtle, Units 3 and 4 Combined License Application, Request for Additional Information Letter No. 061. 8 Page(s)	Letter	Southern Nuclear Operating Co, Inc	NRC/Document Control Desk NRC/NRO	05200025 05200026
09/08/2010	ML102510701	Press Release-10-160: ASLB to	Press Release	NRC/OPA		05200025

Appendix B

Chronology of the Combined License Application for Vogtle Electric Generating Plant Units 3 and 4

This appendix contains a chronological listing of routine licensing correspondence between the staff of the U.S. Nuclear Regulatory Commission (NRC) regarding the review of the Vogtle Electric Generating Plant, Units 3 and 4 plant design under Docket Nos. 052-000025 and 052-000026

Document Date	Accession Number	Title & Estimated Page Count	Document Type	Author Affiliation	Addressee Affiliation	Docket Number
		Hold Sept. 17 Prehearing Conference in Rockville, Md., on Vogtle New Reactor Application. 2 Page(s)				05200026
09/10/2010	ML102570062	Vogtle Electric Generating Plant, Units 3 and 4 - Combined License Application Additional Information for Final Safety Analysis Report Chapter 11. 7 Page(s)	Letter Request for Additional Information (RAI)	Southern Nuclear Operating Co, Inc	NRC/Document Control Desk NRC/NRO	05200025 05200026
09/13/2010	ML102560043	Memo, Vogtle Units 3 & 4, Safety Evaluation Report Input for Quickloc Inservice Inspection for Vogtle Combined Operating License. 2 Page(s)	Memoranda	NRC/NRO/DE/CIB1	NRC/NRO/DNRL/NWE1	05200025 05200026
09/13/2010	ML102560052	Marked Up Copy of Safety Evaluation Report Input for Quickloc Inservice Inspection for Vogtle Combined Operating License. 16 Page(s)	Draft Safety Evaluation Report (DSER)	NRC/NRO/DE/CIB1	NRC/NRO/DNRL/NWE1	05200025 05200026
09/14/2010	ML102571638	2010/09/14 Vogtle RAI for SER – Request for Additional Information Letter No. 062 Related to SRP Section 1.05 for the Vogtle Electric Generating Plant Units 3 and 4 Combined License Application 6 Page(s)	Request for Additional Information (RAI)	NRC/NRO	NRC/NRO/DNRL/NWE1	05200025 05200026

Appendix B

Chronology of the Combined License Application for Vogtle Electric Generating Plant Units 3 and 4

This appendix contains a chronological listing of routine licensing correspondence between the staff of the U.S. Nuclear Regulatory Commission (NRC) regarding the review of the Vogtle Electric Generating Plant, Units 3 and 4 plant design under Docket Nos. 052-000025 and 052-000026

Document Date	Accession Number	Title & Estimated Page Count	Document Type	Author Affiliation	Addressee Affiliation	Docket Number
09/17/2010	ML102590575	Transmittal of Advanced Final Safety Evaluation for Vogtle Electric Generating Plant (VEGP) Units 3 and 4 Combined License Application, Revision 2, Certification Phase B. 9 Page(s)	Memoranda	NRC/NRO/DSRA/SBCV	NRC/NRO/DNRL	05200025 05200026
09/20/2010	ML102450049	Audit Summary - April 13-14, June 7, July 26, and August 23, 2010, Review of Calculations Supporting Evaluation of Control Room Habitability for Vogtle Units 3 and 4 Combined License Application. 4 Page(s)	Meeting Summary Memoranda	NRC/NRO/DNRL/NWE1	NRC/NRO/DNRL/NWE1	05200025 05200026
09/20/2010	ML102650088	Vogtle, Units 3 & 4 - Combined License Application Voluntary Revision to Final Safety Analysis Report Chapter 19. 10 Page(s)	Letter	Southern Nuclear Operating Co, Inc	NRC/Document Control Desk NRC/NRO	05200025 05200026
09/20/2010	ML102650089	Vogtle Units 3 & 4 Combined License Application Voluntary Revision to Final Safety Analysis Report Chapter 3. 6 Page(s)	Final Safety Analysis Report (FSAR) Letter	Southern Nuclear Operating Co, Inc	NRC/Document Control Desk NRC/NRO	05200025 05200026
09/21/2010	ML102070021	10/07/2010 Notice of Forthcoming Public Meeting for the Draft Supplemental Environmental Impact Statement for the Combined Licenses for the Vogtle Electric	Meeting Agenda Meeting Notice Memoranda	NRC/NRO/DSER/RAP1	NRC/NRO/DSER/RAP1	05200025 05200026

Appendix B

Chronology of the Combined License Application for Vogtle Electric Generating Plant Units 3 and 4

This appendix contains a chronological listing of routine licensing correspondence between the staff of the U.S. Nuclear Regulatory Commission (NRC) regarding the review of the Vogtle Electric Generating Plant, Units 3 and 4 plant design under Docket Nos. 052-000025 and 052-000026

Document Date	Accession Number	Title & Estimated Page Count	Document Type	Author Affiliation	Addressee Affiliation	Docket Number
		Generating Plant, Units 3 and 4. 7 Page(s)				
09/21/2010	ML102910448	2010/09/21 Vogtle COL Review - Vogtle Electric Generating Plant Expansion, Burke County, HP-060428-001 3 Page(s)	E-Mail	- No Known Affiliation	NRC/NRO/DNRL/NWE1	05200025 05200026
09/23/2010	ML102780787	2010/09/23 Vogtle COL Review - draft info for response to LTR 62 - 30/40/70 EP 4 Page(s)	E-Mail	- No Known Affiliation	NRC/NRO/DNRL/NWE1	05200025 05200026
09/23/2010	ML102790198	2010/09/23 Vogtle COL Review - draft info for Chapter 14 SUT LC revisions 7 Page(s)	E-Mail	- No Known Affiliation	NRC/NRO/DNRL/NWE1	05200025 05200026
09/27/2010	ML102650632	10/14/2010 - Notice of Meeting with Southern Nuclear Operating Company, to Discuss Plans Regarding Preconstruction Activities Related to the Turbine Building Foundation at Vogtle Units 3 and 4 Site. 7 Page(s)	Meeting Agenda Meeting Notice	NRC/NRO/DNRL/NWE1	NRC/NRO/DNRL/NWE1	05200025 05200026
09/27/2010	ML102790197	2010/09/27 Vogtle COL Review - DRAFT response for Ch15 OI 10 Page(s)	E-Mail	- No Known Affiliation	NRC/NRO/DNRL/NWE1	05200025 05200026

Appendix B

Chronology of the Combined License Application for Vogtle Electric Generating Plant Units 3 and 4

This appendix contains a chronological listing of routine licensing correspondence between the staff of the U.S. Nuclear Regulatory Commission (NRC) regarding the review of the Vogtle Electric Generating Plant, Units 3 and 4 plant design under Docket Nos. 052-000025 and 052-000026

Document Date	Accession Number	Title & Estimated Page Count	Document Type	Author Affiliation	Addressee Affiliation	Docket Number
09/29/2010	ML102700514	Letter to Lucious Abrams: invitation to a Meeting with U.S. Nuclear Regulatory Commission to discuss the Draft Supplemental Environmental Impact Statement for Combined Licenses for Vogtle Electric Generating Plant Units 3 and 4. 5 Page(s)	Letter	NRC/NRO/DSER/RAP1	Burke County, GA	05200025 05200026
09/29/2010	ML102710050	Letter to Commissioner Andrews: Invitation to a Meeting with U.S. Nuclear Regulatory Commission to Discuss the Draft Supplemental Environmetnal Impact Statement for Combined License for Vogtle Electric Generating Plants Units 3 and 4. 5 Page(s)	Letter	NRC/NRO/DSER/RAP1	Burke County, GA	05200025 05200026
09/29/2010	ML102710053	Letter to Commissioner Crockett: Invitation to a Meeting with U.S. Nuclear Regulatory Commission to Discuss the Draft Supplemental Environmetnal Impact Statement for Combined License for Vogtle Electric Generating Plants Units 3 and 4. 5 Page(s)	Letter	NRC/NRO/DSER/RAP1	Burke County, GA	05200025 05200026
09/29/2010	ML102710057	Letter to Commissioner DeLaigle: Invitation to a Meeting with U.S.	Letter	NRC/NRO/DSER/RAP1	Burke County, GA	05200025 05200026

Appendix B

Chronology of the Combined License Application for Vogtle Electric Generating Plant Units 3 and 4

This appendix contains a chronological listing of routine licensing correspondence between the staff of the U.S. Nuclear Regulatory Commission (NRC) regarding the review of the Vogtle Electric Generating Plant, Units 3 and 4 plant design under Docket Nos. 052-000025 and 052-000026

Document Date	Accession Number	Title & Estimated Page Count	Document Type	Author Affiliation	Addressee Affiliation	Docket Number
		Nuclear Regulatory Commission to Discuss the Draft Supplemental Environmetnal Impact Statement for Combined License for Vogtle Electric Generating Plants Units 3 and 4. 5 Page(s)				
09/29/2010	ML102710064	Letter to Commissioner Tinley: Invitation to a Meeting with U.S. Nuclear Regulatory Commission to Discuss the Draft Supplemental Environmetnal Impact Statement for Combined License for Vogtle Electric Generating Plants Units 3 and 4. 5 Page(s)	Letter	NRC/NRO/DSER/RAP1	Burke County, GA	05200025 05200026
09/29/2010	ML102710083	Letter to Mayor Deloach: Invitation to a Meeting with U.S. Nuclear Regulatory Commission to Discuss the Draft Supplemental Environmetnal Impact Statement for Combined License for Vogtle Electric Generating Plants Units 3 and 4. 5 Page(s)	Letter	NRC/NRO/DSER/RAP1	Burke County, GA	05200025 05200026
09/29/2010	ML102710096	Letter to Mr. Herman Brown: Invitation to a Meeting with U.S. Nuclear Regulatory Commission to Discuss the Draft Supplemental	Letter	NRC/NRO/DSER/RAP1	Burke County, GA	05200025 05200026

Appendix B

Chronology of the Combined License Application for Vogtle Electric Generating Plant Units 3 and 4

This appendix contains a chronological listing of routine licensing correspondence between the staff of the U.S. Nuclear Regulatory Commission (NRC) regarding the review of the Vogtle Electric Generating Plant, Units 3 and 4 plant design under Docket Nos. 052-000025 and 052-000026

Document Date	Accession Number	Title & Estimated Page Count	Document Type	Author Affiliation	Addressee Affiliation	Docket Number
		Environmetnal Impact Statement for Combined License for Vogtle Electric Generating Plants Units 3 and 4. 5 Page(s)				
09/29/2010	ML102710115	Letter to Mr. Bill Tinley: Invitation to a Meeting with U.S. Nuclear Regulatory Commission to Discuss the Draft Supplemental Environmetnal Impact Statement for Combined License for Vogtle Electric Generating Plants Units 3 and 4. 5 Page(s)	Letter	NRC/NRO/DSER/RAP1	Burke County, GA	05200025 05200026
09/29/2010	ML102710137	Letter to Mr. James Jones: Invitation to a Meeting with U.S. Nuclear Regulatory Commission to Discuss the Draft Supplemental Environmetnal Impact Statement for Combined License for Vogtle Electric Generating Plants Units 3 and 4. 5 Page(s)	Letter	NRC/NRO/DSER/RAP1	Burke County, GA	05200025 05200026
09/29/2010	ML102710147	Letter to Mr. Richard Byne: Invitation to a Meeting with the U.S. Nuclear Regulatory Commission to Discuss the Draft Supplemental Environmetnal Impact Statement for Combined Licenses for the Vogtle	Letter	NRC/NRO/DSER/RAP1	City of Waynesboro, GA	05200025 05200026

Appendix B

Chronology of the Combined License Application for Vogtle Electric Generating Plant Units 3 and 4

This appendix contains a chronological listing of routine licensing correspondence between the staff of the U.S. Nuclear Regulatory Commission (NRC) regarding the review of the Vogtle Electric Generating Plant, Units 3 and 4 plant design under Docket Nos. 052-000025 and 052-000026

Document Date	Accession Number	Title & Estimated Page Count	Document Type	Author Affiliation	Addressee Affiliation	Docket Number
		Electric Generating Plant units 3 and 4. 5 Page(s)				
09/29/2010	ML102710169	Letter to Portia Lodge Washington: Invitation to a Meeting with the U.S. Nuclear Regulatory Commission to Discuss the Draft Supplemental Environmental Impact Statement for Combined Licenses for the Vogtle Electric Generating Plant Units 3 and 4. 5 Page(s)	Letter	NRC/NRO/DSER/RAP1	City of Waynesboro, GA	05200025 05200026
09/29/2010	ML102710182	Letter to Mr. Willie Williams: Invitation to a Meeting with the U.S. Nuclear Regulatory Commission to Discuss the Draft Supplemental Environmental Impact Statement for Combined Licenses for the Vogtle Electric Generating Plant Units 3 and 4. 5 Page(s)	Letter	NRC/NRO/DSER/RAP1	City of Waynesboro, GA	05200025 05200026

Appendix B

Chronology of the Combined License Application for Vogtle Electric Generating Plant Units 3 and 4

This appendix contains a chronological listing of routine licensing correspondence between the staff of the U.S. Nuclear Regulatory Commission (NRC) regarding the review of the Vogtle Electric Generating Plant, Units 3 and 4 plant design under Docket Nos. 052-000025 and 052-000026

Document Date	Accession Number	Title & Estimated Page Count	Document Type	Author Affiliation	Addressee Affiliation	Docket Number
09/29/2010	ML102710201	Letter to Ms. Linda Bailey: Invitation to a meeting with the U.S. Nuclear Regulatory Commission to discuss the Draft Supplemental Environmental Impact Statement for Combined Licenses for the Vogtle Electric Generating Plant Units 3 and 4. 5 Page(s)	Letter	NRC/NRO/DSER/RAP1	Burke County, GA	05200025 05200026
09/30/2010	ML093560006	VEGP SER- Vogtle Electric Generating Plant Units 3 and 4 Advanced Final Safety Evaulation Report with No Open Items for Chatper 9, "Auxiliary Systems." 84 page(s)	NRO Safety Evaulation Report (SER)–Delayed Safety Evaluation Report	NRC/NRO/DNRL/NWE1		05200025 05200026
09/30/2010	ML100910141	Ltr.-CH9- Vogtle Electric Generating Plant Untis 3 and 4 Advanced Final Safety Evaulation Report with No Open Items for Chapte 8, "Auxiliary Systems". 10 Page(s)	Letter	NRC/NRO/DNRL/NWEQ	NRC/ACRS	05200025 05200026
09/30/2010	ML102940057	2010/09/30 - Comment (4) re: Vogtle COLA DEIS 2 Page(s)	General FR Notice Comment Letter	Public Commenter	NRC/NRO/DSER/RAP1	05200025
09/30/2010	ML102370278	NUREG-1947 DFC, "Draft Supplemental Environmental Impact Statement for Combined License (COLs) for Vogtle Electric Generating Plant Untis 3 and 4"	Letter	Souther Nuclear Operating Co, Inc	NRC/Document Control Desk	05200025 05200026

Appendix B

Chronology of the Combined License Application for Vogtle Electric Generating Plant Units 3 and 4

This appendix contains a chronological listing of routine licensing correspondence between the staff of the U.S. Nuclear Regulatory Commission (NRC) regarding the review of the Vogtle Electric Generating Plant, Units 3 and 4 plant design under Docket Nos. 052-000025 and 052-000026

Document Date	Accession Number	Title & Estimated Page Count	Document Type	Author Affiliation	Addressee Affiliation	Docket Number
10/01/2010	ML100910147	Memo – Vogtle Electric Generating Plant units 3 and 4 Advanced Final Safety Evaluation Report with No Open Items for Chapter 9, "Auxiliary Systems." 3 Page(s) (Draft for Comment). 302 Page(s)	Memoranda	NRC/NRO/DNRL	NRC/ACRS	05200025 05200026
10/01/2010	ML102780279	Vogtle, Units 3 and 4 Combined License Application, Voluntary Revision to Final Safety Analysis Report Chapter 3. 6 Page(s)	Letter	Southern Nuclear Operating Co, Inc	NRC/Document Control Desk NRC/NRO	05200025 05200026
10/04/2010	ML102770201	10/14/10 Revised Notice of Forthcoming SNC Vogtle Meeting, Re: Preconstruction Activities Related to the Turbine Building Foundation at Vogtle Units 3 and 4. 5 Page(s)	Meeting Agenda Meeting Notice Memoranda	NRC/NRO/DNRL/NWE1	NRC/NRO/DNRL/NWE1	05200025 05200026
10/04/2010	ML102810165	10/14/10 Revised Notice of Forthcoming SNC Vogtle Meeting, Re: Preconstruction Activities Related to the Turbine Building Foundation at Vogtle Units 3 and 4. 8 Page(s)	Meeting Agenda Meeting Notice	NRC/NRO/DNRL/NWE1	NRC/NRO/DNRL/NWE1	05200025 05200026
10/05/2010	ML102780789	2010/10/05 Vogtle COL Review - Reissuance of 9/30/10 phone call summary to correct an error and to provide additional discuss regarding	E-Mail	NRC/NRO	NRC/NRO/DNRL/NWE1	05200025 05200026

Appendix B

Chronology of the Combined License Application for Vogtle Electric Generating Plant Units 3 and 4

This appendix contains a chronological listing of routine licensing correspondence between the staff of the U.S. Nuclear Regulatory Commission (NRC) regarding the review of the Vogtle Electric Generating Plant, Units 3 and 4 plant design under Docket Nos. 052-000025 and 052-000026

Document Date	Accession Number	Title & Estimated Page Count	Document Type	Author Affiliation	Addressee Affiliation	Docket Number
		the ACRS action associated with Chapter 18 4 Page(s)				
10/06/2010	ML102790294	2010/10/06 Vogtle RAI for SER – Request for Additional Information Letter No. 063 Related to SRP Section 1.05 for the Vogtle Electric Generating Plant Units 3 and 4 Combined License Application 7 Page(s)	E-Mail	NRC/NRO	NRC/NRO/DNRL/NWE1	05200025 05200026
10/06/2010	ML102790537	2010/10/06 Vogtle COL Review - Summary of October 5, 2010 Public Phone Call to Discuss Issues Associated with the Vogtle Units 3 and 4 Combined License Application Review 18 Page(s)	E-Mail	NRC/NRO	NRC/NRO/DNRL/NWE1	05200025 05200026
10/06/2010	ML102940055	2010/10/06 - Comment (3) re: Vogtle COLA DEIS 2 Page(s)	General FR Notice Comment Letter	Public Commenter	NRC/NRO/DSER/RAP1	05200025
10/07/2010	ML102800090	Press Release-10-179: ASLB to Hold Prehearing Conference Oct. 19 in Rockville, MD., on Vogtle New Reactor Application. 2 Page(s)	Press Release	NRC/OPA		05200025 05200026

Chronology of the Combined License Application for Vogtle Electric Generating Plant Units 3 and 4

This appendix contains a chronological listing of routine licensing correspondence between the staff of the U.S. Nuclear Regulatory Commission (NRC) regarding the review of the Vogtle Electric Generating Plant, Units 3 and 4 plant design under Docket Nos. 052-000025 and 052-000026

Document Date	Accession Number	Title & Estimated Page Count	Document Type	Author Affiliation	Addressee Affiliation	Docket Number
10/07/2010	ML103130523	10/07/2010 Public Meeting List of Attendees for Vogtle Units 3 and 4 DSEIS. 2 Page(s)	- No Document Type Applies	NRC/NRO/DSER/RAP1		05200025 05200026
10/07/2010	ML103130547	10/07/2010 Public Meeting Slides for Vogtle Units 3 and 4 DSEIS. 8 Page(s)	Environmental Impact Statement Slides and Viewgraphs	NRC/NRO/DSER/RAP1		05200025 05200026
10/07/2010	ML103130550	Vogtle Electric Generating Plant Draft EIS - Public Meeting, October 7, 2010, Pages 1-100. 102 Page(s)	Meeting Transcript	NRC/NRO/DSER		05200025 05200026
10/07/2010	ML103130596	10/07/2010 Agenda for Public Meeting to Discuss Draft Supplemental Environmental Impact Statement for Vogtle Units 3 and 4 COL Application. 1 Page(s)	Meeting Agenda	NRC/NRO/DSER/RAP1		05200025 05200026
10/07/2010	ML103140519	10/07/2010 Vogtle Public Meeting Comments- Song Sheet Received. 2 Page(s)	- No Document Type Applies	- No Known Affiliation	NRC/NRO	05200025 05200026
10/07/2010	ML103140538	10/07/2010 Vogtle Public Meeting Comments from Ms. Sara Barczak of Southern Alliance for Clean Energy. 2 Page(s)	Speech	Southern Alliance for Clean Energy	NRC/NRO	05200025 05200026
10/07/2010	ML103140541	10/07/2010 Vogtle Public Meeting Comments- Letter from President Terry D. Elam of Augusta Technical	Letter	Augusta Technical College	NRC/NRO	05200025 05200026

Chronology of the Combined License Application for Vogtle Electric Generating Plant Units 3 and 4

This appendix contains a chronological listing of routine licensing correspondence between the staff of the U.S. Nuclear Regulatory Commission (NRC) regarding the review of the Vogtle Electric Generating Plant, Units 3 and 4 plant design under Docket Nos. 052–000025 and 052-000026

Document Date	Accession Number	Title & Estimated Page Count	Document Type	Author Affiliation	Addressee Affiliation	Docket Number
		Community College. 2 Page(s)				
10/08/2010	ML102810006	2010/10/08 Vogtle COL Review - Summary of September 30, 2010 Public Phone Call to Discuss Issues Associated with the Vogtle Units 3 and 4 Combined License Application Review 7 Page(s)	E-Mail	NRC/NRO	NRC/NRO/DNRL/NWE1	05200025 05200026
10/08/2010	ML102810022	2010/10/08 Vogtle COL Review - Summary of October 7, 2010 Public Phone Call to Discuss Issues Associated with the Vogtle Units 3 and 4 Combined License Application Review 5 Page(s)	E-Mail	NRC/NRO	NRC/NRO/DNRL/NWE1	05200025 05200026
10/08/2010	ML102861726	Vogtle, Units 3 and 4, Combined License Application Response to Request for Additional Information Letter No. 054, Supplement 2, Loss of Large Areas of the Plant Due to Explosions or Fire – Mitigative Strategies Description. 4 Page(s)	Letter	Southern Nuclear Operating Co, Inc	NRC/Document Control Desk NRC/NRO	05200025 05200026
10/08/2010	ML102861727	Vogtle, Units 3 and 4, Revised Response to R-COLA RAI Letter No. 054 RAI No. 19-102, Involving the Loss of Large Areas of the Plant Due to Explosions or Fire Mitigative	- No Document Type Applies	Southern Nuclear Operating Co, Inc	NRC/NRO	05200025 05200026

Appendix B

Chronology of the Combined License Application for Vogtle Electric Generating Plant Units 3 and 4

This appendix contains a chronological listing of routine licensing correspondence between the staff of the U.S. Nuclear Regulatory Commission (NRC) regarding the review of the Vogtle Electric Generating Plant, Units 3 and 4 plant design under Docket Nos. 052-000025 and 052-000026

Document Date	Accession Number	Title & Estimated Page Count	Document Type	Author Affiliation	Addressee Affiliation	Docket Number
10/12/2010	ML102940054	2010/10/12 - Comment (2) re: Vogtle COLA DEIS 3 Page(s)	General FR Notice Comment Letter	Public Commenter	NRC/NRO/DSER/RAP1	05200025
10/14/2010	ML110250411	October 14, 2010 Meeting with Southern Nuclear, Vogtle 3 & 4 Project Turbine Building Foundation. 16 Page(s)	Meeting Briefing Package/Handouts Slides and Viewgraphs	Southern Nuclear Operating Co, Inc	NRC/NRO/DNRL/NWE1	05200025 05200026
10/15/2010	ML102920120	Vogtle, Units 3 and 4 - Combined License Application, Response to Request for Additional Information Letter No. 062. 7 Page(s)	Letter	Southern Nuclear Operating Co, Inc	NRC/Document Control Desk NRC/NRO	05200025 05200026
10/15/2010	ML102920123	Vogtle, Units 3 and 4 - Combined License Application, Response to Request for Additional Information Related to Initial Test Program License Conditions - Voluntary Revision to Part 10. 12 Page(s)	Letter	Southern Nuclear Operating Co, Inc	NRC/Document Control Desk NRC/NRO	05200025 05200026
10/15/2010	ML102920126	Vogtle, Units 3 and 4 - Combined License Application, Voluntary Revision to Final Safety Analysis Report Chapter 3. 6 Page(s)	Letter	Southern Nuclear Operating Co, Inc	NRC/Document Control Desk NRC/NRO	05200025 05200026
10/15/2010	ML102920393	Vogtle, Units 3 and 4 - Combined License Application Voluntary	Letter	Southern Nuclear Operating Co, Inc	NRC/Document Control Desk	05200025 05200026

Appendix B

Chronology of the Combined License Application for Vogtle Electric Generating Plant Units 3 and 4

This appendix contains a chronological listing of routine licensing correspondence between the staff of the U.S. Nuclear Regulatory Commission (NRC) regarding the review of the Vogtle Electric Generating Plant, Units 3 and 4 plant design under Docket Nos. 052-000025 and 052-000026

Document Date	Accession Number	Title & Estimated Page Count	Document Type	Author Affiliation	Addressee Affiliation	Docket Number
		Revision to Final Safety Analysis Report Chapter 8. 9 Page(s)			NRC/NRO	
10/15/2010	ML102920396	Vogtle, Units 3 and 4 - Combined License Application, Second Supplemental Response to Request for Additional Information Letter No. 18. 44 Page(s)	Letter	Southern Nuclear Operating Co, Inc	NRC/Document Control Desk NRC/NRO	05200025 05200026
10/18/2010	ML102920048	G20100644/EDATS: OEDO-2010-0836 - Janet Lepre E-Mail Re: Briefing Packaging for Visit to GE Hitachi Laser Enrichment Facility on November 8, 2010 & to Vogtle on November 11, 2010 4 Page(s)	E-Mail	NRC/OCM/KLS	NRC/EDO	05200025 05200026 07007016
10/20/2010	ML102930188	2010/10/19-Transcript Southern Nuclear Operation Company Vogtle Electric Plant, Units 3 & 4, October 19, 2010, Pages 1 - 121. 122 Page(s)	Legal-Transcript	NRC/ASLBP		05200025 05200026
10/20/2010	ML103010076	Letter from Fish and Wildlife Service Dated October 20, 2010 on Vogtle Units 3 and 4 Review. 3 Page(s)	Letter	US Dept of Interior, Fish & Wildlife Service	NRC/NRO/DSER/RAP1	05200025 05200026
10/22/2010	ML102950025	2010/10/22 Vogtle RAI for SER - RE: Request for Additional Infomration Letter No. 064 Related to SRP Section 1.05 for the Vogtle	Request for Additional Information (RAI)	NRC/NRO	NRC/NRO/DNRL/NWE1	05200025 05200026

B-227

Appendix B

Chronology of the Combined License Application for Vogtle Electric Electric Generating Plant Units 3 and 4

This appendix contains a chronological listing of routine licensing correspondence between the staff of the U.S. Nuclear Regulatory Commission (NRC) regarding the review of the Vogtle Electric Generating Plant, Units 3 and 4 plant design under Docket Nos. 052-000025 and 052-000026

Document Date	Accession Number	Title & Estimated Page Count	Document Type	Author Affiliation	Addressee Affiliation	Docket Number
		Electric Generating Plant Units 3 and 4 Combined License Application 6 Page(s)				
10/22/2010	ML102990047	Vogtle, Units 3 and 4 Combined License Application, Voluntary FSAR Revision to Clarify Implementation Milestones for Security Programs. 8 Page(s)	Final Safety Analysis Report (FSAR) Letter	Southern Nuclear Operating Co, Inc	NRC/Document Control Desk NRC/NRO	05200025 05200026
10/27/2010	ML103000449	IR 05200025-10-001 and Notice of Violation on 09/14/10 - 09/15/10 and 09/22/10 - 09/23/10 for Vogtle, Unit 3 19 Page(s)	Inspection Report Letter Notice of Deviation	NRC/RGN-II/DCP/CPB2	Southern Nuclear Operating Co, Inc	05200025
10/28/2010	ML102990317	Summary of Teleconference with the U.S. Fish and Wildlife Service to Discuss Species List for the Vogtle Combined Licenses Biological Assessment. 2 Page(s)	Meeting Summary Memoranda	NRC/NRO/DSER/RAP1	NRC/NRO/DSER/RAP1	05200025 05200026
10/29/2010	MI102310362	Vogtle Electric Generating Plant Untis 3 and 4 Combined license Application – Revised Review Schedule. 10 Page(s)	Letter Schedule and Calendar	NRC/NRO/DNRL	Southern Nuclear Operating Co, Inc.	05200025 05200026
10/29/2010	ML103060037	Vogtle Electric Generating Plant Untis 3 and 4 Combined License Application Additional Response to	Letter	Southern Nuclear Operating Co, Inc	NRC/Document Control Desk NRC/NRO	05200025 05200026

Appendix B

Chronology of the Combined License Application for Vogtle Electric Generating Plant Units 3 and 4

This appendix contains a chronological listing of routine licensing correspondence between the staff of the U.S. Nuclear Regulatory Commission (NRC) regarding the review of the Vogtle Electric Generating Plant, Units 3 and 4 plant design under Docket Nos. 052-000025 and 052-000026

Document Date	Accession Number	Title & Estimated Page Count	Document Type	Author Affiliation	Addressee Affiliation	Docket Number
		Bellefonte Units 3 and 4 Safety Evaluation Report Open Items for Chapter 15. 13 Page(s)				
11/02/2010	ML103070428	Vogtle, Units 3 and 4, Combined License Application, Response to Request for Additional Information Letter No. 063. 9 Page(s)	Letter	Southern Nuclear Operating Co, Inc	NRC/Document Control Desk NRC/NRO	05200025 05200026
11/02/2010	ML103080090	Vogtle Electric Generating Plant, Units 3 and 4 - Combined License Application Response to ACRS Question on Common Technical Support Center. 8 Page(s)	Letter	Southern Nuclear Operating Co, Inc	NRC/Document Control Desk NRC/NRO	05200025 05200026
11/02/2010	ML103560158	2010/11/02 - Comment (36) re: Vogtle COLA DEIS 4 Page(s)	General FR Notice Comment Letter	Public Commenter	NRC/NRO/DSER/RAP1	05200025
11/03/2010	ML100920051	CH19 – Vogtle Electric Generating Plan Units 3 and 4 Advanced Final Safety Evaulation Report with No Open Items for Chapter 19, "Severe Accidents" 6 Page(s)	Letter	NRC/NRO/DNRL/NWE1	Southern Nuclear Operating Co, Inc.	05200025 05200026
11/04/2010	ML100480787	Memo- CH 5- Vogtle Electric Generating Plant Units 3 and 4 Advanced Final Safety Evaluation Report with No Open Items for Chapter 5, "Reactor Coolant	Memoranda	NRC/NRNO/DNRL	NRC/ACRS	05200025 05200026

Appendix B

Chronology of the Combined License Application for Vogtle Electric Generating Plant Units 3 and 4

This appendix contains a chronological listing of routine licensing correspondence between the staff of the U.S. Nuclear Regulatory Commission (NRC) regarding the review of the Vogtle Electric Generating Plant, Units 3 and 4 plant design under Docket Nos. 052-000025 and 052-000026

Document Date	Accession Number	Title & Estimated Page Count	Document Type	Author Affiliation	Addressee Affiliation	Docket Number
		System and Connected Systems". 3 Page(s)				
11/04/2010	ML100490230	Letter – CH5-Vogtle Electric Generating Plant Units 3 and 4 Advanced Final Safety Evaluation Report with No Open items for Chapte 5, "Reactor Coolant System and Connected Systems." 6 Page(s)	Letter	NRC/NRO/DNRL/NWE1	Southern Nuclear Operating Co, Inc	05200025 05200026
11/04/2010	ML100920066	Memo – CH 19- Vogtle Eelctric Generating Plant Units 3 and 4 Advanced Final Safety Evaluation Report with No Open items for Chapter 19 "Probabilistic Risk Assessment." 3 Page(s)	Memoranda	NRC/NRO/DNRL	NRC/ACRS	05200025 05200026
11/04/2010	ML103020226	Vogtle SE Chapter 5 Concurrence Page "Phase 4 ASE Without Open Items Concurrence Sheet." 1 Page(s)	- No Document Type Applies	NRC/NRO		05200025 05200026
11/04/2010	ML103081103	Vogtle Units 3 & 4 - NRC Generic Fundamental Examination Results - September 2010. 3 Page(s)	Letter	NRC/RGN-II/DRS	Southern Nuclear Operating Co, Inc	05200025 05200026
11/08/2010	ML102861983	Concurrence Page - Vogtle Electric Generating Plant SE Chapter 6 - Engineered Safety Systems. 1 Page(s)	Safety Evaluation	NRC/NRO/DNRL/NWE1	NRC/NRO/DNRL/NWE1	05200025 05200026
11/08/2010	ML100900115	Ltr.- CH8-Vogtle Electric	Letter	NRC/NRO/DNRL/NWE1	Southern Nuclear Operating	05200025

Appendix B

Chronology of the Combined License Application for Vogtle Electric Generating Plant Units 3 and 4

This appendix contains a chronological listing of routine licensing correspondence between the staff of the U.S. Nuclear Regulatory Commission (NRC) regarding the review of the Vogtle Electric Generating Plant, Units 3 and 4 plant design under Docket Nos. 052-000025 and 052-000026

Document Date	Accession Number	Title & Estimated Page Count	Document Type	Author Affiliation	Addressee Affiliation	Docket Number
		Generating Plan Units 3 and 4 Advanced Final Safety Evaluation Report With No Open Items for Chapte 8, "Electric Power" 6 Page(s)			Co, Inc	05200026
11/08/2010	ML100910103	Vogtle Electric Generating Plant Units 3 and 4 Advanced Final Safety Evaluation Report with No Open Items for Chapter 6, "Engineered Safety Features." 6 Page(s)	Letter	NRC/NRO/DNRL/NWE1	Southern Nuclear Operating Co, Inc	05200025 05200026
11/08/2010	MI100920459	SE Chapter 6 – Vogtle Electric Generating Plant Safety Evaluation Report Chapter 6 CLEAN Master. 39 Page(s)	NRO Safety Evaluation Report (SER)-Delayed	NRC/NRO		05200025 05200026
11/08/2010	ML103060013	Vogtle SE Chapter 8 Concurrence Page. 1 Page(s)	- No Document Type Applies	NRC/NRO/DNRL/NWE1		05200025 05200026
11/09/2010	MI100880411	Vogtle Electric Generating Plant, Units 3 and 4 – Combined License Application – Advanced Safety Evaluation without Open items for Chapter 8, "Electric Power". 3 Page(s)	Memoranda	NRC/NRO/DNRL	NRC/ACRS	05200025 05200026
11/10/2010	ML103130579	Summary of Public Meeting for the Draft Supplemental Environmental Impact Statement for the Combined Licenses for Vogtle Electric Generating Plants, Units 3 and 4.	Meeting Summary Memoranda	NRC/NRO/DSER/RAP1	NRC/NRO/DSER/RAP1	05200025 05200026

Appendix B

Chronology of the Combined License Application for Vogtle Electric Generating Plant Units 3 and 4

This appendix contains a chronological listing of routine licensing correspondence between the staff of the U.S. Nuclear Regulatory Commission (NRC) regarding the review of the Vogtle Electric Generating Plant, Units 3 and 4 plant design under Docket Nos. 052-000025 and 052-000026

Document Date	Accession Number	Title & Estimated Page Count	Document Type	Author Affiliation	Addressee Affiliation	Docket Number
11/11/2010	ML103200416	Vogtle, Units 3 & 4 Combined License Application, Voluntary Revision to Final Safety Analysis Report Chapter 1. 10 Page(s)	Letter	Southern Nuclear Operating Co, Inc	NRC/Document Control Desk NRC/NRO	05200025 05200026
11/11/2010	ML103200417	Vogtle, Units 3 and 4, Combined License Application, Voluntary Submittal Regarding COL Information Item 2.5-16. 6 Page(s)	Letter	Southern Nuclear Operating Co, Inc	NRC/Document Control Desk NRC/NRO	05200025 05200026
11/11/2010	ML103200418	Vogtle, Units 3 and 4 Combined License Application, Voluntary Revision to Final Safety Analysis Report Chapter 14. 6 Page(s)	Letter	Southern Nuclear Operating Co, Inc	NRC/Document Control Desk NRC/NRO	05200025 05200026
11/12/2010	ML103200368	Vogtle Electric Generating Plant, Units 3 & 4 - Combined License Application Mitigative Strategies Description, Revision 1. 4 Page(s)	Letter	Southern Nuclear Operating Co, Inc	NRC/Document Control Desk NRC/NRO	05200025 05200026
11/12/2010	ML103200369	Enclosures 1 & 2: Vogtle Electric Generating Plant, Units 3 & 4 Combined License Application Mitigative Strategies Description, Revision 1. 53 Page(s)	Report, Miscellaneous	Southern Nuclear Operating Co, Inc	NRC/NRO	05200025 05200026
11/15/2010	ML103370044	Comment (1) of Heinz J. Mueller on behalf of US Environmental	General FR Notice Comment Letter	US Environmental Protection Agency (EPA)	NRC/ADM/DAS/RDEB	05200025 05200026

Appendix B

Chronology of the Combined License Application for Vogtle Electric Generating Plant Units 3 and 4

This appendix contains a chronological listing of routine licensing correspondence between the staff of the U.S. Nuclear Regulatory Commission (NRC) regarding the review of the Vogtle Electric Generating Plant, Units 3 and 4 plant design under Docket Nos. 052-000025 and 052-000026

Document Date	Accession Number	Title & Estimated Page Count	Document Type	Author Affiliation	Addressee Affiliation	Docket Number
		Protection Agency (EPA) on Draft Supplemental Environmental Impact Statement for Combined Licenses Vogtle, Units 3 & 4, Construction and Operation, Application, NUREG-1947, CEQ No. 20100351. 9 Page(s)				
11/16/2010	ML102730775	Structural Engineering Branch 1 Input to Safety Evaluation Report for Vogtle Nuclear Plant Units 3 and 4 Reference Combined License Application (Phase 4) 34 Page(s)	Memoranda	NRC/NRO/DE/SEB1	NRC/NRO/DNRL/NWE1	05200025 05200026
11/17/2010	ML103210332	12/01/2010, Forthcoming Meeting Southern Nuclear Operating Company to Discuss Licensing Pre-Construction Related to Issues to Combined License Application for Vogtle Units 3 and 4. 7 Page(s)	Meeting Agenda Meeting Notice Memoranda	NRC/NRO/DNRL/NWE1	NRC/NRO/DNRL/NWE1	05200025 05200026
11/18/2010	ML100900291	Ltr.-CH15 – Vogtle Electric Generating Plant Units 3 and 4 Advanced Safety Evaluation Without Open Items for chapter 15, "Accident Analysis." 10 Page(s)	Letter	NRC/NRO/DNRL/NWE1	Southern Nuclear Operating Co, Inc	05200025 05200026
11/18/2010	ML100900320	Memo- CH15-Vogtle Electric Generating Plant Units 3 and 4	Memoranda	NRC/NRO/DNRL	NRC/ACRS	05200025 05200026

Appendix B

Chronology of the Combined License Application for Vogtle Electric Generating Plant Units 3 and 4

This appendix contains a chronological listing of routine licensing correspondence between the staff of the U.S. Nuclear Regulatory Commission (NRC) regarding the review of the Vogtle Electric Generating Plant, Units 3 and 4 plant design under Docket Nos. 052-000025 and 052-000026

Document Date	Accession Number	Title & Estimated Page Count	Document Type	Author Affiliation	Addressee Affiliation	Docket Number
11/18/2010	ML103230427	Advanced Final safety Evaluation Report with No Open Items for Chapter 15, "Accident Analysis". 5 Page(s)				
11/18/2010	ML103230427	11/18/2010 Stop Work Order at Plant Vogtle. 6 Page(s)	Slides and Viewgraphs	NRC/NRO/DCIP/CQVP		05200025 05200026
11/19/2010	MI100880449	Memo – Vogtle Electric Generating Plant Units 3 and 4 Advanced Final Safety Evaluation Report With no Open Items For Chapter 14, Initial Test Programs. 5 Page(s)	Memoranda	NRC/NRO/DNRL	NRC/ACRS	05200025 05200026
11/19/2010	ML100890031	Ltr.- Vogtle Electric Generating Plant Units 3 and 4 Advanced Final Safety Evalution Report with No Open Items for Chapter 14, Initial Test Programs. 10 Page(s)	Letter	NRC/NRO/DNRL/NWE1	Southern Nuclear Operating Co, Inc.	05200025 05200026
11/19/2010	ML100950532	Vogtle Electric Generating Plant, Units 3 and 4 Combined License Application, Advanced Final Safety Evaluation Report with No Open Items for Chapter 3, "Design of Structures, Components, Equipment, and Systems" 3 Page(s)	Memoranda	NRC/NRO/DNRL	NRC/ACRS	05200025 05200026

Appendix B

Chronology of the Combined License Application for Vogtle Electric Generating Plant Units 3 and 4

This appendix contains a chronological listing of routine licensing correspondence between the staff of the U.S. Nuclear Regulatory Commission (NRC) regarding the review of the Vogtle Electric Generating Plant, Units 3 and 4 plant design under Docket Nos. 052-000025 and 052-000026

Document Date	Accession Number	Title & Estimated Page Count	Document Type	Author Affiliation	Addressee Affiliation	Docket Number
11/19/2010	MI100980607	Vogtle Electric Generating Plant, Units 3 and 4 Combined License Application, Advanced Safety Evaluation Without Open Items for Chapter 3, "Design of Structures, Components, Equipment, and Systems" 6 Page(s)	Letter	NRC/NRO/DNRL/NWE1	Southern Nuclear Operating Co, Inc.	05200025 05200026
11/19/2010	ML103060242	Concurrence Page Vogtle SE Chapter 14 "Initial Test Programs". 1 Page(s)	- No Document Type Applies	NRC/NRO/DNRL/NWE1		05200025 05200026
11/23/2010	ML100820408	Vogtle SE Chapter 13 "Conduct of Operations." 187 Page(s)	NRO Safety Evaluation Report (SER)-Delayed	NRC/NRO/DNRL/NWE1		05200025 05200026
11/23/2010	ML100910456	Vogtle Electric Generating Plant Units 3 and 4 Advanced Final Safety Evaluation Report with No Opent Items for Chapter 13, "Conduct of Operations." 6 Page(s)	Letter	NRC/NRO/DNRL/NWE1	Southern Nuclear Operating Co, Inc	05200025 05200026
11/23/2010	ML100910470	Vogtle Electric Generating Plant, Units 3 and 4, Advanced Final Safety Evaluation Report with No Open items for chapter 13, "Conduct of Operations." 3 Page(s)	Memoranda	NRC/NRO/DNRL	NRC/ACRS	05200025 05200026
11/23/2010	ML103300034	Vogtle, Units 3 and 4, Combined License Application, Request for Additional Information Letter No.	Letter	Southern Nuclear Operating Co, Inc	NRC/Document Control Desk NRC/NRO	05200025 05200026

Appendix B

Chronology of the Combined License Application for Vogtle Electric Generating Plant Units 3 and 4

This appendix contains a chronological listing of routine licensing correspondence between the staff of the U.S. Nuclear Regulatory Commission (NRC) regarding the review of the Vogtle Electric Generating Plant, Units 3 and 4 plant design under Docket Nos. 052-000025 and 052-000026

Document Date	Accession Number	Title & Estimated Page Count	Document Type	Author Affiliation	Addressee Affiliation	Docket Number
11/23/2010	ML103300035	064. 22 Page(s) Vogtle, Units 3 and 4, Combined License Application, Comments on Draft Supplemental Environmental Impact Statement. 12 Page(s)	Letter	Southern Nuclear Operating Co, Inc	NRC/Document Control Desk NRC/NRO	05200025 05200026
11/23/2010	ML103330030	2010/11/23 - Comment (5) re: Vogtle COLA DEIS 3 Page(s)	General FR Notice Comment Letter	Public Commenter	NRC/NRO/DSER/RAP1	05200025
11/23/2010	ML103330031	2010/11/23 - Comment (6) re: Vogtle COLA DEIS 3 Page(s)	General FR Notice Comment Letter	Public Commenter	NRC/NRO/DSER/RAP1	05200025
11/23/2010	ML103330032	2010/11/23 - Comment (7) re: Vogtle COLA DEIS 2 Page(s)	General FR Notice Comment Letter	Public Commenter	NRC/NRO/DSER/RAP1	05200025
11/23/2010	ML103330033	2010/11/23 - Comment (8) re: Vogtle COLA DEIS 3 Page(s)	General FR Notice Comment Letter	Public Commenter	NRC/NRO/DSER/RAP1	05200025
11/23/2010	ML103330034	2010/11/23 - Comment (9) re: Vogtle COLA DEIS 3 Page(s)	General FR Notice Comment Letter	Public Commenter	NRC/NRO/DSER/RAP1	05200025
11/23/2010	ML103330035	2010/11/23 - Comment (10) re: Vogtle COLA DEIS 3 Page(s)	General FR Notice Comment Letter	Public Commenter	NRC/NRO/DSER/RAP1	05200025
11/23/2010	ML103330036	2010/11/23 - Comment (11) re: Vogtle COLA DEIS 3 Page(s)	General FR Notice Comment Letter	Public Commenter	NRC/NRO/DSER/RAP1	05200025
11/23/2010	ML103330037	2010/11/23 - Comment (12) re:	General FR Notice	Public Commenter	NRC/NRO/DSER/RAP1	05200025

Appendix B

Chronology of the Combined License Application for Vogtle Electric Generating Plant Units 3 and 4

This appendix contains a chronological listing of routine licensing correspondence between the staff of the U.S. Nuclear Regulatory Commission (NRC) regarding the review of the Vogtle Electric Generating Plant, Units 3 and 4 plant design under Docket Nos. 052-000025 and 052-000026

Document Date	Accession Number	Title & Estimated Page Count	Document Type	Author Affiliation	Addressee Affiliation	Docket Number
		Vogtle COLA DEIS 3 Page(s)	Comment Letter			
11/23/2010	ML103330038	2010/11/23 - Comment (13) re: Vogtle COLA DEIS 3 Page(s)	General FR Notice Comment Letter	Public Commenter	NRC/NRO/DSER/RAP1	05200025
11/23/2010	ML103330069	2010/11/23 - Comment (34) re: Vogtle COLA DEIS 4 Page(s)	General FR Notice Comment Letter	Public Commenter	NRC/NRO/DSER/RAP1	05200025
11/24/2010	ML103210661	12/8/2010 - Notice of Forthcoming Public Meeting to Discuss Licensing Pre-Construction Related Issues To Southern Nuclear Operating Company's Combined License Application for Vogtle Units 3 and 4. 7 Page(s)	Meeting Agenda Meeting Notice Memoranda	NRC/NRO/DNRL/NWE1	NRC/NRO/DNRL/NWE1	05200025 05200026
11/24/2010	ML103330039	2010/11/24 - Comment (14) re: Vogtle COLA DEIS 3 Page(s)	General FR Notice Comment Letter	Public Commenter	NRC/NRO/DSER/RAP1	05200025
11/24/2010	ML103330040	2010/11/24 - Comment (15) re: Vogtle COLA DEIS 3 Page(s)	General FR Notice Comment Letter	Public Commenter	NRC/NRO/DSER/RAP1	05200025
11/24/2010	ML103330043	2010/11/24 - Comment (16) re: Vogtle COLA DEIS 3 Page(s)	General FR Notice Comment Letter	Public Commenter	NRC/NRO/DSER/RAP1	05200025
11/24/2010	ML103330044	2010/11/24 - Comment (17) re: Vogtle COLA DEIS 3 Page(s)	General FR Notice Comment Letter	Public Commenter	NRC/NRO/DSER/RAP1	05200025
11/24/2010	ML103330045	2010/11/24 - Comment (18) re: Vogtle COLA DEIS	General FR Notice Comment Letter	Public Commenter	NRC/NRO/DSER/RAP1	05200025

Appendix B

Chronology of the Combined License Application for Vogtle Electric Generating Plant Units 3 and 4

This appendix contains a chronological listing of routine licensing correspondence between the staff of the U.S. Nuclear Regulatory Commission (NRC) regarding the review of the Vogtle Electric Generating Plant, Units 3 and 4 plant design under Docket Nos. 052-000025 and 052-000026

Document Date	Accession Number	Title & Estimated Page Count	Document Type	Author Affiliation	Addressee Affiliation	Docket Number
11/24/2010	ML103330046	2010/11/24 - Comment (19) re: Vogtle COLA DEIS 3 Page(s)	General FR Notice Comment Letter	Public Commenter	NRC/NRO/DSER/RAP1	05200025
11/24/2010	ML103330047	2010/11/24 - Comment (20) re: Vogtle COLA DEIS 2 Page(s)	General FR Notice Comment Letter	Public Commenter	NRC/NRO/DSER/RAP1	05200025
11/24/2010	ML103330048	2010/11/24 - Comment (21) re: Vogtle COLA DEIS 3 Page(s)	General FR Notice Comment Letter	Public Commenter	NRC/NRO/DSER/RAP1	05200025
11/24/2010	ML103330050	2010/11/24 - Comment (22) re: Vogtle COLA DEIS 3 Page(s)	General FR Notice Comment Letter	Public Commenter	NRC/NRO/DSER/RAP1	05200025
11/24/2010	ML103330052	2010/11/24 - Comment (23) re: Vogtle COLA DEIS 3 Page(s)	General FR Notice Comment Letter	Public Commenter	NRC/NRO/DSER/RAP1	05200025
11/24/2010	ML103330053	2010/11/24 - Comment (24) re: Vogtle COLA DEIS 3 Page(s)	General FR Notice Comment Letter	Public Commenter	NRC/NRO/DSER/RAP1	05200025
11/24/2010	ML103330070	2010/11/24 - Comment (35) re: Vogtle COLA DEIS 23 Page(s)	General FR Notice Comment Letter	Public Commenter	NRC/NRO/DSER/RAP1	05200025
11/25/2010	ML103330054	2010/11/25 - Comment (25) re: Vogtle COLA DEIS 3 Page(s)	General FR Notice Comment Letter	Public Commenter	NRC/NRO/DSER/RAP1	05200025
11/25/2010	ML103330055	2010/11/25 - Comment (26) re: Vogtle COLA DEIS 3 Page(s)	General FR Notice Comment Letter	Public Commenter	NRC/NRO/DSER/RAP1	05200025
11/25/2010	ML103330058	2010/11/25 - Comment (27) re:	General FR Notice	Public Commenter	NRC/NRO/DSER/RAP1	05200025

Appendix B

Chronology of the Combined License Application for Vogtle Electric Generating Plant Units 3 and 4

This appendix contains a chronological listing of routine licensing correspondence between the staff of the U.S. Nuclear Regulatory Commission (NRC) regarding the review of the Vogtle Electric Generating Plant, Units 3 and 4 plant design under Docket Nos. 052-000025 and 052-000026

Document Date	Accession Number	Title & Estimated Page Count	Document Type	Author Affiliation	Addressee Affiliation	Docket Number
		Vogtle COLA DEIS 3 Page(s)	Comment Letter			
11/26/2010	ML103330059	2010/11/26 - Comment (28) re: Vogtle COLA DEIS 3 Page(s)	General FR Notice Comment Letter	Public Commenter	NRC/NRO/DSER/RAP1	05200025
11/26/2010	ML103330060	2010/11/26 - Comment (29) re: Vogtle COLA DEIS 2 Page(s)	General FR Notice Comment Letter	Public Commenter	NRC/NRO/DSER/RAP1	05200025
11/26/2010	ML103330061	2010/11/26 - Comment (30) re: Vogtle COLA DEIS 3 Page(s)	General FR Notice Comment Letter	Public Commenter	NRC/NRO/DSER/RAP1	05200025
11/26/2010	ML103330062	2010/11/26 - Comment (31) re: Vogtle COLA DEIS 3 Page(s)	General FR Notice Comment Letter	Public Commenter	NRC/NRO/DSER/RAP1	05200025
11/27/2010	ML103330063	2010/11/27 - Comment (32) re: Vogtle COLA DEIS 3 Page(s)	General FR Notice Comment Letter	Public Commenter	NRC/NRO/DSER/RAP1	05200025
11/27/2010	ML103330064	2010/11/27 - Comment (33) re: Vogtle COLA DEIS 3 Page(s)	General FR Notice Comment Letter	Public Commenter	NRC/NRO/DSER/RAP1	05200025
12/01/2010	ML103100005	Ltr, Vogtle SE Chapter 1 "Introduction and Interfaces." 6 Page(s)	Letter	NRC/NRO/DNRL/NWE1	Southern Nuclear Operating Co, Inc	05200025 05200026
12/02/2010	ML103100006	Vogtle Electric Generating Plant, Units 3 and 4 - Combined License Application - Advanced Safety Evaluation without Open Items for Chapter 1, "Introduction and interfaces."	Memoranda	NRC/NRO/DNRL/NWE1	NRC/NRO/DNRL/NWE1	05200025 05200026

Appendix B

Chronology of the Combined License Application for Vogtle Electric Generating Plant Units 3 and 4

This appendix contains a chronological listing of routine licensing correspondence between the staff of the U.S. Nuclear Regulatory Commission (NRC) regarding the review of the Vogtle Electric Generating Plant, Units 3 and 4 plant design under Docket Nos. 052-000025 and 052-000026

Document Date	Accession Number	Title & Estimated Page Count	Document Type	Author Affiliation	Addressee Affiliation	Docket Number
12/02/2010	ML103100007	Concurrence Page - Vogtle SE Chapter 1 "Introduction and Interfaces." 1 Page(s) 3 Page(s)	- No Document Type Applies	NRC/NRO		05200025 05200026
12/02/2010	ML103360175	12/15/2010 - Notice of Forthcoming Public Meeting To Discuss Licensing Pre-Construction Related Issues Regarding Southern Nuclear Operating Company's Combined License Application For Vogtle Units 3 And 4. 7 Page(s)	Meeting Agenda Meeting Notice Memoranda	NRC/NRO/DNRL/NWE1	NRC/NRO/DNRL/NWE1	05200025 05200026
12/14/2010	ML103470723	01/05/2011 - Notice of Forthcoming Public Meeting to Discuss Licensing Pre-Construction Related Issues Regarding Southern Nuclear Operating Company's Combined License Application for Vogtle, Units 3 and 4. 7 Page(s)	Meeting Agenda Meeting Notice Memoranda	NRC/NRO/DNRL/NWE1	NRC/NRO/DNRL/NWE1	05200025 05200026
12/14/2010	ML103470724	1/12/2011 - Notice of Forthcoming Public Meeting to Discuss Licensing Pre-Construction Related Issues Regarding Southern Nuclear Operating Company's Combined License Application for Vogtle, Units 3 and 4. 7 Page(s)	Meeting Agenda Meeting Notice Memoranda	NRC/NRO/DNRL/NWE1	NRC/NRO/DNRL/NWE1	05200025 05200026

Appendix B

Chronology of the Combined License Application for Vogtle Electric Electric Generating Plant Units 3 and 4

This appendix contains a chronological listing of routine licensing correspondence between the staff of the U.S. Nuclear Regulatory Commission (NRC) regarding the review of the Vogtle Electric Generating Plant, Units 3 and 4 plant design under Docket Nos. 052-000025 and 052-000026

Document Date	Accession Number	Title & Estimated Page Count	Document Type	Author Affiliation	Addressee Affiliation	Docket Number
12/21/2010	ML103370680	Vogtle Electric Generating Plant, Units 3 And 4 - Combined License Application - Request For Authorization For Installation Of Turbine Building Foundation. 3 Page(s)	Letter	NRC/NRO/DNRL/NWE1	Southern Nuclear Operating Co, Inc	05200025 05200026
12/29/2010	ML103570205	01/19/11 Notice of Meeting with Southern Nuclear Operating Company (SNC) to Discuss Licensing Pre-Construction Related Issues to SNC's Combined License Application for Vogtle Units 3 and 4. 8 Page(s)	Meeting Agenda Meeting Notice	NRC/NRO/DNRL/NWE1	NRC/NRO/DNRL/NWE1	05200025 05200026
12/29/2010	ML103570217	01/12/11 - Notice of Forthcoming Public Meeting To Discuss Licensing Pre-Construction Related Issues To Southern Nuclear Operating Company's Combined License Application For Vogtle Units 3 And 4. 7 Page(s)	Meeting Agenda Meeting Notice Memoranda	NRC/NRO/DNRL/NWE1	NRC/NRO/DNRL/NWE1	05200025 05200026
01/12/2011	ML110140178	Vogtle, Units 3 & 4 - Combined License Application Revision to Final Safety Analysis Report Sections 2.5 and 3.7. 9 Page(s)	Final Safety Analysis Report (FSAR) Letter	Southern Nuclear Operating Co, Inc	NRC/Document Control Desk NRC/NRO	05200025 05200026
01/13/2011	ML110130469	Comments provided to the ACRS Full Committee in Regards to the Final Safety Evaluation Report	E-Mail	- No Known Affiliation	NRC/NRR/DSS	05200025 05200026

Appendix B

Chronology of the Combined License Application for Vogtle Electric Generating Plant Units 3 and 4

This appendix contains a chronological listing of routine licensing correspondence between the staff of the U.S. Nuclear Regulatory Commission (NRC) regarding the review of the Vogtle Electric Generating Plant, Units 3 and 4 plant design under Docket Nos. 052-000025 and 052-000026

Document Date	Accession Number	Title & Estimated Page Count	Document Type	Author Affiliation	Addressee Affiliation	Docket Number
01/24/2011		Associated with the Vogtle Units 3 and 4 Combined License Application Session. 1 Page(s)				
01/24/2011	ML110350282	G20110090/LTR-11-0049/EDATS: SECY-2011-0058 - Ltr: J. S. Armijo re: Report on the Safety Aspects of the Southern Nuclear Operating Company Combined License Application for Vogtle Electric Generating Plant, Units 3 and 4. 10 Page(s)	Letter	NRC/ACRS	NRC/Chairman	05200025 05200026
01/25/2011	ML110240536	2/9/2011 Notice of Forthcoming Public Meeting with Southern Nuclear Operating Company to Discuss SNC's Plans to Submit an Exemption Request from Requirements of 10 CFR 50.10(D) for Its Combined License Application the Vogtle, Units 3 & 4. 8 Page(s)	Meeting Agenda Meeting Notice Memoranda	NRC/NRO/DNRL/NWE1	NRC/NRO/DNRL/NWE1	05200025 05200026
01/26/2011	ML110260399	Vogtle Units 3 & 4 - NRC Generic Fundamentals Examination Results. 3 Page(s)	License-Operator, Part 55 Examination Related Material	NRC/RGN-II/DRS/OLB	Southern Nuclear Operating Co, Inc	05200025 05200026
01/31/2011	ML110240622	Summary Of A Meeting To Discuss Southern Nuclear Operating Company Plans Regarding Preconstruction Activities Related	Meeting Summary	NRC/NRO/DNRL/NWE1		05200025 05200026

Appendix B

Chronology of the Combined License Application for Vogtle Electric Generating Plant Units 3 and 4

This appendix contains a chronological listing of routine licensing correspondence between the staff of the U.S. Nuclear Regulatory Commission (NRC) regarding the review of the Vogtle Electric Generating Plant, Units 3 and 4 plant design under Docket Nos. 052-000025 and 052-000026

Document Date	Accession Number	Title & Estimated Page Count	Document Type	Author Affiliation	Addressee Affiliation	Docket Number
		To The Turbine Building Foundation At The Vogtle Units 3 And 4 Site 3 Page(s)				
01/31/2011	ML110340015	Vogtle, Units 3 and 4 - Combined License Application Voluntary Letter Regarding Cyber Security Scope Clarification for Balance of Plant Systems. 7 Page(s)	Letter	Southern Nuclear Operating Co, Inc	NRC/Document Control Desk NRC/NRO	05200025 05200026
01/31/2011	ML110390329	Vogtle Electric Generating Plant, Units 3 & 4 Combined License Application, Submittal 7 to License Application, Revision. 8 Page(s)	Letter	Southern Nuclear Operating Co, Inc	NRC/Document Control Desk NRC/NRO	05200025 05200026
02/01/2011	ML110320694	2011/02/01 Vogtle COL Review - Telecon Summary--Public Conference call with Vogtle, February 1, 2011 2 Page(s)	E-Mail	NRC/NRO	NRC/NRO/DNRL/NWE1	05200025 05200026
02/08/2011	ML110410187	Vogtle, Units 3 & 4 Combined License Application Information to Address ACRS Questions. 7 Page(s)	Letter	Southern Nuclear Operating Co, Inc	NRC/Document Control Desk NRC/NRO	05200025 05200026
02/08/2011	ML110410188	Vogtle, Units 3 and 4, Combined License Application Voluntary Letter Regarding Conformance with Division 5 Regulatory Guides. 10 Page(s)	Letter	Southern Nuclear Operating Co, Inc	NRC/Document Control Desk NRC/NRO	05200025 05200026
02/11/2011	ML110460538	Vogtle, Units 3 and 4 Combined	Letter	Southern Nuclear Operating	NRC/Document Control	05200025

Appendix B

Chronology of the Combined License Application for Vogtle Electric Generating Plant Units 3 and 4

This appendix contains a chronological listing of routine licensing correspondence between the staff of the U.S. Nuclear Regulatory Commission (NRC) regarding the review of the Vogtle Electric Generating Plant, Units 3 and 4 plant design under Docket Nos. 052-000025 and 052-000026

Document Date	Accession Number	Title & Estimated Page Count	Document Type	Author Affiliation	Addressee Affiliation	Docket Number
02/14/2011	ML110460304	License Application - Submittal No. 7 Roadmap. 25 Page(s)		Co, Inc	Desk NRC/NRO	05200026
02/14/2011	ML110460304	IR 05200011-10-008, IR 05200025-10-002, IR 05200026-10-001; 10/1/2010 through 12/31/2010; Vogtle Electric Generating Plant (VEGP) Units 3 and 4, Routine Integrated Inspection Report & Notice of Violation. 31 Page(s)	Inspection Report	NRC/RGN-II/DCP/CPB4	Southern Nuclear Operating Co, Inc	05200011 05200025 05200026
02/15/2011	ML110460469	2/16/2011 Notice of Forthcoming Closed Meeting With the Southern Nuclear Operating Company (SNC), on the Vogtle COLA Submittal. 8 Page(s)	Meeting Agenda Meeting Notice Memoranda	NRC/NRO/DNRL/NWE1	NRC/NRO/DNRL/NWE1	05200025 05200026
02/24/2011	ML103410229	Biological Assessment for Threatened and Endangered Species and Designated Critical Habitat for the Vogtle Electric Generating Plant, Units 3 and 4 Combined Licenses Application. 7 Page(s)	Letter	NRC/NRO/DSER/RAP1	US Dept of Interior, Fish & Wildlife Service	05200025 05200026

B-244

Appendix B

Chronology of the Combined License Application for Vogtle Electric Generating Plant Units 3 and 4

This appendix contains a chronological listing of routine licensing correspondence between the staff of the U.S. Nuclear Regulatory Commission (NRC) regarding the review of the Vogtle Electric Generating Plant, Units 3 and 4 plant design under Docket Nos. 052-000025 and 052-000026

Document Date	Accession Number	Title & Estimated Page Count	Document Type	Author Affiliation	Addressee Affiliation	Docket Number
02/24/2011	ML103410237	Biological Assessment to FWS on the DEIS for Vogtle Electric Generating Plant, Units 3 and 4 to the COL Review - BA. 40 Page(s)	Report, Miscellaneous	NRC/NRO/DSER/RAP1		05200025 05200026
02/24/2011	ML110480434	Cover Letter, 1st Federal Register Notice Re: Availability of the Combined License Application for Vogtle Units 3 and 4. 6 Page(s)	Letter	NRC/NRO/DNRL/NWE1	Southern Nuclear Operating Co, Inc	05200025 05200026
02/24/2011	ML110600731	Vogtle, Units 3 and 4 - Combined License Application, Voluntary Revision to Final Safety Analysis Report Chapter 1. 7 Page(s)	Final Safety Analysis Report (FSAR) Letter	Southern Nuclear Operating Co, Inc	NRC/Document Control Desk NRC/NRO	05200025 05200026
02/28/2011	ML110600325	Vogtle, Units 3 & 4 Annual Reporting Form for Drug and Alcohol Tests, for Period Ending 2010. 1 Page(s)	License-Fitness for Duty (FFD) Performance Report	Southern Nuclear Operating Co, Inc	NRC/Document Control Desk NRC/NSIR	05200025 05200026
02/28/2011	ML110600328	Vogtle, Units 3 & 4, 10 CFR Part 26, Subpart I - Managing Fatigue, For Period Ending 2010. 1 Page(s)	License-Fitness for Duty (FFD) Performance Report	Southern Nuclear Operating Co, Inc	NRC/Document Control Desk NRC/NSIR	05200025 05200026
02/28/2011	ML110601165	Semiannual Performance Review and Inspection Plan - Vogtle Electric Generating Plant Units 3 & 4. 8 Page(s)	Letter	NRC/RGN-II/DCP/CPB4	Southern Nuclear Operating Co, Inc	05200025 05200026

Appendix B

Chronology of the Combined License Application for Vogtle Electric Generating Plant Units 3 and 4

This appendix contains a chronological listing of routine licensing correspondence between the staff of the U.S. Nuclear Regulatory Commission (NRC) regarding the review of the Vogtle Electric Generating Plant, Units 3 and 4 plant design under Docket Nos. 052-000025 and 052-000026

Document Date	Accession Number	Title & Estimated Page Count	Document Type	Author Affiliation	Addressee Affiliation	Docket Number
03/02/2011	ML110460152	Conference Consultation Letter for the Atlantic Sturgeon for the Vogtle Electric Generating Plant, Units 3 and 4 Combined Licenses Application. 2 Page(s)	Letter	NRC/NRO/DSER/RAP1	State of FL, National Marine Fisheries Services	05200025 05200026
03/02/2011	ML110460166	Analysis Regarding Potential Impacts on Atlantic Sturgeon for the Vogtle Electric Generating Plant Units 3 and 4. 14 Page(s)	Report, Miscellaneous	NRC/NRO/DSER/RAP1		05200025 05200026
03/02/2011	ML110480518	Notification of the Availability of Combined License Application for Vogtle, Units 3 & 4, in Accordance with 10 CFR 50.43 Requirements. 7 Page(s)	Letter	NRC/NRO/DNRL/NWE1	State of GA, Public Service Commission	05200025 05200026
03/02/2011	ML110480540	Notification of the Availability of Combined License Application for Vogtle, Units 3 & 4, in Accordance with 10 CFR 50.43 Requirements. 6 Page(s)	Letter	NRC/NRO/DNRL/NWE1	US Federal Energy Regulatory Commission	05200025 05200026
03/03/2011	ML110480429	G2011090/LTR-11-0049/EDATS: SECY-2011-0058 - Ltr. to J. S. Armijo, ACRS from R. W. Borchardt re: Report on the Safety Aspects of the Southern Nuclear Operating Company Combined License Application for Vogtle Electric Generating Plant, Units 3 and 4.	Letter	NRC/EDO	NRC/ACRS	05200025 05200026

Appendix B

Chronology of the Combined License Application for Vogtle Electric Generating Plant Units 3 and 4

This appendix contains a chronological listing of routine licensing correspondence between the staff of the U.S. Nuclear Regulatory Commission (NRC) regarding the review of the Vogtle Electric Generating Plant, Units 3 and 4 plant design under Docket Nos. 052-000025 and 052-000026

Document Date	Accession Number	Title & Estimated Page Count	Document Type	Author Affiliation	Addressee Affiliation	Docket Number
03/03/2011	ML110660153	Vogtle, Units 3 and 4, Combined License Application, Voluntary Letter Regarding Protection of Special Nuclear Material Prior to 10 CFR 73.55 Implementation. 9 Page(s)	Letter	Southern Nuclear Operating Co, Inc	NRC/Document Control Desk NRC/NRO	05200025 05200026
03/09/2011	ML110680192	03/10/2011 Notice of Forthcoming Closed Meeting with the Southern Nuclear Operating Company on the Vogtle 10 CFR Part 70 Issues. 8 Page(s)	Meeting Agenda Meeting Notice Memoranda	NRC/NRO/DNRL/NWE1	NRC/NRO/DNRL/NWE1	05200025 05200026
03/10/2011	ML110730822	Semiannual Security Performance Review and Inspection Plan Vogtle Electric Generating Plant, Units 3 and 4. 4 Page(s)	Letter	NRC/RGN-II/DRS/PSB2	Southern Nuclear Operating Co, Inc	05200025 05200026
03/15/2011	ML110740658	2011/03/15 Vogtle RAI for SER - RE: Request for Additional Information Letter No. 065 (Revised) Related to SRP Section 13.6 for the Vogtle Electric Plant Units 3 and 4 Combined License Application 9 Page(s)	Request for Additional Information (RAI)	NRC/NRO	NRC/NRO/DNRL/NWE1	05200025 05200026
03/15/2011	ML110760146	Vogtle, Units 3 & 4, Reply to a Notice of Violation Concerning Inspection Report 05200011/2010-008; 05200025/2010-002 and	Letter Licensee Response to Notice of Violation	Southern Nuclear Operating Co, Inc	NRC/Document Control Desk NRC/NRO	05200025 05200026

Appendix B

Chronology of the Combined License Application for Vogtle Electric Generating Plant Units 3 and 4

This appendix contains a chronological listing of routine licensing correspondence between the staff of the U.S. Nuclear Regulatory Commission (NRC) regarding the review of the Vogtle Electric Generating Plant, Units 3 and 4 plant design under Docket Nos. 052-000025 and 052-000026

Document Date	Accession Number	Title & Estimated Page Count	Document Type	Author Affiliation	Addressee Affiliation	Docket Number
		05200026/2010-001. 9 Page(s)				
03/16/2011	ML110770137	Vogtle, Units 3 and 4 Combined License Application, Special Nuclear Material Physical Protection Program Description. 4 Page(s)	Letter	Southern Nuclear Operating Co, Inc	NRC/Document Control Desk NRC/NRO	05200025 05200026
03/16/2011	ML110800088	Vogtle, Units 3 and 4 - Combined License Application Supplemental Information in Support of 10 CFR Part 30 Byproduct Material and 10 CFR Part 40 Source Material License. 12 Page(s)	Letter Report, Miscellaneous	Southern Nuclear Operating Co, Inc	NRC/Document Control Desk NRC/NRO	05200025 05200026
03/18/2011	ML110660362	Final Supplemental Environmental Impact Statement for the Vogtle Electric Generation Plant Units 3 and 4 Combined Licenses Application. 13 Page(s)	Letter	NRC/NRO/DSER/RAP1	US Environmental Protection Agency (EPA)	05200025 05200026
03/18/2011	ML110660397	Notice of Availability of the Final Supplemental Environmental Impact Statement for the Vogtle Electric Generation Plant, Units 3 and 4 Combined Licenses Application. 7 Page(s)	Letter	NRC/NRO/DSER/RAP1	- No Known Affiliation	05200025 05200026

Appendix B

Chronology of the Combined License Application for Vogtle Electric Generating Plant Units 3 and 4

This appendix contains a chronological listing of routine licensing correspondence between the staff of the U.S. Nuclear Regulatory Commission (NRC) regarding the review of the Vogtle Electric Generating Plant, Units 3 and 4 plant design under Docket Nos. 052-000025 and 052-000026

Document Date	Accession Number	Title & Estimated Page Count	Document Type	Author Affiliation	Addressee Affiliation	Docket Number
03/18/2011	ML110660405	FRN: General Notice. Vogtle Electric Generating Plant, Units 3 and 4 FSEIS. 3 Page(s)	Federal Register Notice	NRC/NRO/DSER		05200025 05200026
03/18/2011	ML110670315	Notice of Availability of the Final Supplemental Environmental Impact Statement for Vogtle Electric Generating Plant, Units 3 and 4 Combined Licenses Application Review. 7 Page(s)	Letter	NRC/NRO/DSER/RAP1	US Dept of Interior, Fish & Wildlife Service	05200025 05200026
03/18/2011	ML110670694	Notice of Availability of the Final Environmental Impact Statement for the Vogtle Electric Generating Plant, Units 3 and 4 Combined Licenses Application Review. 2 Page(s)	Letter	NRC/NRO/DSER/RAP1	US Advisory Council On Historic Preservation	05200025 05200026
03/21/2011	ML110670364	Letter to Ms. Rolin Regarding Notice of Availability of the Final Supplemental Environmental Impact Statement for Vogtle Electric Generating Plant, Units 3 and 4 Combined Licenses Application Review. 2 Page(s)	Letter	NRC/NRO/DSER/RAP1	Poarch Band of Creek Nation	05200025 05200026
03/21/2011	ML110770367	Letter to Mr. Tullis Regarding Notice of Availability of the Final Supplemental Environmental Impact Statement for Vogtle Electric	Letter	NRC/NRO/DSER/RAP1	Poarch Band of Creek Nation	05200025 05200026

Appendix B

Chronology of the Combined License Application for Vogtle Electric Generating Plant Units 3 and 4

This appendix contains a chronological listing of routine licensing correspondence between the staff of the U.S. Nuclear Regulatory Commission (NRC) regarding the review of the Vogtle Electric Generating Plant, Units 3 and 4 plant design under Docket Nos. 052-000025 and 052-000026

Document Date	Accession Number	Title & Estimated Page Count	Document Type	Author Affiliation	Addressee Affiliation	Docket Number
		Generating Plant, Units 3 and 4 Combined Licenses Application Review. 7 Page(s)				
03/21/2011	ML110770400	Letter to Ms. Holland: Notice of Availability of the Final Supplemental Environmental Impact Statement for Vogtle Electric Generating Plant, Units 3 and 4 Combined Licenses Application Review. 7 Page(s)	Letter	NRC/NRO/DSER/RAP1	United Keetoowah Band of Cherokee Indians	05200025 05200026
03/21/2011	ML110770406	Letter to Ms. McCoy Regarding Notice of Availability of the Final Supplemental Environmental Impact Statement for Vogtle Electric Generating Plant, Units 3 and 4 Combined Licenses Application Review. 7 Page(s)	Letter	NRC/NRO/DSER/RAP1	Eastern Band of Cherokee Indians	05200025 05200026
03/21/2011	ML110770439	Letter to Mr. Zachary Regarding Notice of Availability of the Final Supplemental Environmental Impact Statement for Vogtle Electric Generating Plant, Units 3 and 4 Combined Licenses Application Review. 7 Page(s)	Letter	NRC/NRO/DSER/RAP1	Coushatta Indian Tribe	05200025 05200026

Appendix B

Chronology of the Combined License Application for Vogtle Electric Generating Plant Units 3 and 4

This appendix contains a chronological listing of routine licensing correspondence between the staff of the U.S. Nuclear Regulatory Commission (NRC) regarding the review of the Vogtle Electric Generating Plant, Units 3 and 4 plant design under Docket Nos. 052-000025 and 052-000026

Document Date	Accession Number	Title & Estimated Page Count	Document Type	Author Affiliation	Addressee Affiliation	Docket Number
03/21/2011	ML110770446	Letter to Ms. Bucktrot Regarding Notice of Availability of the Final Supplemental Environmental Impact Statement for Vogtle Electric Generating Plant, Units 3 and 4 Combined Licenses Application Review. 7 Page(s)	Letter	NRC/NRO/DSER/RAP1	Kialegee Tribal Town	05200025 05200026
03/21/2011	ML110770452	Letter to Mr. Terry Regarding Notice of Availability of the Final Supplemental Environmental Impact Statement for Vogtle Electric Generating Plant, Units 3 and 4 Combined Licenses Application Review. 7 Page(s)	Letter	NRC/NRO/DSER/RAP1	Miccosukee Tribe of Indians of Florida	05200025 05200026
03/21/2011	ML110770456	Letter to Ms. Gale Thrower Regarding Notice of Availability of the Final Supplemental Environmental Impact Statement for Vogtle Electric Generating Plant, Units 3 and 4 Combined Licenses Application Review. 7 Page(s)	Letter	NRC/NRO/DSER/RAP1	Poarch Band of Creek Nation	05200025 05200026

Appendix B

Chronology of the Combined License Application for Vogtle Electric Generating Plant Units 3 and 4

This appendix contains a chronological listing of routine licensing correspondence between the staff of the U.S. Nuclear Regulatory Commission (NRC) regarding the review of the Vogtle Electric Generating Plant, Units 3 and 4 plant design under Docket Nos. 052-000025 and 052-000026

Document Date	Accession Number	Title & Estimated Page Count	Document Type	Author Affiliation	Addressee Affiliation	Docket Number
03/21/2011	ML110770458	Letter to Mr. McGertt Regarding Notice of Availability of the Final Supplemental Environmental Impact Statement for Vogtle Electric Generating Plant, Units 3 and 4 Combined Licenses Application Review. 7 Page(s)	Letter	NRC/NRO/DSER/RAP1	Thlopthlocco Tribal Town	05200025 05200026
03/21/2011	ML110770460	Letter to Chief Ellis Regarding Notice of Availability of the Final Supplemental Environmental Impact Statement for Vogtle Electric Generating Plant, Units 3 and 4 Combined Licenses Application Review. 7 Page(s)	Letter	NRC/NRO/DSER/RAP1	Muscogee (Creek) Nation	05200025 05200026
03/21/2011	ML110770461	Letter to Mr. Allen Regarding Notice of Availability of the Final Supplemental Environmental Impact Statement for Vogtle Electric Generating Plant, Units 3 and 4 Combined Licenses Application Review. 7 Page(s)	Letter	NRC/NRO/DSER/RAP1	Cherokee Nation of Oklahoma	05200025 05200026
03/21/2011	ML110770462	Notice of Availability of the Final Supplemental Environmental Impact Statement for Vogtle Electric Generating Plant, Untis 3 and 4 Combined Licenses Application	Letter	NRC/NRO/DSER/RAP1	Chickasaw Nation	05200025 05200026

Appendix B

Chronology of the Combined License Application for Vogtle Electric Generating Plant Units 3 and 4

This appendix contains a chronological listing of routine licensing correspondence between the staff of the U.S. Nuclear Regulatory Commission (NRC) regarding the review of the Vogtle Electric Generating Plant, Units 3 and 4 plant design under Docket Nos. 052-000025 and 052-000026

Document Date	Accession Number	Title & Estimated Page Count	Document Type	Author Affiliation	Addressee Affiliation	Docket Number
		Review. 7 Page(s)				
03/21/2011	ML110770465	Letter to Mr. Anoatubby Regarding Notice of Availability of the Final Supplemental Environmental Impact Statement for Vogtle Electric Generating Plant, Units 3 and 4 Combined Licenses Application Review. 7 Page(s)	Letter	NRC/NRO/DSER/RAP1	Chickasaw Nation of Oklahoma	05200025 05200026
03/21/2011	ML110770468	Letter to Mr. Thurmond Regarding Notice of Availability of the Final Supplemental Environmental Impact Statement for Vogtle Electric Generating Plant, Units 3 and 4 Combined Licenses Application Review. 7 Page(s)	Letter	NRC/NRO/DSER/RAP1	Georgia Tribe of Eastern Cherokee	05200025 05200026
03/21/2011	ML110770471	Letter to Mr. Yargee Regarding Notice of Availability of the Final Supplemental Environmental Impact Statement for Vogtle Electric Generating Plant, Units 3 and 4 Combined Licenses Application Review. 7 Page(s)	Letter	NRC/NRO/DSER/RAP1	Alabama-Quassarte Tribal Town	05200025 05200026
03/21/2011	ML110770477	Letter to Mr. Bowlegs: Notice of Availability of the Final Supplemental Environmental	Letter	NRC/NRO/DSER/RAP1	Seminole Nation of Oklahoma	05200025 05200026

Appendix B

Chronology of the Combined License Application for Vogtle Electric Generating Plant Units 3 and 4

This appendix contains a chronological listing of routine licensing correspondence between the staff of the U.S. Nuclear Regulatory Commission (NRC) regarding the review of the Vogtle Electric Generating Plant, Units 3 and 4 plant design under Docket Nos. 052-000025 and 052-000026

Document Date	Accession Number	Title & Estimated Page Count	Document Type	Author Affiliation	Addressee Affiliation	Docket Number
		Impact Statement for Vogtle Electric Generating Plant, Units 3 and 4 Combined Licenses Application Review. 7 Page(s)				
03/21/2011	ML110770482	Letter to Chief Hicks Regarding Notice of Availability of the Final Supplemental Environmental Impact Statement for Vogtle Electric Generating Plant, Units 3 and 4 Combined Licenses Application Review. 7 Page(s)	Letter	NRC/NRO/DSER/RAP1	Eastern Band of Cherokee Indians	05200025 05200026
03/21/2011	ML110770487	Letter to Chief Proctor Regarding Notice of Availability of the Final Supplemental Environmental Impact Statement for Vogtle Electric Generating Plant, Units 3 and 4 Combined Licenses Application Review. 7 Page(s)	Letter	NRC/NRO/DSER/RAP1	United Keetoowah Band of Cherokee Indians	05200025 05200026

Appendix B

Chronology of the Combined License Application for Vogtle Electric Generating Plant Units 3 and 4

This appendix contains a chronological listing of routine licensing correspondence between the staff of the U.S. Nuclear Regulatory Commission (NRC) regarding the review of the Vogtle Electric Generating Plant, Units 3 and 4 plant design under Docket Nos. 052-000025 and 052-000026

Document Date	Accession Number	Title & Estimated Page Count	Document Type	Author Affiliation	Addressee Affiliation	Docket Number
03/21/2011	ML110770491	Letter to Ms. Kaniatobe Regarding Notice of Availability of the Final Supplemental Environmental Impact Statement for Vogtle Electric Generating Plant, Units 3 and 4 Combined Licenses Application Review. 7 Page(s)	Letter	NRC/NRO/DSER/RAP1	Absentee-Shawnee Tribe of Oklahoma	05200025 05200026
03/21/2011	ML110770506	Letter to Ms. Thomas: Notice of Availability of the Final Supplemental Environmental Impact Statement for Vogtle Electric Generating Plant, Units 3 and 4 Combined Licenses Application Review. 7 Page(s)	Letter	NRC/NRO/DSER/RAP1	Alabama-Coushatta Tribe of Texas	05200025 05200026
03/21/2011	ML110770517	Letter to Mrs. Bear Regarding Notice of Availability of the Final Supplemental Environmental Impact Statement for Vogtle Electric Generating Plant, Units 3 and 4 Combined Licenses Application Review. 7 Page(s)	Letter	NRC/NRO/DSER/RAP1	Muscogee (Creek) Nation of Oklahoma	05200025 05200026
03/21/2011	ML110770520	Letter to Chief Smith: Notice of Availability of the Final Supplemental Environmental Impact Statement for Vogtle Electric Generating Plant, Units 3 and 4	Letter	NRC/NRO/DSER/RAP1	Cherokee Nation of Oklahoma	05200025 05200026

Appendix B

Chronology of the Combined License Application for Vogtle Electric Generating Plant Units 3 and 4

This appendix contains a chronological listing of routine licensing correspondence between the staff of the U.S. Nuclear Regulatory Commission (NRC) regarding the review of the Vogtle Electric Generating Plant, Units 3 and 4 plant design under Docket Nos. 052-000025 and 052-000026

Document Date	Accession Number	Title & Estimated Page Count	Document Type	Author Affiliation	Addressee Affiliation	Docket Number
		Combined Licenses Application Review. 7 Page(s)				
03/21/2011	ML110770528	Letter to Mr. Steele Regarding Notice of Availability of the Final Supplemental Environmental Impact Statement for Vogtle Electric Generating Plant, Units 3 and 4 Combined Licenses Application Review. 7 Page(s)	Letter	NRC/NRO/DSER/RAP1	Seminole Tribe of Florida	05200025 05200026
03/21/2011	ML110770536	Letter to Mr. Carleton Regarding Notice of Availability of the Final Supplemental Environmental Impact Statement for Vogtle Electric Generating Plant, Units 3 and 4 Combined Licenses Application review. 7 Page(s)	Letter	NRC/NRO/DSER/RAP1	Mississippi Band of Choctaw Indians	05200025 05200026
03/24/2011	ML110690917	Notice of Availability of the Final Supplemental Environmental Impact Statement for Vogtle Electric Generating Plant, Units 3 and 4 Combined Licenses Application Review. 7 Page(s)	Letter	NRC/NRO/DSER/RAP1	State of FL, National Marine Fisheries Services	05200025 05200026
03/24/2011	ML110840708	Vogtle, Units 3 & 4, Nuclear Development Executive Management Addition.	Letter	Southern Co Southern Nuclear Operating Co, Inc	NRC/Document Control Desk NRC/NRO	05200011 05200025 05200026

Appendix B

Chronology of the Combined License Application for Vogtle Electric Generating Plant Units 3 and 4

This appendix contains a chronological listing of routine licensing correspondence between the staff of the U.S. Nuclear Regulatory Commission (NRC) regarding the review of the Vogtle Electric Generating Plant, Units 3 and 4 plant design under Docket Nos. 052-000025 and 052-000026

Document Date	Accession Number	Title & Estimated Page Count	Document Type	Author Affiliation	Addressee Affiliation	Docket Number
03/25/2011	ML110840184	Press Release-11-058: NRC Issues Final Supplemental Environmental Impact Statement for Vogtle New Reactors Application. 2 Page(s)	Press Release	NRC/OPA		05200025 05200026
03/31/2011	ML110880558	Review of Revision 0 to the Physical Security Plan (PSP) for the Vogtle Electric Generating Plant dated March 14, 2011. 2 Page(s)	Memoranda	NRC/NSIR/DSP	NRC/NRO	05200025 05200026
03/31/2011	ML11076A010	NUREG-1947, "Final Supplemental Environmental Impact Statement for Combined Licenses (COLs) for Vogtle Electric Generating Plant, Units 3 and 4." 571 Page(s)	NUREG	NRC/NRO		05200025 05200026
03/31/2011	ML110880599	Review of Revision 0 to the Physical Security Plan for the Vogtle Electric Generating Plant Units 3 & 4 Combined License Application dated March 14, 2011 (TAC RN0265). 4 Page(s)	Safety Evaluation Report	NRC/NSIR/DSP	NRC/NRO	05200025 05200026
03/31/2011	ML110910538	Vogtle Electric, Units 3 and 4 - Combined License Application Summary Identification of Reference Combined License Application, Standard Content	Letter	Southern Nuclear Operating Co, Inc	NRC/Document Control Desk NRC/NRO	05200025 05200026

Appendix B

Chronology of the Combined License Application for Vogtle Electric Generating Plant Units 3 and 4

This appendix contains a chronological listing of routine licensing correspondence between the staff of the U.S. Nuclear Regulatory Commission (NRC) regarding the review of the Vogtle Electric Generating Plant, Units 3 and 4 plant design under Docket Nos. 052-000025 and 052-000026

Document Date	Accession Number	Title & Estimated Page Count	Document Type	Author Affiliation	Addressee Affiliation	Docket Number
		Submittals. 6 Page(s)				
04/05/2011	ML110730475	03/10/11 Summary of Meeting Between The U.S. Nuclear Regulatory Commission Staff And Southern Nuclear Operating Company On The Vogtle 10 CFR Part 70 Issues. 10 Page(s)	Meeting Summary	NRC/NRO/DNRL/NWE1	NRC/NRO/DNRL/NWE1	05200025 05200026
04/08/2011	ML111010362	Vogtle, Units 3 and 4 Combined License Application, 10 CFR 50.46 Annual Report. 4 Page(s)	Annual Operating Report Letter	Southern Nuclear Operating Co, Inc	NRC/Document Control Desk NRC/NRO	05200025 05200026
04/08/2011	ML111010363	Vogtle, Units 3 & 4 Combined License Application, Voluntary Revision to Final Safety Analysis Report Chapter 2. 8 Page(s)	Letter	Southern Nuclear Operating Co, Inc	NRC/Document Control Desk NRC/NRO	05200025 05200026
04/11/2011	ML111030431	Vogtle Units 3 And 4 - NRC Generic Fundamentals Examination Results March 2011. 3 Page(s)	Letter	NRC/RGN-II/DRS/OLB	Southern Nuclear Operating Co, Inc	05200025 05200026
04/12/2011	ML111030348	Southern Nuclear Operating Company Vogtle Electric Generating Plant Units 3 and 4 Acknowledgement of Letter Responding to NRC Integrated Inspection Reports. 7 Page(s)	Letter	NRC/RGN-II/DCP/CPB4	Southern Nuclear Operating Co, Inc	05200011 05200025 05200026

Appendix B

Chronology of the Combined License Application for Vogtle Electric Generating Plant Units 3 and 4

This appendix contains a chronological listing of routine licensing correspondence between the staff of the U.S. Nuclear Regulatory Commission (NRC) regarding the review of the Vogtle Electric Generating Plant, Units 3 and 4 plant design under Docket Nos. 052-000025 and 052-000026

Document Date	Accession Number	Title & Estimated Page Count	Document Type	Author Affiliation	Addressee Affiliation	Docket Number
04/18/2011	ML111080599	Safeguards Evaluation Report, for The Vogtle Electric Generating Plant Units 3 and 4 Combined License Application (TAC RN0265). 5 Page(s)	Safety Evaluation Report	NRC/NSIR/DSP/DDMS	NRC/NRO	05200025 05200026
04/25/2011	ML11115A174	2011/04/25 Vogtle COL Review - RE: Creation of a ADAMS record to track the resolution of VEGP Confirmatory Item 11.3-1 2 Page(s)	E-Mail	NRC/NRO	NRC/NRO/DNRL/NWE1	05200025 05200026
05/03/2011	ML111180204	02/09/2011 Summary Of A Meeting To Discuss Southern Nuclear Operating Company Plans An Exemption Request From Requirements Of 10 CFR 50.10 For Its Combined License Application For The Vogtle Electric Generating Plant Proposed Units 3 And 4. 5 Page(s)	Meeting Summary	NRC/NRO/DNRL/NWE1	Southern Nuclear Operating Co, Inc	05200025 05200026
05/05/2011	ML111250546	Trip Report - Vogtle Units 3 & 4 Construction Site (D. Ayres). 3 Page(s)	Memoranda Trip Report	NRC/RGN-II/DCP/CPB4	NRC/RGN-II/DCI	05200011 05200025 05200026
05/06/2011	ML11129A155	Vogtle, Units 3 and 4, Combined License Application, Voluntary Letter Regarding In-Transit Requirements for New Fuel Shipments. 9 Page(s)	Letter	Southern Nuclear Operating Co, Inc	NRC/Document Control Desk NRC/NRO	05200025 05200026
05/09/2011	ML111300466	Vogtle Electric Generating Plant	Inspection Report	NRC/RGN-II/DCP/CPB4	Southern Nuclear Operating	05200011

Chronology of the Combined License Application for Vogtle Electric Generating Plant Units 3 and 4

This appendix contains a chronological listing of routine licensing correspondence between the staff of the U.S. Nuclear Regulatory Commission (NRC) regarding the review of the Vogtle Electric Generating Plant, Units 3 and 4 plant design under Docket Nos. 052-000025 and 052-000026

Document Date	Accession Number	Title & Estimated Page Count	Document Type	Author Affiliation	Addressee Affiliation	Docket Number
		Units 3 And 4 - NRC Integrated Inspection Report 05200011/2011001 052000025/2011002, 05200026/2011001. 19 Page(s)			Co, Inc	05200025 05200026
05/10/2011	ML111260408	05/24/2011 Notice of Forthcoming Meeting with AP1000 Design-Centered Working Group (DCWG) to Discuss the Closure Plan for the AP1000 Piping Design Acceptance Criteria. 9 Page(s)	Meeting Agenda Meeting Notice Memoranda	NRC/NRO/DNRL/NWE1	NRC/NRO/DNRL/NWE1	05200014 05200015 05200018 05200019 05200022 05200023 05200025 05200026 05200027 05200028 05200029 05200030 05200040 05200041
05/10/2011	ML111390647	05/24/2011 Revised Notice of Meeting With AP1000 Design-Centered Working Group (DCWG). 9 Page(s)	Meeting Agenda Meeting Notice Memoranda	NRC/NRO/DNRL/NWE1	NRC/NRO/DNRL/NWE1	05200014 05200015 05200018 05200019 05200022 05200023 05200025 05200026 05200027

Appendix B

Chronology of the Combined License Application for Vogtle Electric Generating Plant Units 3 and 4

This appendix contains a chronological listing of routine licensing correspondence between the staff of the U.S. Nuclear Regulatory Commission (NRC) regarding the review of the Vogtle Electric Generating Plant, Units 3 and 4 plant design under Docket Nos. 052-000025 and 052-000026

Document Date	Accession Number	Title & Estimated Page Count	Document Type	Author Affiliation	Addressee Affiliation	Docket Number
						05200028 05200029 05200030 05200040 05200041
05/13/2011	ML11138A095	Vogtle, Units 3 and 4 Combined License Application, Voluntary Revision to Combined License Application Part 10. 6 Page(s)	Letter	Southern Co Southern Nuclear Operating Co, Inc	NRC/Document Control Desk NRC/NRO	05200025 05200028
05/19/2011	ML111370271	04/27/2011 Meeting Slide - Vogtle & Summer Licensing: Major Activities and Prerequisites. 2 Page(s)	Meeting Briefing Package/Handouts Slides and Viewgraphs	NRC/NRO/DNRL/NRGA		05200025 05200026 05200027 05200028
05/19/2011	ML111370264	04/27/2011 - Summary of Public Meeting with Industry's New Plant Working Group on Combined License Applications. 13 Page(s)	Meeting Summary	NRC/NRO/DNRL/NRGA	NRC/NRO/DNRL/NRGA	05200025 05200026 05200027 05200028
05/24/2011	ML111460082	5/24/11 - AP1000 DCWG Meeting to Discuss Piping DAC and Initial Test Program License Conditions. 6 Page(s)	Meeting Briefing Package/Handouts Slides and Viewgraphs	NRC/NRO/DNRL/NWE1		05200014 05200015 05200018 05200019 05200022 05200023 05200025 05200026 05200027 05200028

Appendix B

Chronology of the Combined License Application for Vogtle Electric Generating Plant Units 3 and 4

This appendix contains a chronological listing of routine licensing correspondence between the staff of the U.S. Nuclear Regulatory Commission (NRC) regarding the review of the Vogtle Electric Generating Plant, Units 3 and 4 plant design under Docket Nos. 052-000025 and 052-000026

Document Date	Accession Number	Title & Estimated Page Count	Document Type	Author Affiliation	Addressee Affiliation	Docket Number
05/24/2011	ML111460084	Meeting Handouts for 5/24/11 - AP1000 DCWG Meeting to Discuss Piping DAC and Initial Test Program License Conditions. 29 Page(s)	Meeting Briefing Package/Handouts Slides and Viewgraphs	Southern Co Southern Nuclear Operating Co, Inc	NRC/NRO	05200029 05200030 05200040 05200041 05200014 05200015 05200018 05200019 05200022 05200023 05200025 05200026 05200027 05200028 05200029 05200030 05200040 05200041
05/24/2011	ML111460093	AP1000 DCWG Meeting to Discuss Piping DAC and Initial Test Program License Conditions. 3 Page(s)	Meeting Briefing Package/Handouts	NRC/NRO/DNRL/NWE1		05200014 05200015 05200018 05200019 05200022 05200023 05200025 05200026 05200027 05200028 05200029

Appendix B

Chronology of the Combined License Application for Vogtle Electric Generating Plant Units 3 and 4

This appendix contains a chronological listing of routine licensing correspondence between the staff of the U.S. Nuclear Regulatory Commission (NRC) regarding the review of the Vogtle Electric Generating Plant, Units 3 and 4 plant design under Docket Nos. 052-000025 and 052-000026

Document Date	Accession Number	Title & Estimated Page Count	Document Type	Author Affiliation	Addressee Affiliation	Docket Number
05/24/2011	ML111460096	Meeting Handouts for 5/24/11 - AP1000 DCWG Meeting to Discuss Piping DAC and Initial test Program License Conditions - Staff Handouts Draft Inspection Procedure 65001.20. 8 Page(s)	Meeting Briefing Package/Handouts	NRC/NRO/DNRL/NWE1		05200014 05200015 05200018 05200019 05200022 05200023 05200025 05200026 05200027 05200028 05200029 05200030 05200040 05200041
05/27/2011	ML11152A189	White Paper - Requirements for COL and LWA Issuance, Relative to the Finalization of Standard Design Certification Rulemaking. 6 Page(s)	Policy and Program Guidance Technical Paper	Balch & Bingham, LLP	NRC/OGC	05200025 05200026
06/02/2011	ML11151A205	SNC Slides - Environmental and Seismic Qualification Audit Discussion. 6 Page(s)	Slides and Viewgraphs	Southern Nuclear Operating Co, Inc	NRC/NRO	05200025 05200026
06/09/2011	ML111470482	05/24/2011 Summary of Public Meeting with the AP1000 Design Centered Working Group (DCWG)	Meeting Summary Memoranda	NRC/NRO/DNRL/NWE1	NRC/NRO/DNRL/NWE1	05200014 05200015 05200018

Appendix B

Chronology of the Combined License Application for Vogtle Electric Generating Plant Units 3 and 4

This appendix contains a chronological listing of routine licensing correspondence between the staff of the U.S. Nuclear Regulatory Commission (NRC) regarding the review of the Vogtle Electric Generating Plant, Units 3 and 4 plant design under Docket Nos. 052-000025 and 052-000026

Document Date	Accession Number	Title & Estimated Page Count	Document Type	Author Affiliation	Addressee Affiliation	Docket Number
		To Discuss the Closure Plan for AP1000 Piping Design Acceptance Criteria (DAC) and Initial Test Program (ITP) License Conditions. 12 Page(s)				05200019 05200022 05200023 05200026 05200027 05200028 05200029 05200030 05200040 05200041
06/22/2011	ML11175A169	Vogtle, Units 3 and 4, Combined License Application, Supplemental Information in Support of 10 CFR Part 70 Special Nuclear Material License Application. 11 Page(s)	Letter	Southern Nuclear Operating Co, Inc	NRC/Document Control Desk NRC/NRO	05200025 05200026
06/24//2011	ML11180A100	Southern Nuclear VEGP Units 3 & 4 COLA (Final Safety Analysis Report), Rev. 5 - Final Safety Analysis Report 828 Page(s)	Final Safety Analysis Report (FSAR) License-Application for Combined License (COLA)	Southern Nuclear Operating Co, Inc	NRC/Document Control Desk NRC/NRO	05200025 05200026
06/24/2011	ML11180A105	Southern Nuclear VEGP Units 3 & 4 COLA (Departures, Exemptions, and Variances), Rev. 5 - Departures, Exemptions, and Variances 29 Page(s)	Generic DCD Departures Report License-Application for Combined License (COLA)	Southern Nuclear Operating Co, Inc	NRC/Document Control Desk NRC/NRO	05200025 05200026

Appendix B

Chronology of the Combined License Application for Vogtle Electric Generating Plant Units 3 and 4

This appendix contains a chronological listing of routine licensing correspondence between the staff of the U.S. Nuclear Regulatory Commission (NRC) regarding the review of the Vogtle Electric Generating Plant, Units 3 and 4 plant design under Docket Nos. 052-000025 and 052-000026

Document Date	Accession Number	Title & Estimated Page Count	Document Type	Author Affiliation	Addressee Affiliation	Docket Number
06/24/2011	ML11180A111	Southern Nuclear VEGP Units 3 & 4 COLA (Proposed License Conditions (Including ITAAC), Rev. 5 - Proposed License Conditions (Including ITAAC) 27 Page(s)	Inspections, Tests, Analyses, and Acceptance Criteria (ITAAC) License-Application for Combined License (COLA)	Southern Nuclear Operating Co, Inc	NRC/Document Control Desk NRC/NRO	05200025 05200026
06/24/2011	ML11180A086	Vogtle Electric Generating Plant, Units 3 and 4 - Combined License Application, Submittal 8 of the Application. 8 Page(s)	Letter License-Application for Combined License (COLA)	Southern Nuclear Operating Co, Inc	NRC/Document Control Desk NRC/NRO	05200025 05200026
06/24/2011	ML11180A107	Southern Nuclear VEGP Units 3 & 4 COLA (Safeguards and Security Plans), Rev. 3 - Safeguards and Security Plans 3 Page(s)	Letter License-Application for Combined License (COLA)	Southern Nuclear Operating Co, Inc	NRC/Document Control Desk NRC/NRO	05200025 05200026
06/24/2011	ML11180A098	Southern Nuclear VEGP Units 3 & 4 COLA (General and Financial Information), Rev. 4 - General and Financial Information 82 Page(s)	License-Application for Combined License (COLA)	Southern Nuclear Operating Co, Inc	NRC/Document Control Desk NRC/NRO	05200025 05200026
06/24/2011	ML11180A102	Southern Nuclear VEGP Units 3 & 4 COLA (Technical Specifications), Rev. 3 - Technical Specifications 852 Page(s)	License-Application for Combined License (COLA) Technical Specifications	Southern Nuclear Operating Co, Inc	NRC/Document Control Desk NRC/NRO	05200025 05200026

Appendix B

Chronology of the Combined License Application for Vogtle Electric Generating Plant Units 3 and 4

This appendix contains a chronological listing of routine licensing correspondence between the staff of the U.S. Nuclear Regulatory Commission (NRC) regarding the review of the Vogtle Electric Generating Plant, Units 3 and 4 plant design under Docket Nos. 052-000025 and 052-000026

Document Date	Accession Number	Title & Estimated Page Count	Document Type	Author Affiliation	Addressee Affiliation	Docket Number
06/24/2011	ML11180A113	Southern Nuclear VEGP Units 3 & 4 COLA (Part 11 Enclosures Index), Rev. 4 - Part 11 Enclosures Index 3 Page(s)	License-Application for Combined License (COLA) Quality Assurance Program	Southern Nuclear Operating Co, Inc	NRC/Document Control Desk NRC/NRO	05200025 05200026
06/24/2011	ML11180A114	Southern Nuclear VEGP Units 3 & 4 COLA (Part 11 Enclosures Index), Rev. 4 - Part 11 Enclosure A - Nuclear Development Quality Assurance Manual 62 Page(s)	License-Application for Combined License (COLA) Quality Assurance Program	Southern Nuclear Operating Co, Inc	NRC/Document Control Desk NRC/NRO	05200025 05200026
06/24/2011	ML11180A115	Southern Nuclear VEGP Units 3 & 4 COLA (Part 11 Enclosures Index), Rev. 4 - Part 11 Enclosure B - Mitigative Strategies Description and Plans 1 Page(s)	License-Application for Combined License (COLA) Quality Assurance Program	Southern Nuclear Operating Co, Inc	NRC/Document Control Desk NRC/NRO	05200025 05200026
06/24/2011	ML11180A117	Southern Nuclear VEGP Units 3 & 4 COLA (Part 11 Enclosures Index), Rev. 4 - Part 11 Enclosure C - Cyber Security Plan 1 Page(s)	License-Application for Combined License (COLA) Quality Assurance Program	Southern Nuclear Operating Co, Inc	NRC/Document Control Desk NRC/NRO	05200025 05200026
06/24/2011	ML11180A118	Southern Nuclear VEGP Units 3 & 4 COLA (Part 11 Enclosures Index), Rev. 4 - Part 11 Enclosure D - Special Nuclear Material (SNM) Material Control and Accounting	License-Application for Combined License (COLA) Quality Assurance	Southern Nuclear Operating Co, Inc	NRC/Document Control Desk NRC/NRO	05200025 05200026

Appendix B

Chronology of the Combined License Application for Vogtle Electric Generating Plant Units 3 and 4

This appendix contains a chronological listing of routine licensing correspondence between the staff of the U.S. Nuclear Regulatory Commission (NRC) regarding the review of the Vogtle Electric Generating Plant, Units 3 and 4 plant design under Docket Nos. 052-000025 and 052-000026

Document Date	Accession Number	Title & Estimated Page Count	Document Type	Author Affiliation	Addressee Affiliation	Docket Number
		Program Description 10 Page(s)	Program			
06/24/2011	ML11180A119	Southern Nuclear VEGP Units 3 & 4 COLA (Part 11 Enclosures Index), Rev. 4 - Part 11 Enclosure E - New Fuel Shipping Plan 3 Page(s)	License-Application for Combined License (COLA) Quality Assurance Program	Southern Nuclear Operating Co, Inc	NRC/Document Control Desk NRC/NRO	05200025 05200026
06/24/2011	ML11180A120	Southern Nuclear VEGP Units 3 & 4 COLA (Part 11 Enclosures Index), Rev. 4 - Part 11 Enclosure F - Supplemental Information in Support of 10 CFR Part 70 Special Nuclear Material License Application 2 Page(s)	License-Application for Combined License (COLA) Quality Assurance Program	Southern Nuclear Operating Co, Inc	NRC/Document Control Desk NRC/NRO	05200025 05200026
07/01/2011	ML11187A242	Vogtle Units 3 & 4 Combined License Application Submittal 8 Roadmap. 13 Page(s)	Letter	Southern Nuclear Operating Co, Inc	NRC/Document Control Desk NRC/NRO	05200025 05200026
07/08/2011	ML11193A087	Vogtle, Units 3 and 4, Response to Regulatory Issue Summary 2011-04 Preparation and Scheduling of Operator Licensing Examinations. 5 Page(s)	Letter	Southern Nuclear Operating Co, Inc	NRC/Document Control Desk NRC/NRO	05200025 05200026
07/11/2011	ML111920383	07/27/11 - Notice of Forthcoming Public Meeting To Discuss Issues Related To Southern Nuclear Operating Company's Combined	Meeting Agenda Meeting Notice Memoranda	NRC/NRO/DNRL/NWE1	NRC/NRO/DNRL/NWE1	05200025 05200026

Appendix B

Chronology of the Combined License Application for Vogtle Electric Generating Plant Units 3 and 4

This appendix contains a chronological listing of routine licensing correspondence between the staff of the U.S. Nuclear Regulatory Commission (NRC) regarding the review of the Vogtle Electric Generating Plant, Units 3 and 4 plant design under Docket Nos. 052-000025 and 052-000026

Document Date	Accession Number	Title & Estimated Page Count	Document Type	Author Affiliation	Addressee Affiliation	Docket Number
07/12/2011	ML11194A281	License Application. 8 Page(s) Vogtle, Units 3 & 4 - Combined License Application Final Supplemental Environmental Impact Statement Construction Time of Year Limitations. 4 Page(s)	Letter	Southern Nuclear Operating Co, Inc	NRC/Document Control Desk NRC/NRO	05200025 05200026

Appendix C – Requests for Additional Information

Throughout the course of the review of the Vogtle Eclectic Generating Plant (VEGP) Units 3 and 4 combined license (COL) application, the staff requested additional information (RAIs) of Southern Nuclear Operating Company (SNC). The following is a list of these RAIs and the responses.

As noted in Section 1.2.3 of this report, a design-centered review approach (DCRA) was used in the review of the VEGP COL application. The first COL application submitted for NRC staff review in a design center is designated as the reference COL (RCOL), and the subsequent applications in the design center are designated as subsequent COL (SCOL) applications. The Bellefonte Nuclear Plant (BLN) Units 3 and 4 COL application was originally designated as the R-COL application for the AP1000 design center, and the staff issued a safety evaluation report (SER) with open items that documented its review of both standard and site-specific information (for all chapters except Sections 3.7, 3.8, 13.6, 13.7, and 13.8 and Appendix 19A). The RCOL for the AP1000 COL design center switched from the Bellefonte COL application to the Vogtle COL application after issuance of the Bellefonte SER with open items. To effect this transition, SNC responded to all of the open items in the staff's BLN SER that related to standard content on behalf of the AP1000 design center and consistent with its new position as the R-COL for the AP1000 design center. Thus, this FSER documents the staff's review of both standard and site-specific information and is the first complete FSER for a COL application in the AP1000 design center. Therefore, in addition to the list of RAIs that follows that are based on site-specific information, SNC had to endorse RAI responses from the RCOLs (both Bellefonte) that were determined to be standard to the AP1000 COL design center. The endorsement of these standard RAIs can be found in the following letters:

- Endorsement of Bellefonte R-COLA Standard Content Request for Additional Information, dated December 17, 2008, ADAMS accession number ML083570590. This letter provides endorsement of standard responses that were provided in a Tennessee Valley Authority (TVA) letter dated October 24, 2008.

- Endorsement of Bellefonte R-COLA Standard Content Request for Additional Information, dated May 15, 2009, ADAMS accession number ML091390567. This letter provides endorsement of standard responses that were provided in a TVA letter April 15, 2009, which supplements the TVA letter dated October 24, 2008.

- Endorsement of Bellefonte R-COLA Standard Content Request for Additional Information, dated November 20, 2009, ADAMS accession number ML0932807746. This letter provides endorsement of standard responses that were provided in a TVA letter November 16, 2009, which supplements the TVA letter dated October 24, 2008.

The following notes pertain to the table on the proceeding pages:

- It includes questions related to SER Open items in from Bellefonte SER which were addressed by SNC.

- Question numbers were assigned based on the section of the standard review plan that was associated with the question (e.g., question 02.03.04-1 was generated based on the staff's review of the application against section 2.3.4 of the standard review plan).

- The NRC letter number is a unique number that was assigned to the letter that transmitted the RAIs to the applicant. If the RAI was not transmitted to the applicant by an NRC letter (e.g., the applicant provided a "voluntary submittal" to address questions from the staff or to inform the staff of changes to the application as a result of changes to the AP1000 design control document) then the applicant's letter was added to the database for tracking purposes and assigned an RAI number. For these "tracking" RAIs, the RAI letter number is "0" and corresponding Accession Number is characterized as "NA".

Electronic Request for Additional Information Database
Appendix C
(Bellefonte and Vogtle)

Application Title: Bellefonte Units 3 and 4 - Dockets 52-014 and 52-015

Question Number	NRCLetter Number:	SRP Section Title	RAI Issued:	RAI Accession Number	RAI Reponse:	Response Accession No
01-11	142	01 - Introduction and Interfaces	12/16/08	ML083510576	09/21/09	ML092660091
01-17	162	01 - Introduction and Interfaces	06/24/09	ML091750079	07/09/10	ML101940025
01-18	162	01 - Introduction and Interfaces	06/24/09	ML091750079	07/29/09	ML092120064
01-19	162	01 - Introduction and Interfaces	06/24/09	ML091750079	07/29/09	ML092120064
01-15	162	01 - Introduction and Interfaces	06/24/09	ML091750079	07/29/09	ML092120064
01-16	162	01 - Introduction and Interfaces	06/24/09	ML091750079	07/29/09	ML092120064
01-20	0	01 - Introduction and Interfaces	07/29/09	N/A	07/29/09	ml092120064
02.02.03-10	137	02.02.03 - Evaluation of Potential Accidents	11/18/08	ML083230896	10/20/09	ML092990266
03.06.02-1	114	03.06.02 - Determination of Rupture Locations and Dynamic Effects Associated with the Postulated Rupture of Piping	08/07/08	ML082200726	04/23/10	ML101160533
03.09.06-11	7	03.09.06 - Functional Design Qualification and Inservice Testing Programs for Pumps, Valves, and Dynamic Restraints	04/28/08	ML081200506	03/01/10	ML100620826
03.09.06-14	7	03.09.06 - Functional Design Qualification and Inservice Testing Programs for Pumps, Valves, and Dynamic Restraints	04/28/08	ML081200506	01/12/10	ML100141734
03.09.06-16	7	03.09.06 - Functional Design Qualification and Inservice Testing Programs for Pumps, Valves, and Dynamic Restraints	04/28/08	ML081200506	12/14/09	ML093491035
03.09.06-2	7	03.09.06 - Functional Design Qualification and Inservice Testing Programs for Pumps, Valves, and Dynamic Restraints	04/28/08	ML081200506	12/14/09	ML093491035
03.09.06-8	7	03.09.06 - Functional Design Qualification and Inservice Testing Programs for Pumps, Valves, and Dynamic Restraints	04/28/08	ML081200506	05/14/10	ML101380120

	No.	Title	Date	Accession No.	Date	Accession No.
03.09.06-9	7	03.09.06 - Functional Design Qualification and Inservice Testing Programs for Pumps, Valves, and Dynamic Restraints	04/28/08	ML081200506	05/14/10	ML101380120
03.10-1	111	03.10 - Seismic and Dynamic Qualification of Mechanical and Electrical Equipment	08/07/08	ML082200569	04/02/10	ML100960109
03.11-1	48	03.11 - Environmental Qualification of Mechanical and Electrical Equipment	06/20/08	ML081720308	12/14/09	ML093491035
04.04-1	0	04.04 - Thermal and Hydraulic Design	01/08/10	N/A	01/08/10	ML100120291
05.02.04-5	74	05.02.04 - Reactor Coolant Pressure Boundary Inservice Inspection and Testing	07/16/08	ML081980777	08/24/09	ML092390080
05.03.01-1	2	05.03.01 - Reactor Vessel Materials	04/17/08	ML081080069	08/24/09	ML092390080
06.01.02-2	170	06.01.02 - Protective Coating Systems (Paints) - Organic Materials	02/17/10	ML100490030	03/04/10	ML100640639
06.01.02-1	1	06.01.02 - Protective Coating Systems (Paints) - Organic Materials	04/14/08	Ml080990812	05/23/08	ML081490154
06.02.02-1	30	06.02.02 - Containment Heat Removal Systems	06/06/08	ML081580010	06/09/09	ML091610555
06.04-7	168	06.04 - Control Room Habitability System	12/14/09	ML093480250	01/05/10	ML100280941
06.04-8	169	06.04 - Control Room Habitability System	01/07/10	ML100070448	06/17/10	ML101690456
08.01-2	25	08.01 - Electric Power - Introduction	05/23/08	ML081440255	05/15/09	ML091390566
08.02-10	151	08.02 - Offsite Power System	03/11/09	ML090700362	04/23/09	ML091170077
08.03.01-1	138	08.03.01 - AC Power Systems (Onsite)	11/18/08	ML083230706	01/02/09	ML090080183
08.03.01-2	149	08.03.01 - AC Power Systems (Onsite)	03/03/09	ML090620447	04/06/09	ML090980343
09.01.02-1	165	09.01.02 - New and Spent Fuel Storage	08/21/09	ML092330617	04/23/10	ML101160532
09.01.02-2	167	09.01.02 - New and Spent Fuel Storage	12/11/09	ML093451224	04/23/10	ML101160532
09.01.04-1	84	09.01.04 - Light Load Handling System (Related to Refueling)	07/21/08	ML082030667	12/30/09	ML100040142
09.01.05-1	61	09.01.05 - Overhead Heavy Load Handling Systems	07/02/08	ML081840267	12/30/09	ML100040142
09.01.05-2	61	09.01.05 - Overhead Heavy Load Handling Systems	07/02/08	ML081840267	12/30/09	ML100040142
09.02.01-9	0	09.02.01 - Station Service Water System	12/30/09	N/A	12/30/09	ML100040142
09.05.01-16	128	09.05.01 - Fire Protection Program	10/07/08	ML082810021	11/20/08	ML083290334
09.05.04-1	92	09.05.04 - Emergency Diesel Engine Fuel Oil Storage and Transfer System	07/31/08	ML082140290	08/29/08	ML082470060
10.03.06-1	18	10.03.06 - Steam and Feedwater System Materials	05/13/08	ML081340606	07/16/09	ML092010091
10.03.06-2	18	10.03.06 - Steam and Feedwater System Materials	05/13/08	ML081340606	09/09/09	ML092530695
12.01-2	164	12.01 - Assuring that Occupational Radiation Exposures Are As Low as is Reasonably Achievable	07/01/09	MI091820094	10/30/09	ML093070283
12.01-1	0	12.01 - Assuring that Occupational Radiation Exposures Are As Low as is Reasonably Achievable	10/16/09	N/A	10/16/09	ML092930120
12.03-12.04-6	161	12.03-12.04 - Radiation Protection Design Features	06/17/09	ML091680006	07/16/09	ML092010089
12.03-12.04-1	109	12.03-12.04 - Radiation Protection Design Features	08/07/08	ML082200591	07/16/09	ML092010089
12.03-12.04-2	109	12.03-12.04 - Radiation Protection Design Features	08/07/08	ML082200591	10/30/09	ML093070283
12.03-12.04-3	109	12.03-12.04 - Radiation Protection Design Features	08/07/08	ML082200591	07/16/09	ML092010089

12.05-2	109	12.05 - Operational Radiation Protection Program	08/07/08	ML082200591	12/16/08	ML083530554
14.02-5	21	14.02 - Initial Plant Test Program - Design Certification and New License Applicants	05/15/08	ML081360305	06/26/08	ML081790813
14.02-6	21	14.02 - Initial Plant Test Program - Design Certification and New License Applicants	05/15/08	ML081360305	06/26/08	ML081790813
14.02-12	139	14.02 - Initial Plant Test Program - Design Certification and New License Applicants	12/08/08	ML083430755	03/26/09	ML090900181
14.02-13	0	14.02 - Initial Plant Test Program - Design Certification and New License Applicants	07/16/09	N/A	07/16/09	ML092010090
14.02-14	0	14.02 - Initial Plant Test Program - Design Certification and New License Applicants	07/16/09	N/A	07/16/09	ML092010090
14.03-1	27	14.03 - Inspections, Tests, Analyses, and Acceptance Criteria	05/28/08	ML081490339	05/11/09	ML091330116
15-1	0	15 - Introduction - Transient and Accident Analyses	10/29/09	N/A	10/29/10	ML103060037
15.04.06-1	0	15.04.06 - Inadvertent Decrease in Boron Concentration in the Reactor Coolant (PWR)	01/22/10	N/A	01/22/10	ML100260127
16-1	131	16 - Technical Specifications	10/21/08	ML082950444	05/21/10	ML101460202
17.01-2	0	17.01 - Quality Assurance During the Design and Construction Phases 07/1981	12/31/09	N/A	12/31/09	ML100050270
17.5-13	16	17.5 - Quality Assurance Program Description - Design Certification, Early Site Permit and New License Applicants	05/12/08	ML081330640	12/31/09	ML100050270
17.5-14	16	17.5 - Quality Assurance Program Description - Design Certification, Early Site Permit and New License Applicants	05/12/08	ML081330640	12/31/09	ML100050270
17.5-17	16	17.5 - Quality Assurance Program Description - Design Certification, Early Site Permit and New License Applicants	05/12/08	ML081330640	12/31/09	ML100050270
17.5-5	13	17.5 - Quality Assurance Program Description - Design Certification, Early Site Permit and New License Applicants	05/12/08	ML081330515	12/31/09	ML100050270
17.5-6	13	17.5 - Quality Assurance Program Description - Design Certification, Early Site Permit and New License Applicants	05/12/08	ML081330515	12/31/09	ML100050270
17.5-7	13	17.5 - Quality Assurance Program Description - Design Certification, Early Site Permit and New License Applicants	05/12/08	ML081330515	12/31/09	ML100050270
19-20	152	19 - Probabilistic Risk Assessment and Severe Accident Evaluation	03/11/09	ML090700510	04/15/09	ML091100173
19-21	152	19 - Probabilistic Risk Assessment and Severe Accident Evaluation	03/11/09	ML090700510	04/15/09	ML091100173

Application Title: Vogtle Nuclear Site, Units 3 and 4, Dockets 52-0025 and 52-0026

Question Number	NRC Letter Number:	SRP Section Title	RAI Issued:	RAI Accession Number	RAI Reponse:	Response Accession No
01-1	27	01 - Introduction and Interfaces	02/25/09	ML090560655	03/27/09	ML090900179
01-2	37	01 - Introduction and Interfaces	06/16/09	Ml091680204	10/23/09	ML093010572
01-3	0	01 - Introduction and Interfaces	02/24/11	N/A	02/24/11	ML110600731
01-4	0	01 - Introduction and Interfaces	02/08/11	N/A	02/08/11	ML110410188
01.05-1	62	01.05 - Other Regulatory Considerations	09/14/10	ML102571638	10/15/10	ML102920120
01.05-2	63	01.05 - Other Regulatory Considerations	10/06/10	ML102790294	11/02/10	ML103070428
01.05-3	64	01.05 - Other Regulatory Considerations	10/22/10	ML102950025	11/23/10	ML103300034
01.05-4	0	01.05 - Other Regulatory Considerations	03/03/11	N/A	03/03/11	ML110660153
01.05-5	0	01.05 - Other Regulatory Considerations	03/16/11	N/A	03/16/11	ML110800088
01.05-6	0	01.05 - Other Regulatory Considerations	06/22/11	N/A	06/22/11	ML11175A169
02-1	0	02 - Site Characteristics and Site Parameters	07/01/10	N/A	07/01/10	ML101830393
02.02.03-1	19	02.02.03 - Evaluation of Potential Accidents	12/12/08	ML083470948	06/17/10	ML101690456
02.03.01-3	8	02.03.01 - Regional Climatology	10/20/08	ML082940631	11/18/08	ML083250482
02.03.01-4	8	02.03.01 - Regional Climatology	10/20/08	ML082940631	11/18/08	ML083250482
02.03.01-1	2	02.03.01 - Regional Climatology	08/29/08	ML082421137	09/18/08	ML082660545
02.03.01-2	2	02.03.01 - Regional Climatology	08/29/08	ML082421137	09/18/08	ML082660545
02.03.04-1	1	02.03.04 - Short Term Atmospheric Dispersion Estimates for Accident Releases	08/19/08	ML082320837	09/11/08	ML082590051
02.03.05-2	8	02.03.05 - Long-Term Atmospheric Dispersion Estimates for Routine Releases	10/20/08	ML082940631	11/18/08	ML083250482
02.03.05-1	2	02.03.05 - Long-Term Atmospheric Dispersion Estimates for Routine Releases	08/29/08	ML082421137	09/18/08	ML082660545
02.04.02-1	28	02.04.02 - Floods	02/25/09	ML090560817	08/05/09	ML092230147
02.04.02-2	28	02.04.02 - Floods	02/25/09	ML090560817	03/27/09	ML090920480
02.04.02-3	28	02.04.02 - Floods	02/25/09	ML090560817	03/27/09	ML090920480
02.04.02-4	31	02.04.02 - Floods	03/30/09	ML090890314	04/29/09	ML091200579
02.04.13-1	17	02.04.13 - Accidental Releases of Radioactive Liquid Effluents in Ground and Surface Waters	11/26/08	ML083310349	12/23/08	ML083640477
02.05.04-1	13	02.05.04 - Stability of Subsurface Materials and Foundations	11/13/08	ML083180932	12/11/08	ML083530555
02.05.04-2	13	02.05.04 - Stability of Subsurface Materials and Foundations	11/13/08	ML083180932	12/11/08	ML083530555
02.05.04-3	30	02.05.04 - Stability of Subsurface Materials and Foundations	03/10/10	ML090700044	04/09/09	ML091030217
02.05.04-4	30	02.05.04 - Stability of Subsurface Materials and Foundations	03/10/09	ML090700044	04/09/09	ML091030217
02.05.04-5	45	02.05.04 - Stability of Subsurface Materials and Foundations	02/01/10	ML100320769	03/02/10	ML100630208

Application Title: Vogtle Nuclear Site, Units 3 and 4, Dockets 52-0025 and 52-0026 (continued)

ID	Rev	Title	Date	ML	Date	ML
02.05.04-6	0	02.05.04 - Stability of Subsurface Materials and Foundations	07/01/10	N/A	07/01/10	ML101830392
02.05.04-7	0	02.05.04 - Stability of Subsurface Materials and Foundations	11/11/10	N/A	11/11/10	ML103200417
02.05.04-8	0	02.05.04 - Stability of Subsurface Materials and Foundations	01/12/11	N/A	01/12/11	ML110140178
02.05.04-9	0	02.05.04 - Stability of Subsurface Materials and Foundations	04/08/11	N/A	04/08/11	ML111010363
03.03.01-1	0	03.03.01 - Wind Loading	09/20/10	N/A	09/20/10	ML102650089
03.06.01-1	0	03.06.01 - Plant Design for Protection Against Postulated Piping Failures in Fluid Systems Outside Containment	04/23/10	N/A	04/23/10	ML101160533
03.07.01-1	0	03.07.01 - Seismic Design Parameters	10/15/10	N/A	10/15/10	ML102920126
03.07.02-1	18	03.07.02 - Seismic System Analysis	12/15/08	ML083500265	10/15/10	ML102920396
03.07.02-2	18	03.07.02 - Seismic System Analysis	12/15/08	ML083500265	01/14/09	ML090150432
03.07.02-3	36	03.07.02 - Seismic System Analysis	06/01/09	ML091490779	07/01/09	ML092080390
03.08.04-1	22	03.08.04 - Other Seismic Category I Structures	12/18/08	ML083530816	01/16/09	ML090220268
03.08.04-2	0	03.08.04 - Other Seismic Category I Structures	10/01/10	N/A	10/01/10	ML102780279
03.08.05-1	0	03.08.05 - Foundations	08/17/10	N/A	08/17/10	ML102310236
03.08.05-2	0	03.08.05 - Foundations	05/13/11	N/A	05/13/11	ML11138A095
03.09.06-1	56	03.09.06 - Functional Design Qualification and Inservice Testing Programs for Pumps, Valves, and Dynamic Restraints	04/29/10	ML101190566	05/27/10	ML101520183
03.09.06-3	0	03.09.06 - Functional Design Qualification and Inservice Testing Programs for Pumps, Valves, and Dynamic Restraints	03/01/10	N/A	03/01/10	ML100620826
03.12-1	46	03.12 - ASME Code Class 1, 2, and 3 Piping Systems and Piping Components and TheirAssociated Supports	02/01/10	ML100320770	03/02/10	ML100630207
03.12-2	57	03.12 - ASME Code Class 1, 2, and 3 Piping Systems and Piping Components and TheirAssociated Supports	05/18/10	ML101380685	08/06/10	ML102210342
05.03.01-1	0	05.02.04 - Reactor Coolant Pressure Boundary Inservice Inspection and Testing	08/27/10	N/A	08/27/10	ML102430184
05.02.05-1	60	05.02.05 - Reactor Coolant Pressure Boundary Leakage Detection	07/08/10	ML101890438	08/05/10	ML102210129
05.02.05-2	60	05.02.05 - Reactor Coolant Pressure Boundary Leakage Detection	07/08/10	ML101890438	08/05/10	ML102210129
06.01.02-1	0	06.01.02 - Protective Coating Systems (Paints) - Organic Materials	07/02/10	N/A	07/02/10	ML101870659
06.01.02-2	0	06.01.02 - Protective Coating Systems (Paints) - Organic Materials	08/13/10	N/A	08/13/10	ML102290036
06.04-1	12	06.04 - Control Room Habitability System	11/13/08	ML083180071	12/11/08	ML083510077
06.04-2	48	06.04 - Control Room Habitability System	02/05/10	ML100450060	03/05/10	ML100680206

Application Title: Vogtle Nuclear Site, Units 3 and 4, Dockets 52-0025 and 52-0026 (continued)

06.04-3	48	06.04 - Control Room Habitability System	02/05/10	ML100450060	03/05/10	ML100680206
06.04-4	0	06.04 - Control Room Habitability System	06/17/10	N/A	06/17/10	ML101690456
06.04-5	61	06.04 - Control Room Habitability System	07/23/10	ML102040703	09/03/10	ML102500477
06.04-6	0	06.04 - Control Room Habitability System	07/30/10	N/A	07/30/10	ML102150196
07.01-1	0	07.01 - Instrumentation and Controls - Introduction	06/04/10	N/A	06/04/10	ML101590402
07.05-1	6	07.05 - Information Systems Important to Safety	10/16/08	ML082900188	07/06/10	ML101890107
08.02-1	25	08.02 - Offsite Power System	12/19/08	ML083540807	10/23/09	ML093240096
08.02-2	25	08.02 - Offsite Power System	12/19/08	ML083540807	04/28/10	ML101200571
08.02-3	25	08.02 - Offsite Power System	12/19/08	ML083540807	01/16/09	ML090220180
08.02-4	25	08.02 - Offsite Power System	12/19/08	ML083540807	01/16/09	ML090220180
08.02-5	25	08.02 - Offsite Power System	12/19/08	ML083540807	10/23/09	ML093240096
08.02-6	25	08.02 - Offsite Power System	12/19/08	ML083540807	01/16/09	ML090220180
08.02-7	25	08.02 - Offsite Power System	12/19/08	ML083540807	04/28/10	ML101200571
08.02-8	25	08.02 - Offsite Power System	12/19/08	ML083540807	10/23/09	ML093240096
08.02-9	25	08.02 - Offsite Power System	12/19/08	ML083540807	01/16/09	ML090220180
08.02-10	34	08.02 - Offsite Power System	04/22/09	ML091120284	10/23/09	ML093010573
08.02-11	34	08.02 - Offsite Power System	04/22/09	ML091120284	10/23/09	ML093010573
08.02-12	38	08.02 - Offsite Power System	08/04/09	ML092160007	08/31/09	ML092450485
08.02-13	44	08.02 - Offsite Power System	12/08/09	ML093420690	01/07/10	ML100110073
08.02-14	53	08.02 - Offsite Power System	03/25/10	ML100840706	05/06/10	ML101300089
08.03.02-1	0	08.03.02 - DC Power Systems (Onsite)	10/15/10	N/A	10/15/10	ML102920393
09.02.01-1	3	09.02.01 - Station Service Water System	10/06/08	ML082800255	11/04/08	ML083110369
09.02.01-2	3	09.02.01 - Station Service Water System	10/06/08	ML082800255	11/04/08	ML083110369
09.02.01-3	3	09.02.01 - Station Service Water System	10/06/08	ML082800255	11/04/08	ML083110369
09.02.01-4	26	09.02.01 - Station Service Water System	01/26/09	ML090260496	03/12/09	ML090760819
09.02.01-5	26	09.02.01 - Station Service Water System	01/26/09	ML090260496	03/12/09	ML090760819
09.02.01-6	26	09.02.01 - Station Service Water System	01/26/09	ML090260496	03/12/09	ML090760819
09.02.02-1	9	09.02.02 - Reactor Auxiliary Cooling Water Systems	11/04/08	ML083090425	12/02/08	ML083390120
09.02.05-1	4	09.02.05 - Ultimate Heat Sink	10/06/08	ML082800361	11/04/08	ML083110470
09.03.03-1	50	09.03.03 - Equipment and Floor Drainage System	03/16/10	ML100750300	04/15/10	ML101090146
09.05.01-1	15	09.05.01 - Fire Protection Program	11/26/08	ML083310028	12/17/08	ML083570397
09.05.01-2	16	09.05.01 - Fire Protection Program	11/26/08	ML083310029	12/17/08	ML083570588
09.05.02-1	5	09.05.02 - Communications Systems	10/06/08	ML082800382	12/23/08	ML083640476
09.05.02-2	5	09.05.02 - Communications Systems	10/06/08	ML082800382	12/23/08	ML083640476
09.05.02-3	5	09.05.02 - Communications Systems	10/06/08	ML082800382	03/15/10	ML100760095
09.05.02-4	5	09.05.02 - Communications Systems	10/06/08	ML082800382	12/23/08	ML083640476
11.02-1	10	11.02 - Liquid Waste Management System	11/13/08	ML083180069	12/11/08	ML083510076
11.03-1	11	11.03 - Gaseous Waste Management System	11/13/08	ML083180070	12/11/08	ML083510078
11.03-2	0	11.03 - Gaseous Waste Management System	04/25/11	N/A	04/25/11	ML11115A174

ID		Title	Date	ML	Date	ML
11.04-1	39	11.04 - Solid Waste Management System	08/24/09	ML092360579	09/23/09	ML092680023
11.04-2	39	11.04 - Solid Waste Management System	08/24/09	ML092360579	09/23/09	ML092680023
12.03-12.04-1	21	12.03-12.04 - Radiation Protection Design Features	12/19/08	ML083540805	01/29/10	ML100330387
12.03-12.04-2	21	12.03-12.04 - Radiation Protection Design Features	12/19/08	ML083540805	01/29/10	ML100330387
12.03-12.04-3	21	12.03-12.04 - Radiation Protection Design Features	12/19/08	ML083540805	01/16/09	ML090220267
12.03-12.04-4	0	12.03-12.04 - Radiation Protection Design Features	09/10/10	N/A	09/10/10	ML102570062
13.01.01-1	23	13.01.01 - Management and Technical Support Organization	12/19/08	ML083540133	01/16/09	ML090220269
13.01.01-2	23	13.01.01 - Management and Technical Support Organization	12/19/08	ML083540133	01/16/09	ML090220269
13.01.02-13.01.01.03-1	24	13.01.02-13.01.03 - Operating Organization	12/19/08	ML083540135	01/16/09	ML090220179
13.01.02-13.01.01.03-2	24	13.01.02-13.01.03 - Operating Organization	12/19/08	ML083540135	01/16/09	ML090220179
13.01.02-13.01.01.03-3	24	13.01.02-13.01.03 - Operating Organization	12/19/08	ML083540135	01/16/09	ML090220179
13.03-1	29	13.03 - Emergency Planning	03/06/09	ML090650536	06/18/09	ML091750106
13.03-2	29	13.03 - Emergency Planning	03/06/09	ML090650536	06/18/09	ML091750106
13.03-3	29	13.03 - Emergency Planning	03/06/09	ML090650536	04/28/10	ML101200570
13.03-4	29	13.03 - Emergency Planning	03/06/09	ML090650536	06/18/09	ML091750106
13.03-5	32	13.03 - Emergency Planning	04/17/09	ML091070271	05/15/09	ML091390050
13.03-6	35	13.03 - Emergency Planning	05/13/09	ML091330816	06/26/09	ML091810095
13.03-7	35	13.03 - Emergency Planning	05/13/09	ML091330816	06/26/09	ML091810095
13.03-8	0	13.03 - Emergency Planning	07/09/10	N/A	07/09/10	ML101940029
13.03-9	0	13.03 - Emergency Planning	11/02/10	N/A	11/02/10	ML103080090
13.03-10	0	13.03 - Emergency Planning	05/27/10	N/A	05/27/10	ML101520182
13.05.02.01-1	0	13.05.02.01 - Operating and Emergency Operating Procedures	05/06/11	N/A	05/06/11	ML11129A155
13.06-1	41	13.06 - Physical Security	09/17/09	ML092600660	10/16/09	ML092930117
13.06-2	41	13.06 - Physical Security	09/17/09	ML092600660	10/16/09	ML092930117
13.06-3	41	13.06 - Physical Security	09/17/09	ML092600660	10/16/09	ML092930117
13.06-4	41	13.06 - Physical Security	09/17/09	ML092600660	10/16/09	ML092930117
13.06-5	41	13.06 - Physical Security	09/17/09	ML092600660	10/16/09	ML092930117
13.06-6	41	13.06 - Physical Security	09/17/09	ML092600660	05/21/10	ML101460298
13.06-7	41	13.06 - Physical Security	09/15/09	ML092600660	10/16/09	ML092930117
13.06-8	41	13.06 - Physical Security	09/15/09	ML092600660	10/16/09	ML092930117
13.06-9	41	13.06 - Physical Security	09/15/09	ML092600660	10/16/09	ML092930117
13.06-10	41	13.06 - Physical Security	09/15/09	ML092600660	10/16/09	ML092930117
13.06-11	41	13.06 - Physical Security	09/15/09	ML092600660	10/16/09	ML092930117
13.06-12	41	13.06 - Physical Security	09/15/09	ML092600660	10/16/09	ML092930117
13.06-13	41	13.06 - Physical Security	09/15/09	ML092600660	10/16/09	ML092930117
13.06-14	41	13.06 - Physical Security	09/15/09	ML092600660	10/16/09	ML092930117

Application Title: Vogtle Nuclear Site, Units 3 and 4, Dockets 52-0025 and 52-0026 (continued)

13.06-15	41	13.06 - Physical Security	09/15/09	ML092600660	10/16/09	ML092930117
13.06-16	41	13.06 - Physical Security	09/15/09	ML092600660	10/16/09	ML092930117
13.06-17	41	13.06 - Physical Security	09/15/09	ML092600660	10/16/09	ML092930117
13.06-18	41	13.06 - Physical Security	09/15/09	ML092600660	10/16/09	ML092930117
13.06-19	41	13.06 - Physical Security	09/15/09	ML092600660	05/21/10	ML101460204
13.06-20	41	13.06 - Physical Security	09/15/09	ML092600660	10/16/09	ML092930117
13.06-21	43	13.06 - Physical Security	10/21/09	ML112140298	06/14/10	ML101670137
13.06-22	43	13.06 - Physical Security	10/21/09	ML112140298	06/14/10	ML101670137
13.06-23	43	13.06 - Physical Security	10/21/09	ML112140298	06/14/10	ML101670137
13.06-24	43	13.06 - Physical Security	10/21/09	ML112140298	06/14/10	ML101670137
13.06-25	43	13.06 - Physical Security	10/21/09	ML112140298	06/14/10	ML101670137
13.06-26	43	13.06 - Physical Security	10/21/09	ML112140298	06/14/10	ML101670137
13.06-27	47	13.06 - Physical Security	02/04/10	ML100350179	03/05/10	ML100680210
13.06-28	47	13.06 - Physical Security	02/04/10	ML100350179	03/05/10	ML100680210
13.06-29	47	13.06 - Physical Security	02/04/10	ML100350179	03/05/10	ML100680210
13.06-30	47	13.06 - Physical Security	02/04/10	ML100350179	03/05/10	ML100680210
13.06-31	47	13.06 - Physical Security	02/04/10	ML100350179	05/28/10	ML101530057
13.06-32	47	13.06 - Physical Security	02/04/10	ML100350179	03/05/10	ML100680210
13.06-33	49	13.06 - Physical Security	02/05/10	ML100360823	03/05/10	ML100680207
13.06-34	49	13.06 - Physical Security	02/05/10	ML100360823	03/05/10	ML100680207
13.06-35	49	13.06 - Physical Security	02/05/10	ML100360823	06/22/10	ML101740487
13.06-36	55	13.06 - Physical Security	04/28/10	ML101180327	05/14/10	ML101380352
13.06-37	65	13.06 - Physical Security	03/14/11	ML110740658	03/16/11	ML110770137
13.06.01-1	0	13.06.01 - Physical Security - Combined License	10/22/10	N/A	10/22/10	ML102990047
13.06.06-1	51	13.06.06 - Cyber Security (Future SRP Section)	03/17/10	ML100760357	10/22/10	ML102990047
13.06.06-10	51	13.06.06 - Cyber Security (Future SRP Section)	03/17/10	ML100760357	10/22/10	ML102990047
13.06.06-11	51	13.06.06 - Cyber Security (Future SRP Section)	03/17/10	ML100760357	10/22/10	ML102990047
13.06.06-12	51	13.06.06 - Cyber Security (Future SRP Section)	03/17/10	ML100760357	10/22/10	ML102990047
13.06.06-13	51	13.06.06 - Cyber Security (Future SRP Section)	03/17/10	ML100760357	10/22/10	ML102990047
13.06.06-14	51	13.06.06 - Cyber Security (Future SRP Section)	03/17/10	ML100760357	10/22/10	ML102990047
13.06.06-2	51	13.06.06 - Cyber Security (Future SRP Section)	03/17/10	ML100760357	10/22/10	ML102990047
13.06.06-3	51	13.06.06 - Cyber Security (Future SRP Section)	03/17/10	ML100760357	10/22/10	ML102990047
13.06.06-4	51	13.06.06 - Cyber Security (Future SRP Section)	03/17/10	ML100760357	10/22/10	ML102990047
13.06.06-5	51	13.06.06 - Cyber Security (Future SRP Section)	03/17/10	ML100760357	10/22/10	ML102990047
13.06.06-6	51	13.06.06 - Cyber Security (Future SRP Section)	03/17/10	ML100760357	10/22/10	ML102990047
13.06.06-7	51	13.06.06 - Cyber Security (Future SRP Section)	03/17/10	ML100760357	10/22/10	ML102990047
13.06.06-8	51	13.06.06 - Cyber Security (Future SRP Section)	03/17/10	ML100760357	10/22/10	ML102990047
13.06.06-9	51	13.06.06 - Cyber Security (Future SRP Section)	03/17/10	ML100760357	10/22/10	ML102990047
13.06.06-15	0	13.06.06 - Cyber Security (Future SRP Section)	06/04/10	N/A	06/14/10	ml101670137

Section	Rev	Title (Future SRP Section)	Date	ML Number	Date	ML Number
13.06.06-16	0	13.06.06 - Cyber Security (Future SRP Section)	01/31/11	N/A	01/31/11	ML110340015
14.02-1	58	14.02 - Initial Plant Test Program - Design Certification and New License Applicants	05/26/10	ML101520046	06/18/10	ML101720578
14.02-2	59	14.02 - Initial Plant Test Program - Design Certification and New License Applicants	05/26/10	ML101520053	06/18/10	ML101720640
14.02-3	0	14.02 - Initial Plant Test Program - Design Certification and New License Applicants	11/11/10	N/A	11/11/10	ML103200418
14.02-4	0	14.02 - Initial Plant Test Program - Design Certification and New License Applicants	10/15/10	N/A	10/15/10	ML102920123
14.03-1	0	14.03 - Inspections, Tests, Analyses, and Acceptance Criteria	08/06/10	N/A	08/06/10	ML102210342
14.03.12-1	47	14.03.12 - Physical Security Hardware - Inspections, Tests, Analyses, and Acceptance Criteria	02/04/10	ML100350179	06/11/10	ML101660063
14.03.12-2	47	14.03.12 - Physical Security Hardware - Inspections, Tests, Analyses, and Acceptance Criteria	02/04/10	ML100350179	03/05/10	ML100680212
14.03.12-3	47	14.03.12 - Physical Security Hardware - Inspections, Tests, Analyses, and Acceptance Criteria	02/04/10	ML100350179	03/05/10	ML100680212
15-1	0	15 - Introduction - Transient and Accident Analyses	02/08/11	N/A	02/08/11	ML110410187
15.00.03-1	7	15.00.03 - Design Basis Accidents Radiological Consequence Analyses for Advanced Light Water Reactors	10/17/08	ML082910046	11/14/08	ML083230221
17.5-1	14	17.5 - Quality Assurance Program Description - Design Certification, Early Site Permit and New License Applicants	11/25/08	ML083300157	12/17/08	ML083570594
17.5-2	14	17.5 - Quality Assurance Program Description - Design Certification, Early Site Permit and New License Applicants	11/25/08	ML083300157	12/17/08	ML083570594
17.5-3	14	17.5 - Quality Assurance Program Description - Design Certification, Early Site Permit and New License Applicants	11/25/08	ML083300157	12/17/08	ML083570594
17.5-4	14	17.5 - Quality Assurance Program Description - Design Certification, Early Site Permit and New License Applicants	11/25/08	ML083300157	12/17/08	ML083570594
17.5-5	14	17.5 - Quality Assurance Program Description - Design Certification, Early Site Permit and New License Applicants	11/25/08	ML083300157	12/17/08	ML083570594
17.5-6	14	17.5 - Quality Assurance Program Description - Design Certification, Early Site Permit and New License Applicants	11/25/08	ML083300157	12/17/08	ML083570594

Application Title: Vogtle Nuclear Site, Units 3 and 4, Dockets 52-0025 and 52-0026 (continued)

17.5-7	14	17.5 - Quality Assurance Program Description - Design Certification, Early Site Permit and New License Applicants	11/25/08	ML083300157	12/17/08	ML083570594
17.5-8	14	17.5 - Quality Assurance Program Description - Design Certification, Early Site Permit and New License Applicants	11/25/08	ML083300157	12/17/08	ML083570594
17.5-9	0	17.5 - Quality Assurance Program Description - Design Certification, Early Site Permit and New License Applicants	12/31/09	N/A	12/31/09	ML100050270
17.5-10	0	17.5 - Quality Assurance Program Description - Design Certification, Early Site Permit and New License Applicants	01/29/10	N/A	01/29/10	ML100330720
17.5-11	0	17.5 - Quality Assurance Program Description - Design Certification, Early Site Permit and New License Applicants	12/31/09	N/A	12/31/09	ML100050270
17.5-12	0	17.5 - Quality Assurance Program Description - Design Certification, Early Site Permit and New License Applicants	12/31/09	N/A	12/31/09	ML100050270
18-1	0	18 - Human Factors Engineering	07/27/10	N/A	07/27/10	ML102100204
19-1	20	19 - Probabilistic Risk Assessment and Severe Accident Evaluation	12/17/08	ML083520028	02/10/09	ML090490095
19-2	20	19 - Probabilistic Risk Assessment and Severe Accident Evaluation	12/17/08	ML083520028	02/10/09	ML090490095
19-3	33	19 - Probabilistic Risk Assessment and Severe Accident Evaluation	04/22/09	ML091120011	05/22/09	ML091470574
19-4	33	19 - Probabilistic Risk Assessment and Severe Accident Evaluation	04/22/09	ML091120011	10/23/09	ML093010571
19-5	33	19 - Probabilistic Risk Assessment and Severe Accident Evaluation	04/22/09	ML091120011	05/22/09	ML091470574
19-6	33	19 - Probabilistic Risk Assessment and Severe Accident Evaluation	04/22/09	ML091120011	05/22/09	ML091470574
19-7	33	19 - Probabilistic Risk Assessment and Severe Accident Evaluation	04/22/09	ML091120011	05/22/09	ML091470574
19-8	33	19 - Probabilistic Risk Assessment and Severe Accident Evaluation	04/22/09	ML091120011	05/22/09	ML091470574
19-9	33	19 - Probabilistic Risk Assessment and Severe Accident Evaluation	04/22/09	ML091120011	05/22/09	ML091470574
19-10	42	19 - Probabilistic Risk Assessment and Severe Accident Evaluation	10/01/09	ML092750349	10/29/09	ML093070086
19-11	42	19 - Probabilistic Risk Assessment and Severe Accident Evaluation	10/01/09	ML092750349	10/29/09	ML093070086

Application Title: Vogtle Nuclear Site, Units 3 and 4, Dockets 52-0025 and 52-0026 (continued)

19-12	42	19 - Probabilistic Risk Assessment and Severe Accident Evaluation	10/01/09	ML092750349	12/23/09	ML093630679
19-13	42	19 - Probabilistic Risk Assessment and Severe Accident Evaluation	10/01/09	ML092750349	10/29/09	ML093070086
19-14	42	19 - Probabilistic Risk Assessment and Severe Accident Evaluation	10/01/09	ML092750349	10/29/09	ML093070086
19-15	42	19 - Probabilistic Risk Assessment and Severe Accident Evaluation	10/01/09	ML092750349	12/23/09	ML093630679
19-16	42	19 - Probabilistic Risk Assessment and Severe Accident Evaluation	10/01/09	ML092750349	02/05/10	ML100481140
19-17	42	19 - Probabilistic Risk Assessment and Severe Accident Evaluation	10/01/09	ML092750349	12/23/09	ML093630679
19-18	42	19 - Probabilistic Risk Assessment and Severe Accident Evaluation	10/01/09	ML092750349	11/13/09	ML093210475
19-19	42	19 - Probabilistic Risk Assessment and Severe Accident Evaluation	10/01/09	ML092750349	11/13/09	ML093210475
19-20	42	19 - Probabilistic Risk Assessment and Severe Accident Evaluation	10/01/09	ML092750349	10/29/09	ML093070086
19-21	42	19 - Probabilistic Risk Assessment and Severe Accident Evaluation	10/01/09	ML092750349	12/23/09	ML093630679
19-22	42	19 - Probabilistic Risk Assessment and Severe Accident Evaluation	10/01/09	ML092750349	10/29/09	ML093070086
19-23	42	19 - Probabilistic Risk Assessment and Severe Accident Evaluation	10/01/09	ML092750349	11/13/09	ML093210475
19-24	42	19 - Probabilistic Risk Assessment and Severe Accident Evaluation	10/01/09	ML092750349	10/29/09	ML093070086
19-25	42	19 - Probabilistic Risk Assessment and Severe Accident Evaluation	10/01/09	ML092750349	12/23/09	ML093630679
19-26	42	19 - Probabilistic Risk Assessment and Severe Accident Evaluation	10/01/09	ML092750349	12/23/09	ML093630679
19-27	42	19 - Probabilistic Risk Assessment and Severe Accident Evaluation	10/01/09	ML092750349	11/13/09	ML093210475
19-28	42	19 - Probabilistic Risk Assessment and Severe Accident Evaluation	10/01/09	ML092750349	12/23/09	ML093630679
19-29	42	19 - Probabilistic Risk Assessment and Severe Accident Evaluation	10/01/09	ML092750349	10/29/09	ML093070086
19-30	42	19 - Probabilistic Risk Assessment and Severe Accident Evaluation	10/01/09	ML092750349	10/29/09	ML093070086
19-31	42	19 - Probabilistic Risk Assessment and Severe Accident Evaluation	10/01/09	ML092750349	11/13/09	ML093210475
19-32	42	19 - Probabilistic Risk Assessment and Severe Accident Evaluation	10/01/09	ML092750349	11/13/09	ML093210475

Application Title: Vogtle Nuclear Site, Units 3 and 4, Dockets 52-0025 and 52-0026 (continued)

19-33	42	19 - Probabilistic Risk Assessment and Severe Accident Evaluation	10/01/09	ML092750349	12/23/09	ML093630679
19-34	42	19 - Probabilistic Risk Assessment and Severe Accident Evaluation	10/01/09	ML092750349	11/13/09	ML093210475
19-35	42	19 - Probabilistic Risk Assessment and Severe Accident Evaluation	10/01/09	ML092750349	11/13/09	ML093210475
19-37	42	19 - Probabilistic Risk Assessment and Severe Accident Evaluation	10/01/09	ML092750349	10/29/09	ML093070086
19-38	42	19 - Probabilistic Risk Assessment and Severe Accident Evaluation	10/01/09	ML092750349	12/23/09	ML093630679
19-39	42	19 - Probabilistic Risk Assessment and Severe Accident Evaluation	10/01/09	ML092750349	12/23/09	ML093630679
19-40	42	19 - Probabilistic Risk Assessment and Severe Accident Evaluation	10/01/09	ML092750349	11/13/09	ML093210475
19-41	42	19 - Probabilistic Risk Assessment and Severe Accident Evaluation	10/01/09	ML092750349	11/13/09	ML093210475
19-42	42	19 - Probabilistic Risk Assessment and Severe Accident Evaluation	10/01/09	ML092750349	11/13/09	ML093210475
19-43	42	19 - Probabilistic Risk Assessment and Severe Accident Evaluation	10/01/09	ML092750349	11/13/09	ML093210475
19-44	42	19 - Probabilistic Risk Assessment and Severe Accident Evaluation	10/01/09	ML092750349	02/05/10	ML100481140
19-45	42	19 - Probabilistic Risk Assessment and Severe Accident Evaluation	10/01/09	ML092750349	12/23/09	ML093630679
19-46	42	19 - Probabilistic Risk Assessment and Severe Accident Evaluation	10/01/09	ML092750349	12/23/09	ML093630679
19-47	42	19 - Probabilistic Risk Assessment and Severe Accident Evaluation	10/01/09	ML092750349	12/23/09	ML093630679
19-48	42	19 - Probabilistic Risk Assessment and Severe Accident Evaluation	10/01/09	ML092750349	10/29/09	ML093070086
19-49	42	19 - Probabilistic Risk Assessment and Severe Accident Evaluation	10/01/09	ML092750349	10/29/09	ML093070086
19-50	42	19 - Probabilistic Risk Assessment and Severe Accident Evaluation	10/01/09	ML092750349	10/29/09	ML093070086
19-51	42	19 - Probabilistic Risk Assessment and Severe Accident Evaluation	10/01/09	ML092750349	10/29/09	ML093070086
19-52	42	19 - Probabilistic Risk Assessment and Severe Accident Evaluation	10/01/09	ML092750349	10/29/09	ML093070086
19-53	42	19 - Probabilistic Risk Assessment and Severe Accident Evaluation	10/01/09	ML092750349	10/29/09	ML093070086
19-54	42	19 - Probabilistic Risk Assessment and Severe Accident Evaluation	10/01/09	ML092750349	10/29/09	ML093070086

Application Title: Vogtle Nuclear Site, Units 3 and 4, Dockets 52-0025 and 52-0026 (continued)

19-55	42	19 - Probabilistic Risk Assessment and Severe Accident Evaluation	10/01/09	ML092750349	10/29/09	ML093070086
19-56	42	19 - Probabilistic Risk Assessment and Severe Accident Evaluation	10/01/09	ML092750349	12/23/09	ML093630679
19-57	42	19 - Probabilistic Risk Assessment and Severe Accident Evaluation	10/01/09	ML092750349	12/23/09	ML093630679
19-58	42	19 - Probabilistic Risk Assessment and Severe Accident Evaluation	10/01/09	ML092750349	10/29/09	ML093070086
19-59	42	19 - Probabilistic Risk Assessment and Severe Accident Evaluation	10/01/09	ML092750349	10/29/09	ML093070086
19-60	42	19 - Probabilistic Risk Assessment and Severe Accident Evaluation	10/01/09	ML092750349	10/29/09	ML093070086
19-61	42	19 - Probabilistic Risk Assessment and Severe Accident Evaluation	10/01/09	ML092750349	10/29/09	ML093070086
19-62	42	19 - Probabilistic Risk Assessment and Severe Accident Evaluation	10/01/09	ML092750349	11/13/09	ML093210475
19-63	42	19 - Probabilistic Risk Assessment and Severe Accident Evaluation	10/01/09	ML092750349	12/23/09	ML093630679
19-64	42	19 - Probabilistic Risk Assessment and Severe Accident Evaluation	10/01/09	ML092750349	12/23/09	ML093630679
19-65	42	19 - Probabilistic Risk Assessment and Severe Accident Evaluation	10/01/09	ML092750349	02/05/10	ML100481140
19-66	42	19 - Probabilistic Risk Assessment and Severe Accident Evaluation	10/01/09	ML092750349	11/13/09	ML093210475
19-67	42	19 - Probabilistic Risk Assessment and Severe Accident Evaluation	10/01/09	ML092750349	11/13/09	ML093210475
19-68	42	19 - Probabilistic Risk Assessment and Severe Accident Evaluation	10/01/09	ML092750349	08/13/10	ML102290038
19-69	42	19 - Probabilistic Risk Assessment and Severe Accident Evaluation	10/01/09	ML092750349	02/05/10	ML100481140
19-70	42	19 - Probabilistic Risk Assessment and Severe Accident Evaluation	10/01/09	ML092750349	10/29/09	ML093070086
19-71	42	19 - Probabilistic Risk Assessment and Severe Accident Evaluation	10/01/09	ML092750349	10/29/09	ML093070086
19-72	42	19 - Probabilistic Risk Assessment and Severe Accident Evaluation	10/01/09	ML092750349	02/05/10	ML100481140
19-74	42	19 - Probabilistic Risk Assessment and Severe Accident Evaluation	10/01/09	ML092750349	10/29/09	ML093070086
19-75	42	19 - Probabilistic Risk Assessment and Severe Accident Evaluation	10/01/09	ML092750349	12/23/09	ML093630679
19-76	42	19 - Probabilistic Risk Assessment and Severe Accident Evaluation	10/01/09	ML092750349	10/29/09	ML093070086

Application Title: Vogtle Nuclear Site, Units 3 and 4, Dockets 52-0025 and 52-0026 (continued)

19-77	19 - Probabilistic Risk Assessment and Severe Accident Evaluation	10/01/09	ML092750349	12/23/09	ML093630679	42
19-78	19 - Probabilistic Risk Assessment and Severe Accident Evaluation	10/01/09	ML092750349	10/29/09	ML093070086	42
19-79	19 - Probabilistic Risk Assessment and Severe Accident Evaluation	10/01/09	ML092750349	12/23/09	ML093630679	42
19-10	19 - Probabilistic Risk Assessment and Severe Accident Evaluation	09/30/09	ML092730351	10/30/09	ML093070284	40
19-81	19 - Probabilistic Risk Assessment and Severe Accident Evaluation	03/19/10	ML100850120	05/28/10	ML101530059	52
19-82	19 - Probabilistic Risk Assessment and Severe Accident Evaluation	03/19/10	ML100850120	05/05/10	ML101270075	52
19-83	19 - Probabilistic Risk Assessment and Severe Accident Evaluation	03/19/10	ML100850120	05/05/10	ML101270075	52
19-84	19 - Probabilistic Risk Assessment and Severe Accident Evaluation	03/19/10	ML100850120	05/05/10	ML101270075	52
19-85	19 - Probabilistic Risk Assessment and Severe Accident Evaluation	03/19/10	ML100850120	08/13/10	ML102290038	52
19-86	19 - Probabilistic Risk Assessment and Severe Accident Evaluation	03/19/10	ML100850120	05/05/10	ML101270075	52
19-87	19 - Probabilistic Risk Assessment and Severe Accident Evaluation	03/19/10	ML100850120	05/05/10	ML101270075	52
19-88	19 - Probabilistic Risk Assessment and Severe Accident Evaluation	03/19/10	ML100850120	05/05/10	ML101270075	52
19-89	19 - Probabilistic Risk Assessment and Severe Accident Evaluation	03/19/10	ML100850120	05/05/10	ML101270075	52
19-90	19 - Probabilistic Risk Assessment and Severe Accident Evaluation	03/19/10	ML100850120	05/05/10	ML101270075	52
19-91	19 - Probabilistic Risk Assessment and Severe Accident Evaluation	03/19/10	ML100850120	05/28/10	ML101530059	52
19-92	19 - Probabilistic Risk Assessment and Severe Accident Evaluation	03/19/10	ML100850120	05/05/10	ML101270075	52
19-93	19 - Probabilistic Risk Assessment and Severe Accident Evaluation	03/19/10	ML100850120	05/28/10	ML101530059	52
19-94	19 - Probabilistic Risk Assessment and Severe Accident Evaluation	03/19/10	ML100850120	05/05/10	ML101270075	52
19-100	19 - Probabilistic Risk Assessment and Severe Accident Evaluation	04/09/10	ML100980600	06/04/10	ML101590403	54
19-101	19 - Probabilistic Risk Assessment and Severe Accident Evaluation	04/09/10	ML100980600	05/24/10	ML101460301	54
19-102	19 - Probabilistic Risk Assessment and Severe Accident Evaluation	04/09/10	ML100980600	10/08/10	ML102861726	54

Application Title: Vogtle Nuclear Site, Units 3 and 4, Dockets 52-0025 and 52-0026 (continued)

19-103	54	19 - Probabilistic Risk Assessment and Severe Accident Evaluation	04/09/10	ML100980600	05/24/10	ML101460301
19-95	54	19 - Probabilistic Risk Assessment and Severe Accident Evaluation	04/09/10	ML100980600	05/24/10	ML101460301
19-96	54	19 - Probabilistic Risk Assessment and Severe Accident Evaluation	04/09/10	ML100980600	05/24/10	ML101460301
19-98	54	19 - Probabilistic Risk Assessment and Severe Accident Evaluation	04/09/10	ML100980600	05/24/10	ML101460301
19-99	54	19 - Probabilistic Risk Assessment and Severe Accident Evaluation	04/09/10	ML100980600	05/24/10	ML101460301
19-104	0	19 - Probabilistic Risk Assessment and Severe Accident Evaluation	09/20/10	N/A	09/20/10	ML102650088
19-105	0	19 - Probabilistic Risk Assessment and Severe Accident Evaluation	11/12/10	N/A	11/12/10	ML103200368

APPENDIX D. REFERENCES

American Concrete Institute (ACI)

— — — — —, ACI-349, "Code Requirements for Nuclear Safety Related Concrete Structures"

American National Standards Institute (ANSI)

— — — — —, ANSI 15.8-2009, "Material Control Systems – Special Nuclear Material Control and Accounting Systems for Nuclear Power Plants"

— — — — —, ANSI B30.2, "Overhead and Gantry Cranes"

— — — — —, ANSI B30.9, "Slings"

— — — — —, ANSI N13.5-1972, "Performance Specifications for Direct Reading and Indirect Reading Pocket Dosimeters for X- and Gamma Radiation"

— — — — —, ANSI N13.27-1981, "Performance Specifications for Pocket-Sized Alarming Dosimeters/Ratemeters"

— — — — —, ANSI N14.6, "Special Lifting Devices for Shipping Containers Weighing 10,000 Pounds or More"

— — — — —, ANSI N18.7, "Administrative Controls and Quality Assurance for the Operational Phase of Nuclear Power Plants"

— — — — —, ANSI N42.17A-1989, "Performance Specifications for Health Physics Instrumentation–Portable Instrumentation for Use in Normal Environmental Conditions"

— — — — —, ANSI N42.18, "Specification and Performance of On-Site Instrumentation for Continuously Monitoring Radioactivity in Effluents"

— — — — —, ANSI N322-1997, "ANSI Test, Construction, and Performance Requirements for Direct Reading Electrostatic/Electroscope Type Dosimeters"

— — — — —, ANSI N323A-1997, "Radiation Protection Instrumentation Test and Calibration, Portable Survey Instruments"

— — — — —, ANSI N545-1975, "Performance, Testing, and Procedural Specification for TLD, Environmental Application"

American National Standards Institute/American Nuclear Society (ANSI/ANS)

— — — — —, ANSI/ANS 3.1-1993, "American National Standard for Selection, Qualification, and Training of Personnel for Nuclear Power Plants"

— — — — —, ANSI/ANS 57.1-1992, "Design Requirements for LWR Fuel Handling Systems"

American National Standards Institute/Health Physics Society (ANSI/HPS)

— — — — —, ANSI/HPS N13.1, "Sampling and Monitoring Releases of Airborne Radioactive Substances from the Stacks and Ducts of Nuclear Facilities"

American National Standards Institute/Instrument Society of America (ANSI/ISA)

— — — — —, ANSI/ISA Standard 67.04-2000, "Setpoints for Nuclear Safety-Related Instrumentation"

American Nuclear Society/International Standardization Organization/International Electrotechnical Commission (ANS/ISO/IEC)

— — — — —, ANS/ISO/IEC 17025, "General Requirements for the Competence of Testing and Calibration Laboratories"

American Society of Civil Engineers (ASCE)

— — — — —, ASCE 7-98, "Minimum Design Loads for Buildings and Other Structures"

American Society of Mechanical Engineers (ASME)

ASME Code Cases

— — — — —, ASME Code Case N-729-1, "Alternative Examination Requirements for Pressurized-Water Reactor (PWR) Vessel Upper Heads With Nozzles Having Pressure-Retaining Partial-Penetration Welds"

— — — — —, ASME Operation and Maintenance (OM) Code Case OMN-1, "Alternative Rules for the Preservice and Inservice Testing of Certain Electric Motor-Operated Valve Assemblies in Light Water Reactor Power Plants"

— — — — —, ASME Operation and Maintenance (OM) Code Case OMN-11, "Risk-Informed Testing of Motor-Operated Valves"

Other ASME Documents

— — — — —, ASME AG-1-1997, "Code on Nuclear Air and Gas Treatment"

— — — — —, ASME B31.1, "Power Piping"

— — — — —, ASME N509-1989, "Nuclear Power Plant Air-Cleaning Units and Components"

— — — — —, ASME N510-1989, "Testing of Nuclear Air-Treatment Systems"

— — — — —, ASME NOG-1, "Rules for Construction of Overhead and Gantry Cranes (Top Running Bridge, Multiple Girder)"

— — — — —, ASME Operation and Maintenance (OM) Code (ASME OM Code)

— — — — —, ASME Standard NQA-1-1994, "Quality Assurance Requirements for Nuclear Facility Applications"

— — — — —, ASME Standard QME-1-2007, "Qualification of Active Mechanical Equipment Used in Nuclear Power Plants"

American Society for Testing and Materials (ASTM)

— — — — —, ASTM D3359, "Test Methods for Measuring Adhesion by Tape Test"

— — — — —, ASTM D3911-03, "Standard Test Method for Evaluating Coatings Used in Light Water Nuclear Power Plants at Simulated Design Basis Accident (DBA) Conditions"

— — — — —, ASTM D4176, "Standard Test Method for Free Water and Particulate Contamination in Distillate Fuels (Visual Inspection Procedures)"

— — — — —, ASTM D5144-08, "Standard Guide for Use of Protective Coating Standards in Nuclear Power Plants"

— — — — —, ASTM D5163-05a, "Standard Guide for Establishing Procedures to Monitor the Performance of Coating Service Level I Coating Systems in an Operating Nuclear Power Plant"

— — — — —, ASTM D7167-05, "Standard Guide for Establishing Procedures to Monitor the Performance of Safety-Related Coating Service Level III Lining

— — — — —, ASTM D975, "Standard Specification for Diesel Fuel Oils"

— — — — —, ASTM E-185 Annual Book of ASTM Standards, Part 30

American Water Works Association (AWWA)

— — — — —, AWWA C906, "Polyethylene (PE) Pressure Pipe and Fittings, 4 in (100mm) through 63 in (1,575mm), for Water Distribution and Transmission"

Electric Power Research Institute (EPRI)

————, EPRI NP-4354, "Large Scale Hydrogen Burn Equipment Experiments"

————, EPRI NP-5930, "A Criterion for Determining Exceedance of the Operating Basis Earthquake"

————, EPRI NP-6695, "Guidelines for Nuclear Plant Response to an Earthquake"

————, EPRI NSAC-202L, "Recommendations for an Effective Flow-Accelerated Corrosion Program"

————, EPRI TR-100082, "Standardization of the Cumulative Absolute Velocity"

————, EPRI TR-1002884, "Pressurized Water Reactor Primary Water Chemistry Guidelines: Volume 1"

————, EPRI TR-102134-R5, "PWR Secondary Water Chemistry Guidelines"

Institute of Electrical and Electronics Engineers (IEEE)

————, IEEE Standard 80, "Guide for Safety in AC Substation Grounding"

————, IEEE Standard 323, "IEEE Standard for Qualifying Class 1E Equipment for Nuclear Power Generating Stations"

————, IEEE Standard 336-1985, "IEEE Standard Installation, Inspection, and Testing Requirements for Power, Instrumentation, and Control Equipment at Nuclear Facilities"

————, IEEE Standard 384, "IEEE Standard Criteria for Independence of Class 1E Equipment and Circuits"

————, IEEE Standard 450, "Recommended Practice for the Maintenance, Testing, and Replacement of Vented Lead-Acid Batteries for Stationary Applications"

————, IEEE Standard 498-1985, "IEEE Standard Requirements for the Calibration and Control of Measuring and Test Equipment Used in Nuclear Facilities"

————, IEEE Standard 603-1980, "IEEE Standard Criteria for Safety Systems for Nuclear Power Generating Stations"

————, IEEE Standard 665, "Guide for Generating Station Grounding"

U.S. Code of Federal Regulations

— — — — —, *Title 10, Energy*, 2.390, "Public inspections, exemptions, requests for withholding"

— — — — —, *Title 10, Energy*, Part 11, "Criteria and procedures for determining eligibility for access to or control over special nuclear material"

— — — — —, *Title 10, Energy*, 11.11, "General requirements"

— — — — —, *Title 10, Energy*, Part 19, "Notices, instructions and reports to workers: inspection and investigations"

— — — — —, *Title 10, Energy*, 19.12, "Instructions to workers"

— — — — —, *Title 10, Energy*, Part 20, "Standards for protection against radiation"

— — — — —, *Title 10, Energy*, Part 20, Appendix B, "Annual Limits on Intake (ALIs) and Derived Air Concentrations (DACs) of Radionuclides for Occupational Exposure; Effluent Concentrations; Concentrations for Release to Sewerage"

— — — — —, *Title 10, Energy*, 20.1003, "Definitions"

— — — — —, *Title 10, Energy*, 20.1004, "Units of radiation dose"

— — — — —, *Title 10, Energy*, 20.1101, "Radiation protection programs"

— — — — —, *Title 10, Energy*, 20.1204, "Determination of internal exposure"

— — — — —, *Title 10, Energy*, 20.1301, "Dose limits for individual members of the public"

— — — — —, *Title 10, Energy*, 20.1302, "Compliance with dose limits for individual members of the public"

— — — — —, *Title 10, Energy*, 20.1406, "Minimization of contamination"

— — — — —, *Title 10 Energy, 20.1501*, "General"

— — — — —, *Title 10, Energy*, 20.1502, "Conditions requiring individual monitoring of external and internal occupational dose"

— — — — —, *Title 10, Energy*, 20.1601, "Control of access to high radiation areas"

— — — — —, *Title 10, Energy*, 20.1602, "Control of access to very high radiation areas"

— — — — —, *Title 10, Energy*, 20.1801, "Security of stored material"

— — — — —, *Title 10, Energy*, 20.1802, "Control of material not in storage"

— — — — —, *Title 10, Energy*, Part 21, "Reporting of defects and noncompliance"

—————, *Title 10, Energy*, Part 26, "Fitness for duty programs"

—————, *Title 10, Energy*, 26.3, "Scope"

—————, *Title 10, Energy*, 26.4, "FFD program applicability to categories of individuals"

—————, *Title 10, Energy*, 26.205, "Work hours"

—————, *Title 10, Energy*, Part 30, "Rules of general applicability to domestic licensing of byproduct material"

—————, *Title 10, Energy*, 30.18, "Exempt quantities"

—————, *Title 10, Energy*, 30.32, "Application for specific licenses"

—————, *Title 10, Energy*, 30.72, "Schedule C-Quantities of radioactive materials requiring consideration of the need for an emergency plan for responding to a release"

—————, *Title 10, Energy*, Part 31, "General domestic licenses for byproduct material"

—————, *Title 10, Energy*, Part 32, "Specific domestic licenses to manufacture or transfer certain items containing byproduct material"

—————, *Title 10, Energy*, Part 33, "Specific domestic licenses of broad scope for byproduct material"

—————, *Title 10, Energy*, Part 34, "Licenses for industrial radiography and radiation safety requirements for industrial radiographic operations"

—————, *Title 10, Energy*, Part 40, "Domestic licensing of source material"

—————, *Title 10, Energy*, 40.31, "Application for specific licenses"

—————, *Title 10, Energy*, Part 50, "Domestic licensing of production and utilization facilities"

—————, *Title 10, Energy*, Part 50, Appendix A, "General Design Criteria for Nuclear Power Plants"

—————, *Title 10, Energy*, Part 50, Appendix A, GDC 1, "Quality Standards and Records"

—————, *Title 10, Energy*, Part 50, Appendix A, GDC 2, "Design Bases for Protection Against Natural Phenomena"

—————, *Title 10, Energy*, Part 50, Appendix A, GDC 3, "Fire Protection"

—————, *Title 10, Energy*, Part 50, Appendix A, GDC 4, "Environmental and Dynamic Effects Design Bases"

—————, *Title 10, Energy*, Part 50, Appendix A, GDC 5, "Sharing of Structures, Systems, and Components"

—————, *Title 10, Energy*, Part 50, Appendix A, GDC 13, "Instrumentation and Control"

—————, *Title 10, Energy*, Part 50, Appendix A, GDC 14, "Reactor Coolant Pressure Boundary"

—————, *Title 10, Energy*, Part 50, Appendix A, GDC 17, "Electric Power Systems"

—————, *Title 10, Energy*, Part 50, Appendix A, GDC 18, "Inspection and Testing of Electrical Power Systems"

—————, *Title 10, Energy*, Part 50, Appendix A, GDC 19, "Control Room"

—————, *Title 10, Energy*, Part 50, Appendix A, GDC 26, "Reactivity Control System Redundancy and Capability"

—————, *Title 10, Energy*, Part 50, Appendix A, GDC 27, "Combined Reactivity Control Systems Capability"

—————, *Title 10, Energy*, Part 50, Appendix A, GDC 29, "Protection Against Anticipated Operational Occurrences"

—————, *Title 10, Energy*, Part 50, Appendix A, GDC 32, "Inspection of Reactor Coolant Pressure Boundary"

—————, *Title 10, Energy*, Part 50, Appendix A, GDC 44, "Cooling Water"

—————, *Title 10, Energy*, Part 50, Appendix A, GDC 45, "Inspection of Cooling Water System"

—————, *Title 10, Energy*, Part 50, Appendix A, GDC 52, "Capability for Containment Leakage Rate Testing"

—————, *Title 10, Energy*, Part 50, Appendix A, GDC 53, "Provisions for Containment Testing and Inspection"

—————, *Title 10, Energy*, Part 50, Appendix A, GDC 54, "Piping System Penetrating Containment"

—————, *Title 10, Energy*, Part 50, Appendix A, GDC 60, "Control of Releases of Radioactive Materials to the Environment"

—————, *Title 10, Energy*, Part 50, Appendix A, GDC 61, "Fuel Storage and Handling and Radioactivity Control"

—————, *Title 10, Energy*, Part 50, Appendix A, GDC 64, "Monitoring Radioactivity Releases"

—————, *Title 10, Energy*, Part 50, Appendix B, "Quality Assurance Criteria for Nuclear Power Plants and Fuel Processing Plants"

—————, *Title 10, Energy*, Part 50, Appendix E, "Emergency Planning and Preparedness for Production and Utilization Facilities"

—————, *Title 10, Energy*, Part 50, Appendix G, "Fracture Toughness Requirements"

—————, *Title 10, Energy*, Part 50, Appendix H, "Reactor Vessel Material Surveillance Program Requirements"

—————, *Title 10, Energy*, Part 50, Appendix I, "Numerical Guides for Design Objectives and Limiting Conditions for Operation to Meet the Criterion 'As Low as is Reasonably Achievable' for Radioactive Material in Light-Water-Cooled Nuclear Power Reactor Effluents"

—————, *Title 10, Energy*, Part 50, Appendix J, "Primary Reactor Containment Leakage Testing for Water-Cooled Power Reactors"

—————, *Title 10, Energy*, Part 50, Appendix K, "ECCS Evaluation Models"

—————, *Title 10, Energy*, Part 50, Appendix S, "Earthquake Engineering Criteria for Nuclear Power Plants"

—————, *Title 10, Energy*, 50.2, "Definitions"

—————, *Title 10, Energy*, 50.9, "Completeness and accuracy of information"

—————, *Title 10, Energy*, 50.10, "License required; limited work authorization"

—————, *Title 10, Energy*, 50.12, "Specific exemptions"

—————, *Title 10, Energy*, 50.33, "Contents of applications; general information"

—————, *Title 10, Energy*, 50.34, "Contents of applications; technical information"

—————, *Title 10, Energy*, 50.34a, "Design objectives for equipment to control releases of radioactive material in effluents—nuclear power reactors"

—————, *Title 10, Energy*, 50.34(a), "Preliminary safety analysis report"

—————, *Title 10, Energy*, 50.34(b), "Final safety analysis report"

—————, *Title 10, Energy*, 50.36, "Technical specifications"

—————, *Title 10, Energy*, 50.36a, "Technical specifications on effluents from nuclear power reactors"

—————, *Title 10, Energy*, 50.40, "Common Standards"

—————, *Title 10, Energy*, 50.43, "Additional standards and provisions affecting class 103 licenses and certifications for commercial power"

—————, *Title 10, Energy*, 50.47, "Emergency plans"

— — — — —, *Title 10, Energy*, 50.48, "Fire protection"

— — — — —, *Title 10, Energy*, 50.49, "Environmental qualification of electric equipment important to safety for nuclear power plants"

— — — — —, *Title 10, Energy*, 50.54, "Conditions of licenses"

— — — — —, *Title 10, Energy*, 50.55, "Conditions of construction permits, early site permits, combined licenses, and manufacturing licenses"

— — — — —, *Title 10, Energy*, 50.55a, "Codes and standards"

— — — — —, *Title 10, Energy*, 50.59, "Changes, tests and experiments"

— — — — —, *Title 10, Energy*, 50.60, "Acceptance criteria for fracture prevention measures for lightwater nuclear power reactors for normal operation"

— — — — —, *Title 10, Energy*, 50.61, "Fracture toughness requirements for protection against pressurized thermal shock events"

— — — — —, *Title 10, Energy*, 50.63, "Loss of all alternating current power"

— — — — —, *Title 10, Energy*, 50.65, "Requirements for monitoring the effectiveness of maintenance at nuclear power plants"

— — — — —, *Title 10, Energy*, 50.68, "Criticality accident requirements"

— — — — —, *Title 10, Energy*, 50.71, "Maintenance of records, making of reports"

— — — — —, *Title 10, Energy*, 50.72, "Immediate notification requirements for operating nuclear power reactors"

— — — — —, *Title 10, Energy*, 50.73, "Licensee event report system"

— — — — —, *Title 10, Energy*, 50.75, "Reporting and recordkeeping for decommissioning planning"

— — — — —, *Title 10, Energy*, 50.90, "Application for amendment of license, construction permit, or early site permit"

— — — — —, *Title 10, Energy*, 50.120, "Training and qualification of nuclear power plant personnel"

— — — — —, *Title 10, Energy*, Part 51, "Environmental protection regulations for domestic licensing and related regulatory functions"

— — — — —, *Title 10, Energy*, 51-49, "Environmental report – limited work Authorization"

— — — — —, *Title 10, Energy*, 51.50, "Environmental report-construction permit, early site permit, or combined license stage"

— — — — —, *Title 10, Energy*, 51.75, "Draft environmental impact statement—construction permit, early site permit, or combined license"

— — — — —, *Title 10, Energy*, 51.92, "Supplement to the final environmental impact statement"

— — — — —, *Title 10, Energy*, Part 52, "Licenses, certifications and approvals for nuclear power plants"

— — — — —, *Title 10, Energy*, Part 52, Appendix D, "Design Certification Rule for the AP1000 Design"

— — — — —, *Title 10, Energy*, 52.6, "Completeness and accuracy of information"

— — — — —, *Title 10, Energy*, 52.7, "Specific exemptions"

— — — — —, *Title 10, Energy*, 52.17, "Contents of applications; technical information"

— — — — —, *Title 10, Energy*, 52.34, "Contents of applications; technical information"

— — — — —, *Title 10, Energy*, 52.39, "Finality of early site permit determinations"

— — — — —, *Title 10, Energy*, 52.47, "Contents of applications; technical information"

— — — — —, *Title 10, Energy*, 52.63, "Finality of standard design certifications"

— — — — —, *Title 10, Energy*, 52.77, "Contents of applications; general information"

— — — — —, *Title 10, Energy*, 52.79, "Contents of applications; technical information in final safety analysis report"

— — — — —, *Title 10, Energy*, 52.80, "Contents of applications; additional technical information"

— — — — —, *Title 10, Energy*, 52.81, "Standards for review of applications"

— — — — —, *Title 10, Energy*, 52.83, "Finality of referenced NRC approvals; partial initial decision on site suitability"

— — — — —, *Title 10, Energy*, 52.85, "Administrative review of applications; hearings"

— — — — —, *Title 10, Energy*, 52.87, "Referral to the Advisory Committee on Reactor Safeguards (ACRS)"

— — — — —, *Title 10, Energy*, 52.93, "Exemptions and variances"

— — — — —, *Title 10, Energy*, 52.97, "Issuance of combined licenses"

— — — — —, *Title 10, Energy*, 52.98, "Finality of combined licenses; information requests"

— — — — —, *Title 10, Energy*, 52.99, "Inspection during construction"

—————, *Title 10, Energy*, 52.103, "Operation under a combined license"

—————, *Title 10, Energy*, Part 54, Requirements for renewal of operating licenses for nuclear power plants"

—————, *Title 10, Energy*, Part 55, "Operator's licenses"

—————, *Title 10, Energy*, 55.13, "General exemptions"

—————, *Title 10, Energy*, 55.31, "How to apply"

—————, *Title 10, Energy*, 55.41, "Written examinations: Operators"

—————, *Title 10, Energy*, 55.43, "Written examinations: Senior operators"

—————, *Title 10, Energy*, 55.45, "Operating tests"

—————, *Title 10, Energy*, 55.59, "Requalification"

—————, *Title 10, Energy*, Part 61, "Licensing requirements for land disposal of radioactive waste"

—————, *Title 10, Energy*, 61.55, "Waste classification"

—————, *Title 10, Energy*, 61.56, "Waste characteristics"

—————, *Title 10, Energy*, Part 70, "Domestic licensing of special nuclear material"

—————, *Title 10, Energy*, 70.17, "Specific exemptions"

—————, *Title 10, Energy*, 70.22, "Contents of applications"

_ _ _ _ _ _, *Title 10, Energy*, 70.24, "Criticality accident requirements"

—————, *Title 10, Energy*, 70.32, "Conditions of licenses"

—————, *Title 10, Energy*, Part 71, "Packaging and transportation of radioactive material"

—————, *Title 10, Energy*, Part 73, "Physical protection of plants and materials"

—————, *Title 10, Energy*, Part 73, Appendix B, "General Criteria for Security Personnel"

—————, *Title 10, Energy*, Part 73, Appendix C, "Nuclear Power Plant Safeguards Contingency Plans"

—————, *Title 10, Energy*, Part 73, Appendix G, "Reportable Safeguards Events"

—————, *Title 10, Energy*, 73.1, "Purpose and scope"

—————, *Title 10, Energy*, 73.2, "Definitions"

—————, *Title 10, Energy*, 73.21, "Protection of safeguards information: performance requirements"

—————, *Title 10, Energy*, 73.45, "Performance capabilities for fixed site physical protection systems"

—————, *Title 10, Energy*, 73.46, "Fixed site physical protection systems, subsystem, components, and procedures"

—————, *Title 10, Energy*, 73.54, "Protection of digital computer and communication systems and networks"

—————, *Title 10, Energy*, 73.55, "Requirements for physical protection of licensed activities in nuclear power reactors against radiological sabotage"

—————, *Title 10, Energy*, 73.55, Appendix B, "General Criteria for Security Personnel"

—————, *Title 10, Energy*, 73.55, Appendix C, "Nuclear Power Plant Safeguards Contingency Plans"

—————, *Title 10, Energy*, 73.55, Appendix G, "Reportable Safeguards Events"

—————, *Title 10, Energy*, 73.55, Appendix H, "Weapons Qualification Criteria"

—————, *Title 10, Energy*, 73.56, "Personnel access authorization requirements for nuclear power plants"

—————, *Title 10, Energy*, 73.57, "Requirements for criminal history records checks of individuals granted unescorted access to a nuclear power facility or access to safeguards information"

—————, *Title 10, Energy*, 73.58, "Safety/security interface requirements for nuclear power reactors"

—————, *Title 10, Energy*, 73.67, "Licensee fixed site and in-transit requirements for the physical protection of special nuclear material of moderate and low strategic significance"

—————, *Title 10, Energy*, 73.70, "Records"

—————, *Title 10, Energy*, 73.71, "Reporting of safeguards events"

—————, *Title 10, Energy*, Part 74, "Material control and accounting of special nuclear material"

—————, *Title 10, Energy*, 74.4, "Definitions"

—————, *Title 10, Energy*, 74.7, "Specific exemptions"

—————, *Title 10, Energy*, 74.11, "Reports of loss or theft or attempted theft or unauthorized production of special nuclear material"

— — — — —, *Title 10, Energy*, 74.13, "Material status reports"

— — — — —, *Title 10, Energy*, 74.15, "Nuclear material transaction reports"

— — — — —, *Title 10, Energy*, 74.19, "Recordkeeping"

— — — — —, *Title 10, Energy*, 74.31, "Nuclear material control and accounting for special nuclear material of low strategic significance"

— — — — —, *Title 10, Energy*, 74.33, "Nuclear material control and accounting for uranium enrichment facilities authorized to produce special nuclear material of low strategic significance"

— — — — —, *Title 10, Energy*, 74.41, "Nuclear material control and accounting for special nuclear material of moderate strategic significance"

— — — — —, *Title 10, Energy*, 74.51, "Nuclear material control and accounting for strategic special nuclear material"

— — — — —, *Title 10, Energy*, Part 100, "Reactor site criteria"

— — — — —, *Title 10, Energy*, 100.20, "Factors to be considered when evaluating sites"

— — — — —, *Title 10, Energy*, 100.21, "Non-seismic site criteria"

— — — — —, *Title 10, Energy*, 100.23, "Geologic and seismic siting criteria"

— — — — —, *Title 10, Energy*, Part 140, "Financial protection requirements and indemnity agreements"

— — — — —, *Title 40, Energy*, Part 190, "Environmental Radiation Protection Standards for Nuclear Power Operations"

— — — — —, *Title 44, Energy*, Part 353, "Memorandum of Understanding (MOU) Between Federal Emergency Management Agency and Nuclear Regulatory Commission Relating to Radiological Emergency Planning and Preparedness"

— — — — —, *Title 49, Energy*, Part 173, "Shippers—General Requirements for Shipments and Packagings"

U.S. Environmental Protection Agency (EPA)

— — — — —, "ALOHA (Areal Location of Hazardous Atmospheres)," Version 5.4.1, February 2007

U.S. Nuclear Regulatory Commission (NRC)

Generic Communications

Bulletin

— — — — —, 80-15, "Possible Loss of Emergency Notification System (ENS) with Loss of Offsite Power"

— — — — —, 88-11, "Pressurizer Surge Line Thermal Stratification"

— — — — —, 2003-01, "Potential Impact of Debris Blockage on Emergency Sump Recirculation at Pressurized-Water Reactors"

Commission Papers

— — — — —, CMWCO-10-0001, "Regulation of Cyber Systems in an Operating Nuclear Power Plant"

— — — — —, SECY-05-0197, "Review of Operational Programs in a Combined License Application and Generic Emergency Planning Inspections, Tests, Analyses, and Acceptance Criteria"

— — — — —, SECY-06-0187, "Semiannual Update of The Status of New Reactor Licensing Activities and Future Planning for New Reactors," dated November 16, 2006

— — — — —, SECY-93-087, "Policy, Technical, and Licensing Issues Pertaining to Evolutionary and Advanced Light-Water Reactor (ALWR) Designs"

— — — — —, SECY-94-084, "Policy and Technical Issues Associated with the Regulatory Treatment of Non-Safety Systems in Passive Plant Designs"

— — — — —, SECY-95-132, "Policy and Technical Issues Associated with the Regulatory Treatment of Non-safety Systems in Passive Plant Designs"

Generic Letter

— — — — —, GL 80-009, "Low Level Radioactive Waste Disposal"

— — — — —, GL 81-038, "Storage of Low-Level Radioactive Wastes at Power Reactor Sites"

— — — — —, GL 81-039, "NRC Volume Reduction Policy"

— — — — —, GL 85-05, "Inadvertent Boron Dilution Events"

— — — — —, GL 88-05, "Staff Position on Boric Acid Corrosion of Carbon Steel Reactor Pressure Boundary Components in PWR Plants"

— — — — —, GL 89-02, "Actions to Improve the Detection of Counterfeit and Fraudulently Marked Products"

— — — — —, GL 89-08, "Erosion/Corrosion-Induced Pipe Wall Thinning"

— — — — —, GL 91-05, "Licensee Commercial-Grade Procurement and Dedication Programs"

— — — — —, GL 92-01, "Reactor Vessel Structural Integrity"

— — — — —, GL 96-03, "Relocation of the Pressure Temperature Limit Curves and Low Temperature Overpressure Protection System Limits"

— — — — —, GL 96-05, "Periodic Verification of Design-Basis Capability of Safety-Related Motor-Operated Valves"

— — — — —, GL 97-06, "Degradation of Steam Generator Internals"

— — — — —, GL 2004-02, "Potential Impact of Debris Blockage on Emergency Recirculation during Design Basis Accidents at Pressurized-Water Reactors"

— — — — —, GL 2006-2, "Grid Reliability and the Impact on Plant Risk and the Operability of Offsite Power"

Generic Safety Issue

— — — — —, GSI-43, "Reliability of Air Systems"

— — — — —, GSI-83, "Control Room Habitability"

— — — — —, GSI-163, "Multiple Steam Generator Tube Leakage"

— — — — —, GSI-191, "Assessment of Debris Accumulation on PWR Sump Performance"

Information Notice

— — — — —, IN 86-83, "Underground Pathways into Protected Areas, Vital Areas, and Controlled Access Areas," September 19, 1986

Interim Staff Guidance

— — — — —, DC/COL-ISG-3, "Probabilistic Risk Assessment Information to Support Design Certification and Combined License Applications"

— — — — —, DC/COL-ISG-7, "Interim Staff Guidance on Assessment of Normal and Extreme Winter Precipitation Loads on the Roofs of Seismic Category I Structures"

— — — — —, DC/COL-ISG-8, "Necessary Content of Plant-Specific Technical Specifications When a Combined License is Issued"

— — — — —, DC/COL-ISG-16, "Compliance with 10 CFR 50.54(hh)(2) and 10 CFR 52.80(d) Loss of Large Areas of the Plant due to Explosions or Fires from a Beyond-Design Basis Event" **(Not Publicly Available)**

— — — — —, DC/COL-ISG-20, "Implementation of a Probabilistic Risk Assessment-Based Seismic Margin Analysis for New Reactors"

— — — — —, ISG-1, "Interim Staff Guidance on Seismic Issues Associated with High Frequency Ground Motion in Design Certification and Combined License Applications"

— — — — —, ISG-15, "Final Interim Staff Guidance on the Post-Combined License Commitments"

NUREG-Series Reports

— — — — —, NUREG-0570, "Toxic Vapor Concentrations in the Control Room Following a Postulated Accidental Release"

— — — — —, NUREG-0588, "Interim Staff Position on Environmental Qualification of Safety-Related Electrical Equipment"

— — — — —, NUREG-0612, "Control of Heavy Loads at Nuclear Power Plants"

— — — — —, NUREG-0654/FEMA-REP-1, "Criteria for Preparation and Evaluation of Radiological Emergency Response Plans and Preparedness in Support of Nuclear Power Plants," Revision 1

— — — — —, NUREG-0696, "Functional Criteria for Emergency Response Facilities"

— — — — —, NUREG-0711, "Human Factors Engineering Program Review Model," Revision 2

— — — — —, NUREG-0737, "Clarification of TMI Action Plan Requirements"

— — — — —, NUREG-0800, "Standard Review Plan for the Review of Safety Analysis Reports for Nuclear Power Plants (LWR Edition)"

— — — — —, NUREG-0927, "Evaluation of Water Hammer Occurrence in Nuclear Power Plants," Revision 1

— — — — —, NUREG-0933, "Resolution of Generic Safety Issues (Formerly entitled 'A Prioritization of Generic Safety Issues')"

— — — — —, NUREG-1021, "Operator Licensing Examination Standards for Power Reactors"

— — — — —, NUREG-1022, "Event Reporting Guidelines: 10 CFR 50.72 and 50.73," Revision 2

— — — — —, NUREG-1307, "Report on Waste Burial Charges: Changes in Decommissioning Waste Disposal Costs at Low-Level Waste Burial Facilities," Revision 12

— — — — —, NUREG-1407, "Procedural and Submittal Guidance for the Individual Plant Examination of External Events (IPEEE) for Severe Accident Vulnerabilities"

— — — — —, NUREG-1431, "Standard Technical Specifications — Westinghouse Plants"

— — — — —, NUREG-1482, "Guidelines for Inservice Testing at Nuclear Power Plants"

— — — — —, NUREG-1555, "Standard Review Plans for Environmental Reviews for Nuclear Power Plants," Supplement 1

— — — — —, NUREG-1556, "Consolidated Guidance about Materials Licenses"

— — — — —, NUREG-1577, "Standard Review Plan on Power Reactor Licensee Financial Qualifications and Decommissioning Funding Assurance"

— — — — —, NUREG-1736, "Consolidated Guidance: 10 CFR Part 20 – Standards for Protection Against Radiation"

— — — — —, NUREG/CR-1748, "Hazards to Nuclear Power Plants from Nearby Accidents Involving Hazardous Materials-A Preliminary Assessment"

— — — — —, NUREG-1793, "Final Safety Evaluation Report Related to Certification of the AP1000 Standard Design"

— — — — —, NUREG-1801, "Generic Aging Lessons Learned (GALL) Report," Volume 2, Revision 2

— — — — —, NUREG-1872, "Final Environmental Impact Statement for an Early Site Permit (ESP) at the Vogtle Electric Generating Plant Site"

— — — — —, NUREG-1923, "Safety Evaluation Report for an Early Site Permit (ESP) at the Vogtle Electric Generating Plant (VEGP) ESP Site"

— — — — —, NUREG-1947, "Final Supplemental Environmental Impact Statement for Combined Licenses (COLs) for Vogtle Electric Generating Plant Units 3 and 4"

— — — — —, NUREG/CR-2858, "PAVAN: An Atmospheric Dispersion Program for Evaluating Design Basis Accidental Releases of Radioactive Materials from Nuclear Power Stations"

— — — — —, NUREG/CR-6190, "Update of NUREG/CR-6190 Material to Reflect Postulated Threat Requirements" **(Includes security-related or safeguards information and is not publicly available)**

— — — — —, NUREG/CR-6331, "Atmospheric Relative Concentrations in Building Wakes," PNNL-10521, Revision 1

— — — — —, NUREG/CR-7000, "Essential Elements of an Electric Cable Condition Monitoring Program"

Regulatory Guide

—————, RG 1.8, "Qualification and Training of Personnel for Nuclear Power Plants," Revision 3

—————, RG 1.12, "Nuclear Power Plant Instrumentation for Earthquakes," Revision 2

—————, RG 1.16, "Reporting of Operating Information"

—————, RG 1.21, "Measuring, Evaluating, and Reporting Radioactive Material in Liquid and Gaseous Effluents and Solid Waste," Revision 2

—————, RG 1.23, "Meteorological Monitoring Programs for Nuclear Power Plants," Revision 1

—————, RG 1.26, "Quality Group Classification and Standards for Water-, Steam-, and Radioactive-Waste-Containing Components of Nuclear Power Plants," Revision 4

—————, RG 1.27, "Ultimate Heat Sink for Nuclear Power Plants," Revision 2

—————, RG 1.28, "Quality Assurance Program Criteria (Design and Construction)," Revision 4

—————, RG 1.29, "Seismic Design Classification," Revision 4

—————, RG 1.30, "Quality Assurance Requirements for the Installation, Inspection, and Testing of Instrumentation and Electric Equipment (Safety Guide 30)"

—————, RG 1.31, "Control of Ferrite Content in Stainless Steel Weld Metal," Revision 3

—————, RG 1.33, "Quality Assurance Program Requirements (Operation)," Revision 2

—————, RG 1.37, "Quality Assurance Requirements for Cleaning of Fluid Systems and Associated Components of Water-Cooled Nuclear Power Plants," Revision 1

—————, RG 1.38, "Quality Assurance Requirements for Packaging, Shipping, Receiving, Storage, and Handling of Items for Water-Cooled Nuclear Power Plants," **(Withdrawn -- See 75 FR 54921, 09/09/2010)**

—————, RG 1.39, "Housekeeping Requirements for Water-Cooled Nuclear Power Plants," **(Withdrawn -- See 75 FR 70044, 11/16/2010)**

—————, RG 1.44, "Control of the Use of Sensitized Steel," Revision 0

—————, RG 1.45, "Guidance on Monitoring and Responding to Reactor Coolant System Leakage," Revision 1

—————, RG 1.52, "Design, Inspection, and Testing Criteria for Air Filtration and Adsorption Units of Post Accident Engineered Safety Feature Atmosphere Cleanup Systems in Light Water Cooled Nuclear Power Plants," Revision 3

— — — — —, RG 1.54, "Service Level I, II, and III Protective Coatings Applied to Nuclear Power Plants," Revision 1

— — — — —, RG 1.59, "Design Basis Floods for Nuclear Power Plants," Revision 2

— — — — —, RG 1.60, "Design Response Spectra for Seismic Design of Nuclear Power Plants," Revision 1

— — — — —, RG 1.61, "Damping Values for Seismic Design of Nuclear Power Plants," Revision 1

— — — — —, RG 1.63, "Electric Penetration Assemblies in Containment Structures for Nuclear Power Plants," Revision 3

— — — — —, RG 1.65, "Materials and Inspections for Reactor Vessel Closure Studs," Revision 1

— — — — —, RG 1.68, "Initial Test Program for Water-Cooled Nuclear Power Plants," Revision 3

— — — — —, RG 1.70, "Standard Format and Content of Safety Analysis Reports for Nuclear Power Plants (LWR Edition)," Revision 3

— — — — —, RG 1.75, "Physical Independence of Electrical Systems," Revision 3

— — — — —, RG 1.76, "Design-Basis Tornado and Tornado Missiles for Nuclear Power Plants," Revision 1

— — — — —, RG 1.78, "Evaluating the Habitability of a Nuclear Power Plant Control Room During a Postulated Hazardous Chemical Release," Revision 1

— — — — —, RG 1.82, "Potential Impact of Debris Blockage on Emergency Recirculation during Design Basis Accidents at Pressurized-Water Reactors," Revision 3

— — — — —, RG 1.84, "Design and Fabrication Code Case Acceptability, ASME Section III, Division 1"

— — — — —, RG 1.89, "Environmental Qualification of Certain Electric Equipment Important to Safety for Nuclear Power Plants," Revision 1

— — — — —, RG 1.91, "Evaluations of Explosions Postulated to Occur at Transportation Routes Near Nuclear Power Plants," Revision 1

— — — — —, RG 1.94, "Quality Assurance Requirements for Installation, Inspection, and Testing of Structural Concrete and Structural Steel During the Construction Phase of Nuclear Power Plants," **(Withdrawn -- See 75 FR 54321, 09/09/2010)**

— — — — —, RG 1.97, "Instrumentation for Light-Water-Cooled Nuclear Power Plants to Assess Plant Conditions During and Following an Accident," Revision 4

— — — — —, RG 1.99, "Radiation Embrittlement of Reactor Vessel Materials," Revision 2

—————, RG 1.100, "Seismic Qualification of Electric and Active Mechanical Equipment and Functional Qualification of Active Mechanical Equipment for Nuclear Power Plants," Revision 3

—————, RG 1.101, "Emergency Planning and Preparedness for Nuclear Power Reactors," Revision 5

—————, RG 1.105, "Setpoints for Safety-Related Instrumentation," Revision 2

—————, RG 1.109, "Calculation of Annual Doses to Man from Routine Releases of Reactor Effluents for the Purpose of Evaluating Compliance with 10 CFR Part 50, Appendix I," Revision 1

—————, RG 1.110, "Cost-Benefit Analysis for Radwaste Systems for Light-Water-Cooled Nuclear Power Reactors"

—————, RG 1.111, "Methods for Estimating Atmospheric Transport and Dispersion of Gaseous Effluents in Routine Releases from Light-Water-Cooled Reactors," Revision 1

—————, RG 1.112, "Calculation of Releases of Radioactive Materials in Gaseous and Liquid Effluents from Light-Water-Cooled Power Reactors," Revision 1

—————, RG 1.113, "Estimating Aquatic Dispersion of Effluents form Accidental and Routine Reactor Releases for the Purpose of Implementing Appendix I," Revision 1

—————, RG 1.115, "Protection Against Low-Trajectory Turbine Missiles," Revision 1

—————, RG 1.116, "Quality Assurance Requirements for Installation, Inspection, and Testing of Mechanical Equipment and Systems," **(Withdrawn -- See 75 FR 54921, 09/09/2010)**

—————, RG 1.117, "Design Basis Tornado and Tornado Missiles for Nuclear Power Plants," Revision 1

—————, RG 1.121, "Bases for Plugging Degraded PWR Steam Generator Tubes," (for Comment)

—————, RG 1.125, "Physical Models for Design and Operation of Hydraulic Structures and Systems for Nuclear Power Plants," Revision 2

—————, RG 1.129, "Maintenance, Testing, and Replacement of Large Lead Storage Batteries for Nuclear Power Plants," Revision 2

—————, RG 1.132, "Site Investigations for Foundations of Nuclear Power Plants," Revision 2

—————, RG 1.133, "Loose-Part Detection Program for the Primary System of Light-Water-Cooled Reactors," Revision 1

—————, RG 1.138, "Laboratory Investigations of Soils and Rocks for Engineering Analysis and Design of Nuclear Power Plants," Revision 2

— — — — —, RG 1.140, "Design, Inspection, and Testing Criteria for Air Filtration and Adsorption Units of Normal Atmosphere Cleanup Systems in Light-Water-Cooled Nuclear Power Plants," Revision 2

— — — — —, RG 1.143, "Design Guidance for Radioactive Waste Management Systems, Structures, and Components Installed in Light-Water-Cooled Nuclear Power Plants," Revision 2

— — — — —, RG 1.145, "Atmospheric Dispersion Models for Potential Accident Consequence Assessments at Nuclear Power Plants," Revision 1

— — — — —, RG 1.147, "Inservice Inspection Code Case Acceptability, ASME Section XI, Division 1"

— — — — —, RG 1.149, "Nuclear Power Plant Simulation Facilities for Use in Operator Training and License Examinations"

— — — — —, RG 1.150, "Ultrasonic Testing of Reactor Vessel Welds During Preservice and Inservice Examinations" **(Withdrawn-- See 73 FR 7766, 02/11/2008)**

— — — — —, RG 1.152, "Criteria for Digital Computers in Safety Systems of Nuclear Power Plants"

— — — — —, RG 1.155, "Station Blackout"

— — — — —, RG 1.160, "Monitoring the Effectiveness of Maintenance at Nuclear Power Plants," Revision 2

— — — — —, RG 1.163, "Performance-Based Containment Leak-Test Program"

— — — — —, RG 1.165, "Identification and Characterization of Seismic Sources and Determination of Safe Shutdown Earthquake Ground Motion"

— — — — —, RG 1.166, "Pre-Earthquake Planning and Immediate Nuclear Power Plant Operator Postearthquake Actions"

— — — — —, RG 1.167, "Restart of a Nuclear Power Plant Shut Down by a Seismic Event"

— — — — —, RG 1.182, "Assessing and Managing Risk Before Maintenance Activities at Nuclear Power Plants"

— — — — —, RG 1.183, "Alternative Radiological Source Terms for Evaluating Design Basis Accidents at Nuclear Power Reactors"

— — — — —, RG 1.189, "Fire Protection for Nuclear Power Plants," Revision 2

— — — — —, RG 1.192, "Operation and Maintenance Code Case Acceptability, ASME OM Code"

— — — — —, RG 1.194, "Atmospheric Relative Concentrations for Control Room Radiological Habitability Assessments at Nuclear Power Plants"

— — — — —, RG 1.196, "Control Room Habitability at Light Water Nuclear Power Reactors"

— — — — —, RG 1.198, "Procedures and Criteria for Assessing Seismic Soil Liquefaction at Nuclear Power Plant Sites"

— — — — —, RG 1.200, "An Approach for Determining the Technical Adequacy of Probabilistic Risk Assessment Results for Risk-Informed Activities," Revision 1

— — — — —, RG 1.204, "Guidelines for Lightning Protection of Nuclear Power Plants"

— — — — —, RG 1.206, "Combined License Applications for Nuclear Power Plants (LWR Edition)"

— — — — —, RG 1.208, "A Performance-Based Approach to define the Site-Specific Earthquake Ground Motion"

— — — — —, RG 1.214, "Response Strategies for Potential Aircraft Threats"

— — — — —, RG 4.7, "General Site Suitability Criteria for Nuclear Power Stations," Revision 2

— — — — —, RG 4.13, "Performance, Testing, and Procedural Specifications for Thermoluminescence Dosimetry: Environmental Applications," Revision 1

— — — — —, RG 4.15, "Quality Assurance for Radiological Monitoring Programs (Inception through Normal Operations to License Termination) – Effluent Streams and the Environment," Revision 2

— — — — —, RG 4.21, "Minimization of Contamination and Radioactive Waste Generation: Life-Cycle Planning"

— — — — —, RG 5.7, "Entry/Exit Control for Protected Areas, Vital Areas, and Material Access Areas," Revision 1

— — — — —, RG 5.12, "General Use of Locks in the Protection and Control of Facilities and Special Nuclear Materials"

— — — — —, RG 5.29, "Material Control and Accounting for Nuclear Power Reactors"

— — — — —, RG 5.44, "Perimeter Intrusion Alarm Systems," Revision 3

— — — — —, RG 5.62, "Reporting of Safeguards Events," Revision 1

— — — — —, RG 5.65, "Vital Area Access Controls, Protection of Physical Protection System Equipment and Key and Lock Controls"

— — — — —, RG 5.66, "Access Authorization Program for Nuclear Power Plants," Revision 1

— — — — —, RG 5.68, "Protection Against Malevolent Use of Vehicles at Nuclear Power Plants"

—————, RG 5.69, "Guidance for the Application of Radiological Sabotage Design Basis Threat in the Design, Development, and Implementation of a Physical Security Protection Program that Meets 10 CFR 73.55 Requirements" **(Includes security-related or safeguards information and is not publicly available)**

—————, RG 5.71, "Cyber Security Programs for Nuclear Facilities"

—————, RG 5.74, "Managing the Safety/Security Interface"

—————, RG 5.75, "Training and Qualification of Security Personnel at Nuclear Power Reactor Facilities"

—————, RG 5.76, "Physical Protection Programs at Nuclear Power Reactors" **(Includes security-related or safeguards information and is not publicly available)**

—————, RG 5.77, "Insider Mitigation Program"

—————, RG 8.2, "Guide for Administrative Practices in Radiation Monitoring"

—————, RG 8.4, "Direct Reading and Indirect Reading Pocket Dosimeters"

—————, RG 8.6, "Standard Test Procedures for Gieger-Muller Counters"

—————, RG 8.7, "Instructions for Recording and Reporting Occupational Radiation Exposure Data," Revision 2

—————, RG 8.8, "Information Relevant to Ensuring that Occupational Radiation Exposures at Nuclear Power Stations Will Be ALARA," Revision 3

—————, RG 8.9, "Acceptable Concepts, Models, Equations, and Assumptions for a Bioassay Program," Revision 1

—————, RG 8.10, "Operating Philosophy for Maintaining Occupational Radiation Exposures ALARA," Revision 1-R

—————, RG 8.13, "Instruction Concerning Prenatal Radiation Exposure," Revision 3

—————, RG 8.15, "Acceptable Programs for Respiratory Protection," Revision 1

—————, RG 8.20, "Applications of Bioassay for I-125 and I-131," Revision 1

—————, RG 8.25, "Air Sampling in the Workplace," Revision 1

—————, RG 8.26, "Applications of Bioassay for Fission and Activation Products"

—————, RG 8.27, "Radiation Protection Training for Personnel at Light-Water-Cooled Nuclear Power Plants"

—————, RG 8.28, "Audible-Alarm Dosimeters"

— — — — —, RG 8.29, "Instruction Concerning Risks from Occupational Radiation Exposure," Revision 1

— — — — —, RG 8.32, "Criteria for Establishing a Tritium Bioassay Program"

— — — — —, RG 8.34, "Monitoring Criteria and Methods To Calculate Occupational Radiation Doses"

— — — — —, RG 8.35, "Planned Special Exposures," Revision 1

— — — — —, RG 8.36, "Radiation Dose to the Embryo/Fetus"

— — — — —, RG 8.38, "Control of Access to High and Very High Radiation Areas in Nuclear Power Plants," Revision 1

Regulatory Issue Summary

— — — — —, RIS 00-011, "NRC Emergency Telecommunications System"

— — — — —, RIS 2000-03, "Resolution of Generic Safety Issue 158: Performance of Safety-Related Power-Operated Valves Under Design Basis Conditions"

— — — — —, RIS 2000-18, "Guidance on Managing Quality Assurance Records in Electronic Media"

— — — — —, RIS 2002-22, "Use of EPRI/NEI Joint Task Force Report, 'Guideline on Licensing Digital Upgrades: EPRI TR-102348, Revision 1, NEI 01-01: A Revision of EPRI TR-102348 to Reflect Changes to the 10 CFR 50.59 Rule'"

— — — — —, RIS 2005-02, "Clarifying the Process for Making Emergency Plan Changes"

— — — — —, RIS 2005-04, "Guidance on the Protection of Unattended Openings that Intersect a Security Boundary or Area," **(Exempt from public disclosure in accordance with 10 CFR 2.390)**

— — — — —, RIS 2005-022, "Inadequate Criticality Safety Analysis of Ventilation Systems at Fuel Cycle Facilities"

— — — — —, RIS 2005-026, "Control of Sensitive Unclassified Nonsafeguards Information Related to Nuclear Power Reactors"

— — — — —, RIS 2006-06, "New Reactor Standardization Needed to Support the Design-Centered Licensing Review Approach"

Other NRC Documents

— — — — —, Atomic Safety and Licensing Board (ASLB) pleading (Vogtle Electric Generating Plant (VEGP) Early Site Permit (ESP) application): *NRC Staff Response to the Licensing Board's Questions Regarding Safety Matters* (at 41 through 46), January 16, 2009, Docket No. 52-011-ESP

— — — — —, Atomic Safety and Licensing Board (ASLB) pleading (Vogtle Electric Generating Plant (VEGP) Early Site Permit (ESP) application): *NRC Staff Response to the Licensing Board's Questions Regarding Safety Matters* (at 46 through 48), January 16, 2009, Docket No. 52-011-ESP

— — — — —, First Revised Order, EA-03-009, "Interim Inspection Requirements for Reactor Pressure Vessel Heads at Pressurized Water Reactors"

— — — — —, letter dated April 9, 2009, NRC Staff Review of NEI 03-12, "Template for Security Plan, Training and Qualification, Safeguards Contingency Plan, [and Independent Spent Fuel Storage Installation Security Program]" (Revision 6) (ML090920528)

— — — — —, letter, dated September 15, 2009, entitled *Safety Evaluation Report with Open Items for Chapter 13, Not Including Section 13.6, Titled "Conduct of Operations,"* of NUREG-1793, Supplement 2 – AP1000 Design Certification Amendment, Agencywide Documents Access and Management System (ADAMS) Accession No. ML092540088

National Fire Protection Association (NFPA)

— — — — —, NFPA 25, "Standard for the Inspection, Testing, and Maintenance of Water-Based Fire Protection Systems"

— — — — —, NFPA 72, "National Fire Alarm and Signaling Code"

— — — — —, NFPA 780, "Standard for the Installation of Lightening Protection"

— — — — —, NFPA 804, "Standard for Fire Protection for Advanced Light Water Reactor Electric Generating Plants"

U.S. Army Corps of Engineers

— — — — —, 2008a. "HEC-RAS River Analysis System: Hydraulic Reference Manual, Version 4.0." March. Hydrologic Engineering Center, Davis, California

— — — — —, 2008b. "HEC-HMS Hydrologic Modeling System User's Manual, Version 3.3." Hydrologic Engineering Center, Davis, California

U.S. Nuclear Energy Institute (NEI)

— — — — —, "Consistent Site-Response/Soil-Structure Interaction Analysis and Evaluation"

— — — — —, "White Paper in Support of New Plant Applications"

— — — — —, NEI 03-12, "Template for the Security Plan, Training and Qualification Plan, Safeguards Contingency Plan, and Independent Spent Fuel Installation Security Program," Revision 6 **(Includes security-related or safeguards information and is not publicly available)**

— — — — —, NEI 04-07, "Pressurized Water Reactor Sump Performance Evaluation Methodology" Revision 0, Volume 1, as supplemented by the NRC in the "Safety Evaluation by The Office of Nuclear Reactor Regulation Related to NRC Generic Letter 2004-02," in NEI 04-07, Revision 0, Volume 2

— — — — —, NEI 06-06, "Fitness for Duty Program Guidance for New Nuclear Power Plant Construction Sites"

— — — — —, NEI 06-12, "B.5.b Phase 2 & 3 Submittal Guideline," Revision 3 **(Not Publicly Available)**

— — — — —, NEI 06-13, "Template for an Industry Training Program Description"

— — — — —, NEI 06-13A, "Template for an Industry Training Program Description," Revision 1

— — — — —, NEI 06-14A, "Quality Assurance Program Description," Revision 7

— — — — —, NEI 07-01, "Methodology for Development of Emergency Action Levels Advanced Passive Light Water Reactors"

— — — — —, NEI 07-02A, "Generic FSAR Template Guidance for Maintenance Rule Program Description for Plants Licensed Under 10 CFR Part 52"

— — — — —, NEI 07-03A, "Generic FSAR Template Guidance for Radiation Protection Program Description"

— — — — —, NEI 07-08A, "Generic FSAR Template Guidance for Ensuring That Occupational Radiation Exposures Are As Low As Is Reasonably Achievable (ALARA)," Revision 0

— — — — —, NEI 07-09, "Generic FSAR Template Guidance for Offsite Dose Calculation Manual (ODCM) Program Description"

— — — — —, NEI 07-11, "FSAR Template Guidance for Process Control Program (PCP) Description"

— — — — —, NEI 08-08, "Generic FSAR Template Guidance for Life Cycle Minimization of Contamination," Revision 0

— — — — —, NEI 08-08A, "Generic FSAR Template Guidance for Life Cycle Minimization of Contamination"

— — — — —, NEI 94-01, "Industry Guideline for Implementing the Performance-Based Option of 10 CFR Part 50, Appendix J"

— — — — —, NEI 97-06, "Steam Generator Program Guidelines"

— — — — —, NEI 99-04, "Guidelines for Managing NRC Commitment Changes," Revision 0

— — — — —, NUMARC 87-00, "Guidelines and Technical Bases for NUMARC Initiatives Addressing Station Blackout at Light Water Reactors"

— — — — —, NUMARC 93-01, "Industry Guidance for Monitoring the Effectiveness of Maintenance at Nuclear Power Plants"

Nuclear Information and Records Management Association (NIRMA)

— — — — —, NIRMA Guidelines TG 11-1998, "Authentication of Records and Media"

— — — — —, NIRMA Guidelines TG 15-1998, "Management of Electronic Records"

— — — — —, NIRMA Guidelines TG 16-1998, "Software Configuration Management and Quality Assurance"

— — — — —, NIRMA Guidelines TG 21-1998, "Electronic Records Protection and Restoration"

Vogtle Electric Generating Plant (VEGP)

— — — — —, Vogtle Electric Generating Plant (VEGP) Combined License (COL) Application, Revision 4

— — — — —, Vogtle Electric Generating Plant (VEGP) Early Site Permit (ESP) Application Site Safety Analysis Report (SSAR), Revision 5

Westinghouse

— — — — —, Westinghouse Calculation Note, APP-PGS-M3C-011, "AP1000 Gas Spill or Release Effects on Control Room Habitability," Revision 0 and Revision 1

— — — — —, Westinghouse Calculation Note, APP-VES-M3C-006, "Main Control Room Emergency Habitability from Toxic Chemical Effluents," Revision 0 and Revision 1

— — — — —, Westinghouse Commercial Atomic Power (WCAP)-14655, "Designer's Input to the Training of the Human Factors Engineering Verification and Validation Personnel"

— — — — —, Westinghouse Commercial Atomic Power (WCAP)-15985, "AP1000 Implementation of the Regulatory Treatment of Nonsafety-Related System Process"

— — — — —, Westinghouse Commercial Atomic Power (WCAP)-16361, APP-PMS-JEP-001, Revision 0, May 2006, "Westinghouse Setpoint Methodology for Protection Systems – AP1000"

— — — — —, Westinghouse Design Guideline, APP-OCS-J1-002, "AP1000 Human System Interface Design Guidelines"

— — — — —, Westinghouse Technical Report, (TR)-3, APP-GW-S2R-010, "Extension of Nuclear island Seismic Analyses to Soil Sites"

— — — — —, Westinghouse Technical Report (TR)-49, "AP1000 Enhancement Report

— — — — —, Westinghouse Technical Report (TR)-68, APP-GW-GLR-069, "Equipment Survivability Assessment"

— — — — —, Westinghouse Technical Report (TR)-70, APP-GW-GLR-040, "Plant Operations, Surveillance, and Maintenance Procedures"

— — — — —, Westinghouse Technical Report (TR)-94, APP-GW-GLR-066, "AP1000 Safeguards Assessment Report"

— — — — —, Westinghouse Technical Report (TR)-96, "Interim Compensatory Measures Report"

— — — — —, Westinghouse Technical Report (TR)-101, APP-GW-GLR-101, "AP1000 Probabilistic Risk Assessment Site-Specific Considerations"

— — — — —, Westinghouse Technical Report (TR)-107, APP-GW-GLR-107, "AP1000 Technical Support Center (TR107)," Revision 1, June 14, 2007

— — — — —, Westinghouse Technical Report (TR)-134, APP-GW-GLR-134, "AP1000 DCD Impacts to Support COLA Standardization"

— — — — —, Westinghouse Technical Report (TR)-136, APP-GW-GLR-136, "AP1000 Human Factors Program Implementation for the Emergency Operations Facility and Technical Support Center," Revision 1

Other References

— — — — —, Advisory Committee on Reactor Safeguards (ACRS) Early Site Permit (ESP) Subcommittee Transcript, December 3, 2008 (Tr. at 156 to 189)

— — — — —, AP1000 Design Control Document (DCD), Revision 18

— — — — —, Atomic Energy Act, Sections 103 and 185(b)

— — — — —, Atomic Safety and Licensing Board (ASLB) Vogtle Electric Generating Plant (VEGP) Early Site Permit (ESP) Hearing Transcript, March 24, 2009 (Tr. at M 2084)

— — — — —, Atomic Safety and Licensing Board (ASLB) Vogtle Electric Generating Plant (VEGP) Early Site Permit (ESP) Hearing Transcript, March 24, 2009 (Tr. at M 2172 to M 2186)

— — — — —, Caldon Topical Report, ER-157P, Revision 8, "Supplement to Topical Report ER-80P: Basis for a Power Uprate with the LEFM Check or Checkplus™ System"

— — — — —, Murphy, K.G., and K.M. Campe, "Nuclear Power Plant Control Room Ventilation System Design for Meeting General Criterion 19," U.S. Atomic Energy Commission, 13th Air Cleaning Conference, 1974

— — — — —, NuStart Technical Report, AP-TR-NS01-A, Revision 2, "Containment Leak Rate Test Program," April 4, 2007

— — — — —, Technical Specification Task Force (TSTF)-449, "Steam Generator Tube Integrity," Revision 4

— — — — —, Technical Specification Task Force (TSTF)-511, "Eliminate Working Hour Restrictions from TS 5.2.2 to Support Compliance with 10 CFR Part 26 ['Fitness for Duty Programs']," Revision 0

— — — — —, Three Mile Island (TMI) Action Plan, Item III.D.3.4, "Control Room Habitability"

APPENDIX E. PRINCIPAL CONTRIBUTORS

Name	Responsibility
Ahmed, Sardar	Engineering Mechanics
Ahn, Hosung	Hydrology
Anderson, Brian	Project Management
Barss, Daniel	Emergency Planning
Bavol, Bruce	Project Management
Bongarra, James	Operator Training
Budzynski, John	Reactor Systems
Caruso, Mark	PRA/Severe Accidents
Caverly, Jill	Hydrology
Chalk, Wayne	Fitness for Duty
Chapman, Travis	Technical Specifications
Chen, Pie-Ying	Engineering Mechanics
Chien, Nan	Containment Systems
Chopra, Om	Electrical Engineering
Chuang, Jerry	Plant Systems
Chuang, Tze-Jer	Structural Engineering
Clinton, Rebecca	Special Nuclear Material Security
Comar, Manny	Project Management
Concepcion-Robles, Milton	Initial Test Programs
Curran, Gordon	Plant Systems
Dinh, Thinh	Fire Protection
Downey, Steven	Materials Engineering
Dusaniwskyj, Michael	Financial
Frost, John	Plant Security
Galletta, Thomas	Project Management
Goetz, Sue	Project Management
Goldstein, Kay	Licensing assistant
Grady, Anne-Marie	Containment Systems
Green, Sharon	Licensing Assistant
Habib, Donald	Project Management
Haggerty, Sharon	Management Analyst
Harris, Larry	Special Nuclear Material Security
Harris, Paul	Fitness for Duty
Hart, Michelle	Accident Analysis

Name	Responsibility
Harvey, Brad	Meteorology
Hayes, Michelle	Containment Systems
Hernandez, Raul	Plant Systems
Hinson, Charles	Health Physics
Honcharik, John	Component Integrity Performance Testing; Materials Engineering
Hood, Tanya	Project Management
Hsii, Yi-Hsiung	Reactor Systems
Hsu, Kaihwa	Engineering Mechanics
Huang, Jason	Engineering Mechanics
Jackson, Christopher	Containment Systems
Jenkins, Joel	Materials Engineering
Joshi, Ravindra	Project Management
Kavanagh, Kerri	Quality Assurance
Kellum, James	Operator Training
Kelly, Glen	PRA/Severe Accidents
Kleeh, Edmund	Technical Specifications
Law, Yiu	Engineering Mechanics
Le, Tuan	Engineering Mechanics
Lee, Eric	Cyber Security
Lee, Jay	Radiological Dose Analysis
Li, Chang-Yang	Plant Systems
Li, Yueh-Li (Renne)	Engineering Mechanics
Lintz, Mark	Operator Training
Lyon, Warren	Reactor Systems
Makar, Gregory	Materials Engineering
Martinez-Navedo, Tania	Electrical Engineering
McGovern, Denise	Project Management
McNally, Richard	Engineering Mechanics
Minarik, Anthony	Project Management
Misenhimer, David	Project Management
Moody, Robert	Emergency Planning
Morton, Wendell	Plant Systems
Musico, Bruce	Emergency Planning
Nakanishi, Tony	Reactor Systems
Oesterle, Eric	Project Management

Name	Responsibility
Patel, Pravin	Structural Engineering
Patterson, Malcolm	Probabilistic Risk Assessment (PRA)/Severe Accidents
Pelton, Richard	Operator Training
Peng, Shie-Jeng	Containment Systems
Peralta, Juan	Quality Assurance/Initial Test Program
Pieringer, Paul	Human Factors
Poehler, Jeffrey	Material Engineering
Powell, Eric	PRA/Severe Accidents
Radlinski, Robert	Fire Protection
Ray, Nihar	Materials Engineering
Reddy, Devender	Plant Systems
Reichelt, Eric	Component Integrity Performance Testing
Roach, Edward	Health Physics
Rodriquez, Rafael	Special Nuclear Material Safety Analysis
Rycyna, John	Cyber Security
Sastre, Eduardo	Chemical Engineering
Scarbrough, Thomas	Component Integrity Performance Testing
Schaffer, Steve	Waste Management/Process & Effluent Monitoring Systems
Schroer, Suzanne	PRA/Severe Accidents
Sebrosky, Joseph	Project Management
Sekerak, Patrick	Engineering Mechanics
Shum, David	Plant Systems
Spicher, Terri	Project Management
Steingass, Timothy	Materials Engineering
Stubbs, Angelo	Plant Systems
Tabatabai, Sarah	Geology & Seismology
Tamara, Seshagiri Rao	Site Hazards
Tegeler, Bret	Structural Engineering
Tjader, Theodore	Technical Specifications
Truong, Tung	Instrumentation & Controls
Ullrich, Elizabeth	Byproduct and Source Material Safety Analysis
Valentin-Olmeda, Milton	Structural Engineering
Van Wert, Christopher	Reactor Systems
Walker, Jacquam	Operator Training
Wang, Weijun	Geotechnical Engineering
Wentzel, Michael	Project Management

Name	Responsibility
Wheeler, Larry	Plant Systems
White, Duncan	Byproduct and Source Material Safety Analysis
Wilson, Joshua	Plant Systems
Wong, Yuken	Engineering Mechanics
Wray, Barry	Special Nuclear Material Security
Wu, Cheng-Ih	Engineering Mechanics

APPENDIX F

REPORT BY THE ADVISORY COMMITTEE ON REACTOR SAFEGUARDS

UNITED STATES
NUCLEAR REGULATORY COMMISSION
ADVISORY COMMITTEE ON REACTOR SAFEGUARDS
WASHINGTON, DC 20555 - 0001

January 24, 2011

The Honorable Gregory B. Jaczko
Chairman
U.S. Nuclear Regulatory Commission
Washington, DC 20555-0001

Subject: REPORT ON THE SAFETY ASPECTS OF THE SOUTHERN NUCLEAR
             OPERATING COMPANY COMBINED LICENSE APPLICATION FOR VOGTLE
             ELECTRIC GENERATING PLANT, UNITS 3 AND 4

Dear Chairman Jaczko:

During the 579th meeting of the Advisory Committee on Reactor Safeguards (ACRS), January
13-15, 2011, we reviewed the NRC staff's Advanced Safety Evaluation Report (ASER) for the
pending Southern Nuclear Operating Company (SNC) Combined License Application (COLA)
for Vogtle Electric Generating Plant (VEGP), Units 3 and 4. This COLA incorporates by
reference the Westinghouse Electric Company (WEC) AP1000 Design Certification Amendment
(DCA) application and SNC VEGP Early Site Permit (ESP). Our AP1000 subcommittee also
held four meetings (June 24-25, July 21-22, September 20-21, and December 15-16, 2010) to
review various chapters of the COLA and the staff's ASER. During these meetings, we had the
benefit of discussions with representatives of the NRC staff, NuStart Energy Development, LLC
(NuStart)[1], SNC, SNC's supporting vendors, and the public. We also had the benefit of the
documents referenced. This report fulfills the requirement of 10 CFR 52.53 that the ACRS
report on those portions of the application which concern safety.

CONCLUSION AND RECOMMENDATIONS

1. There is reasonable assurance that VEGP, Units 3 and 4, can be built and operated
    without undue risk to the health and safety of the public. The SNC COLA for VEGP
    should be approved following its final revision.

2. The containment interior cleanliness limits on latent debris should be included in the
    Technical Specifications.

---

[1]NuStart is a multi-utility consortium group. Each of the current and planned combined license applicants
referencing the AP1000 reactor design is a member of NuStart.

3. A regulatory requirement focused on the development of an operational in-service inspection/in-service testing (ISI/IST) program for squib valves should be established, including a review of the lessons-learned from the design and qualification process for these valves.

4. An explicit requirement should be established to assure the accuracy of the feedwater flow measurement by in-plant testing.

5. The staff should review with us the changes in design or commitments that are not yet incorporated in the COLA or referenced in the Design Control Document (DCD), which significantly deviate from those presented during our review.

BACKGROUND

By letter dated March 28, 2008, SNC submitted an application to the U.S. Nuclear Regulatory Commission (NRC) for a combined license for VEGP, Units 3 and 4, in accordance with the requirements of 10 CFR Part 52, "Licenses, Certifications, and Approvals for Nuclear Power Plants." In the application, SNC stated that VEGP, Units 3 and 4, would be two Westinghouse AP1000 advanced passive pressurized water reactors and would be located adjacent to the sites of the operating reactors (VEGP, Units 1 and 2). By letter dated April 28, 2009, NuStart informed the NRC that the AP1000 Design-Centered Work Group has designated the SNC COLA for VEGP, Units 3 and 4, as the AP1000 reference plant.

DISCUSSION

Containment Vessel Exterior Surface

The containment vessel (CV) exterior is subject to a continual flow of outside air, which is an inherent passive safety feature of the AP1000 design. The annular space between the CV and the surrounding shield building includes a baffle to direct air flow; water distribution weirs and associated dams, distribution boxes, and supports; and structures to provide personnel access for inspection and maintenance of the CV exterior. The inorganic zinc exterior coating of the 1.75 in. thick steel CV is of particular interest due to its importance in protecting the pressure boundary from corrosion.

The potential for airborne debris to accumulate on surfaces and in crevices to facilitate undetected corrosion of the CV was reviewed. SNC described the CV exterior coating inspection and maintenance program, which complies with 10 CFR Part 50 Appendix B, applicable ASTM standards, and regulatory guidance. This program is acceptable and is expected to ensure against undetected corrosion of the CV pressure boundary.

Also, the potential for debris to accumulate and impede the performance of the CV exterior water distribution system and cooling during an accident was reviewed. Protective screens and grates are provided in the design which, in combination with in-service inspection of the containment exterior, will ensure acceptable performance.

-3-

Containment Interior Debris Limitation

In our December 20, 2010, letter we concluded that the long-term core cooling requirements were adequately met, provided that the stringent cleanliness requirements specified for the containment interior is maintained. These requirements should not be relaxed without additional analyses, a much wider range of experiments at prototypical conditions, and NRC review.

The cleanliness requirements during operation, limiting latent debris to not more than 59 kg of which not more than 3 kg is fiber, are challenging but achievable. In order to ensure that they are not relaxed during plant life without consideration by the NRC staff of the provisions stated in our letter and to make them highly visible to both the plant operators and to the NRC staff, we recommend that the requirements be included in the plant Technical Specifications. We make this recommendation due to the importance these limits have in this instance, recognizing that debris limits are normally not part of the Technical Specifications.

ISI/IST Program Requirements for Squib Valves

The Automatic Depressurization System (ADS) ADS-4 squib valves must operate to achieve post loss-of-coolant accident (LOCA) passive long-term cooling. They are actuated by an explosive charge and are one-time-use valves until the internals are replaced. The development of an effective ISI/IST program to assure operability of the valves is needed. Periodic removal and firing of the explosive charge that initiates operation of the valve may not be sufficient for these critical components. SNC stated that, jointly with Westinghouse, it will develop ISI/IST procedures based on the final valve design and lessons-learned from the valve qualification process. While the AP1000 DCD includes Inspections, Tests, Analyses, and Acceptance Criteria (ITAAC) to confirm squib valve qualification, we recommend that a regulatory requirement be established focused on the development of the ISI/IST program, including a review of the lessons-learned from the valve design and qualification process.

Seismic Margin Analysis

The VEGP site-specific safe shutdown earthquake (SSE) design response spectra are the site-specific ground motion response spectra (GMRS) approved in the ESP. The GMRS slightly exceeded the certified seismic design response spectrum (CSDRS) in the lower frequency range. Therefore, in accordance with provisions in the DCD, plant-specific seismic evaluations were performed to demonstrate that the AP1000 plant designed for the CSDRS was acceptable for the VEGP site.

SNC performed an alternative site-specific analysis of soil-structure interaction using a three-dimensional model that uses the operating basis earthquake damping values of 4% specified in Regulatory Guide (RG) 1.61. The result indicated that the VEGP GMRS excitation will not compromise structures, systems, and components (SSC) under design-basis loads.

In response to a request for additional information, SNC provided additional seismic margin analyses confirming that the AP1000 certified design meets the 1.67 margin specified in SECY-93-087 at the VEGP site. A review-level earthquake equal to 0.5g was set for the seismic margin analysis and used to demonstrate the specified margin over the SSE of 0.3g. SNC also conducted a seismic margin analysis demonstrating that site-specific high confidence of low probability of failure values are equal to or greater than 1.67 times the GMRS of the design-basis SSE. Further, SNC completed a site-specific analysis of phenomena with the potential to reduce seismic margin. Evaluations were made of the potential for soil liquefaction and its effect on bearing capacity as well as nuclear island demand and seismic stability. The results of these additional analyses also demonstrated an adequate seismic margin of 1.67 times the VEGP GMRS, in accordance with SECY-93-087.

Technical Support Center

In a departure from the certified design, the SNC COLA provides for the Technical Support Center (TSC) for the new Units 3 and 4 to be combined with that for the existing Units 1 and 2 in a central Communication Support Center located between the power blocks for Units 2 and 3. This was reflected in the approved ESP, and human factors considerations for the combined TSC were discussed in the COLA review. However, insufficient detail is available at this time to evaluate how the TSC will function to assure that the four units, of two different designs, will be effectively supported in an emergency affecting one or more units. The COLA includes an ITAAC to demonstrate the capability of the TSC equipment and data displays to clearly identify and reflect the affected unit.

The staff should review with us the need for generic design guidance to assure adequate display of information at a multi-unit TSC.

During our review of the VEGP cyber security plan (CSP), we noted that the level of protection designated for the TSC (Level 2) was less than that for the respective units (Level 3 or 4). While it is recognized that control function decisions will be made only in the plant, and that the TSC is limited to advisory and management functions, this difference raised a concern as to the possible consequences during an emergency response if the information displayed in the TSC was corrupted as a result of the lower level of cyber security assigned. Since the CSP is consistent with RG 5.71 guidance, this is a potentially generic concern. The staff stated that this would be addressed in an ACRS Digital I&C Systems subcommittee meeting planned in the near future. We look forward to this further review of the appropriate level of protection for the TSC.

Power Measurement Uncertainty

The amended DCD states that the combined license holder will calculate the primary power calorimetric uncertainty using, "...an NRC acceptable method and confirm that the safety analysis primary power calorimetric uncertainty bounds the calculated values." The initial reactor power for a large-break LOCA, as well as for certain mass and energy release calculations, is assumed to be within 1% of the licensed power. To measure power, SNC proposes to use a secondary side heat balance which requires measurement of certain pressures, temperatures, and flow rates. The largest contributor to uncertainty in the estimate of power is the measurement of the feedwater flow rate.

The Caldon Check Plus™ Leading Edge Flow Meter (LEFM), which is an ultrasonic flow measurement system, will be used to measure feedwater flow rate. The staff has approved this device to support a 1% power measurement uncertainty, provided two criteria for a newly constructed system are met. SNC proposes to address these criteria using an ITAAC to confirm that the instrumentation has been installed correctly, a License Condition to provide confirmation that the administrative controls are in place, and some COLA changes to be incorporated in a future application revision.

One of the criteria allows for use of a calibrated LEFM, where calibration was performed off-site at a lower Reynolds number than would exist in the plant, provided that acceptable justification is provided. Part of this justification is provided by confirmatory in-plant tests following installation. These tests assure that actual performance is within the uncertainty bounds established for the instrumentation.

The NRC should require that SNC make an explicit commitment to perform calibrations with representative piping configurations and conduct in-plant confirmatory tests.

Site-Specific PRA

We expected the COLA PRAs to be revised to include all available plant and site-specific information. This is not the case for the SNC COLA because Chapter 19 of the AP1000 DCD provides guidance to combined license applicants to identify plant-specific information and compare it with specified interface requirements. If the interface requirements are satisfied, the DCD PRA results will be conservative and are considered adequate for the COLA PRA. We find such a bounding approach acceptable at the combined license stage, given that substantial plant-specific, as-built information is not yet available.

NRC regulations require a full-scope, plant-specific PRA before fuel load. This PRA should meet the criteria of RG 1.200, providing a realistic picture of the plant risk, including uncertainty. The passive safety features of the AP1000 design were developed to eliminate or greatly reduce many of the more important contributors to plant risk. However, this improvement in risk comes via a replacement of active high pressure, high flow cooling systems with gravity driven systems.

Possible upsets to adequate performance of the passive phenomena relied upon in the design could be important contributors to risk and should be incorporated into the PRA, if it is to be considered a complete calculation of the risk and used for risk-informed applications or in Reactor Oversight Program (ROP) evaluations. For example, if an inspection should find many times the allowed inventory of fibrous material inside containment, the PRA must be able to show the potential impact of that finding, if it is to be useful in the ROP. (The DCD PRA acknowledges that core damage frequency would increase by a factor of 6,000 if failures of containment recirculation and in-containment refueling water storage tank screens occur, but uses only a "conservative" screen failure rate, rather than a model that would account for debris.) Another example would be the discovery of deposits, grease, or unauthorized paint on the exterior of the containment vessel; again, the DCD PRA is not structured to account for such departures from the assumptions of the passive design.

At this time, it is not as important that such possibilities be fully amenable to engineering analysis as it is to include the possible failure modes and uncertainties in the PRA. For example, they could be addressed using an expert elicitation of the likelihood of failure in the presence of the best available experimental, theoretical, and analytical information.

Incorporation of DCD Changes

The SNC COLA review was conducted in parallel with the review of the AP1000 DCA application. As a consequence, the SNC COLA references Revision 17 of the DCD, whereas the current version is Revision 18, and there may be a further revision prior to certification rulemaking. The staff has described the licensing steps needed to complete the COLA Final Safety Evaluation Report. These include a revision to the COLA following the final DCD revision prior to rulemaking. As described, the process does not provide for further ACRS review of either the DCD or COLA revisions that incorporate changes in design and commitments made by applicants during our review. The staff should review with us the changes and commitments which deviate significantly from those presented during our review.

In summary, we agree with the staff's resolution of all of the open items for the SNC COLA for VEGP, Units 3 and 4, with respect to the specific safety issues. We conclude that there is reasonable assurance that VEGP, Units 3 and 4, can be built and operated without undue risk to the health and safety of the public. The SNC COLA for VEGP, Units 3 and 4, should be approved following its final revision.

Dr. Said Abdel-Khalik did not participate in the Committee's deliberations regarding this matter.

Sincerely,

/RA/

J. S. Armijo
Vice-Chairman

REFERENCES

1. Letter to U.S. Nuclear Regulatory Commission, "Southern Nuclear Operating Company Application for Combined License for Vogtle Electric Generating Plant, Units 3 and 4," March 28, 2008 (ML081050133)

2. During the course of ACRS review, the staff provided the following ASER chapters:

Chapter	Chapter Title	Transmittal Memo to ACRS (Accession Numbers)	ASER (Accession Numbers)
1	Introduction and Interfaces	ML103100006	ML092810005
2	Sites Characteristics	ML100950499	ML100320032
3	Design of Structures, Components, Equipment, and Systems	ML100950532	ML093210002
4	Reactor	ML100331243	ML092610415
5	Reactor Coolant System and Connected Systems	ML100480787	ML092610460
6	Engineered Safety Features	ML100910118	ML100920459
7	Instrumentation and Controls	ML100360236	ML093230696
8	Electric Power	ML100880411	ML092870782
9	Auxiliary Systems	ML100910147	ML093560006
10	Steam and Power Conversion Systems	ML100540758	ML092720790
11	Radioactive Waste Management	ML100340674	ML092610470
12	Radiation Protection	ML100890389	ML092650039
13	Conduct of Operations	ML100910470	ML100820408
14	Initial Test Programs	ML100880449	ML092650048
15	Accident Analysis	ML100900320	ML100130006
16	Technical Specifications	ML100900407	ML092650055
17	Quality Assurance	ML100890413	ML092650063
18	Human Factors Engineering	ML100910031	ML093000107
19	Probabilistic Risk Assessment	ML100920066	ML092650121
19 Appendix 19.A	Loss of Large Areas of the Plant due to Explosions or Fires (LOLA)	ML103090198	ML103260024 (Public Version), ML101810029 (Non-Public Version)
Appendix A	License Conditions, ITAAC, and FSAR Commitments	ML103100006	ML103330312

NRC FORM 335
(12-2010)
NRCMD 3.7

U.S. NUCLEAR REGULATORY COMMISSION

BIBLIOGRAPHIC DATA SHEET

(See instructions on the reverse)

1. REPORT NUMBER
(Assigned by NRC, Add Vol., Supp., Rev., and Addendum Numbers, if any.)

NUREG-2124
Volume 3

2. TITLE AND SUBTITLE

Final Safety Evaluation Report for Combined Licenses for Vogtle Electric Generating Plant (VEGP), Units 3 and 4, Volume 3

3. DATE REPORT PUBLISHED

MONTH	YEAR
September	2012

4. FIN OR GRANT NUMBER

5. AUTHOR(S)

6. TYPE OF REPORT

Technical

7. PERIOD COVERED (Inclusive Dates)

8. PERFORMING ORGANIZATION - NAME AND ADDRESS (If NRC, provide Division, Office or Region, U. S. Nuclear Regulatory Commission, and mailing address; if contractor, provide name and mailing address.)

Division of New Reactor Licensing
Office of New Reactors
U.S. Nuclear Regulatory Commission
Washington, D.C. 20555-0001

9. SPONSORING ORGANIZATION - NAME AND ADDRESS (If NRC, type "Same as above". if contractor, provide NRC Division, Office or Region, U. S. Nuclear Regulatory Commission, and mailing address.)

same as above

10. SUPPLEMENTARY NOTES
Dockets 52-025 and 52-026

11. ABSTRACT (200 words or less)

This final safety evaluation report (FSER) documents the U.S. Nuclear Regulatory Commission (NRC) staff's technical review of the combined license (COL) application submitted by Southern Nuclear Operating Company (SNC or the applicant), for the Vogtle Electric Generating Plant (VEGP) Units 3 and 4. The FSER also documents the NRC staff's technical review of the limited work authorization (LWA) activities for which SNC has requested approval.

By letter dated March 28, 2008, SNC submitted its application to the NRC for COLs for two AP1000 advanced passive pressurized water reactors (PWRs) pursuant to the requirements of Sections 103 and 185(b) fo the Atomic Energy Act of 1954, as amended; Title 10 of the Code of Federal Regulations (10 CFR) Part 52, "Licenses, certifications and approvals for nuclear power plants"; and the associated material licenses under 10 CFR Part 30, "Rules of general applicability to domestic licensing of byproduct material"; 10 CFR Part 40, "Domestic licensing of source material"; and 10 CFR Part 70, "Domestic licensing of special nuclear material." These reactors are identified as VEGP Units 3 and 4, and will be located on the existing VEGP site in Burke County, Georgia. In October 2009, SNC supplemented its COL application to include a request for an LWA.

This FSER presents the results of the staff's review of information submitted in conjunction with the COL and LWA application.

12. KEY WORDS/DESCRIPTORS (List words or phrases that will assist researchers in locating the report.)

Vogtle
Combined License (COL)
Final Safety Evaluation Report (FSER)
Inspections, Tests, Analyses and Acceptance Criteria (ITAAC)

13. AVAILABILITY STATEMENT
unlimited

14. SECURITY CLASSIFICATION

(This Page)
unclassified

(This Report)
unclassified

15. NUMBER OF PAGES

16. PRICE

NRC FORM 335 (12-2010)

NUREG-2124
Volume 3

Final Safety Evaluation Report Related to the Combined Licenses for
Vogtle Electric Generating Plant, Units 3 and 4

September 2012

www.ingramcontent.com/pod-product-compliance
Lightning Source LLC
Chambersburg PA
CBHW080232180526

45167CB00006B/2248